Encyclopedia of Thermodynamics: Scientific Theory and Applications

Volume I

Encyclopedia of Thermodynamics: Scientific Theory and Applications
Volume I

Edited by **Barney Tyler**

NY RESEARCH PRESS

New York

Published by NY Research Press,
23 West, 55th Street, Suite 816,
New York, NY 10019, USA
www.nyresearchpress.com

Encyclopedia of Thermodynamics: Scientific Theory and Applications
Volume I
Edited by Barney Tyler

International Standard Book Number: 978-1-63238-172-9 (Hardback)

Printed in the United States of America.

Contents

Preface

The book contains a collection of numerous high level contributions regarding thermodynamics. It begins with the fundamentals and extends up to several applications in distinct scientific fields. The book covers topics like Classical Thermodynamics, Statistical Thermodynamics and Property Prediction in Thermodynamics. The book aims at providing the readers with in-depth knowledge and serves as a state of the art reference book in the field of thermodynamics.

Significant researches are present in this book. Intensive efforts have been employed by authors to make this book an outstanding discourse. This book contains the enlightening chapters which have been written on the basis of significant researches done by the experts.

Finally, I would also like to thank all the members involved in this book for being a team and meeting all the deadlines for the submission of their respective works. I would also like to thank my friends and family for being supportive in my efforts.

Editor

Section 1

Classical Thermodynamics

Chapter 1

Useful Work and Gibbs Energy

Nikolai Bazhin

Additional information is available at the end of the chapter

1. Introduction

Devices for performing chemical reactions are widely used to produce heat and work. Heat, in turn, produces work, e.g., in the form of electric energy in the so-called heat engines. It is the well-known fact that the efficiency of heat engines is restricted by Carnot principle. Therefore, it is generally recognized that heat cannot be fully converted to work. The efficiency of electric energy production due to the burning of fossil fuel of various kinds varies from 30 to 50 %.

On the other hand, there are galvanic and fuel cells whose efficiency can reach theoretically unity if it implies the ratio between the electric energy produced and the value of a change in Gibbs energy during chemical reactions occurring in a cell. These devices operate at constant temperature and pressure. It is concluded then that the devices, similar to a galvanic cell, cannot work at constant and uniform temperature according to the principle of heat engine. These devices assumed to operate only due to the direct transformation of chemical reaction energy, described by a change in the Gibbs energy, into work [1]. This viewpoint causes, however, numerous contradictions. The goal of this work is to analyze in detail the mechanism of useful work and heat production in chemical systems functioning at constant temperature and pressure.

2. Fundamental functions

In this Section, the fundamental functions of thermodynamics will be characterized, as the fundamental functions play the important role in the description of the process of the energy transformation. This notion includes four functions, i.e., internal energy, enthalpy, Helmholtz energy, and Gibbs energy. All of them are the state functions of energy dimension. It is generally believed that the value with energy dimension describes energy but this is by no means always the case. Below, the fundamental functions will elucidate whether these are energy values or not.

2.1. Internal energy

According to IUPAC [2], "**internal energy** U is the quantity the change in which is equal to the sum of heat, q, brought to the system and work, w, performed on it, $\Delta U = q + w$ ". Because of various transformations, the internal energy can be converted to other kinds of energy. However, the initial quantity of internal energy should be conserved due to the law of energy conservation. Conservation is the most important characteristic of energy. Hence U is energy.

2.2. Enthalpy

According to IUPAC, "**enthalpy,** $H = U + pV$ *is the internal energy* of a system plus the product of pressure and volume. Its change in a system is equal to the heat, brought to the system at constant pressure" [2]. It is worth noting that the change in the value describing energy always corresponds to the change in energy and not only in some special cases. Generally, a change in enthalpy may be inconsistent with the change in real physical values. Let us consider, e.g., the process of ideal gas heating at constant volume. The quantity of heat, taken in by ideal gas from the heat bath, q, equivalently changes only the internal gas energy, $\Delta U = q$, and simultaneously causes changes in gas enthalpy, $\Delta H = \Delta U + \Delta pV > q$. However, this change fails to reflect the changes in any physically significant values. Thus, the enthalpy is not energy but a function of state having the dimension of energy. It is easier and more correct to assume that the enthalpy is the part of a calculating means used to describe thermodynamic processes.

2.3. Helmholtz energy

According to IUPAC, "**Helmholtz energy (function),** A, *is the internal energy* minus the product of thermodynamic temperature and entropy: $A = U - TS$. It was formerly called free energy" [2]. Let us see why this quantity was called free energy. According to the first law [3]

$$\Delta U = q + w, \tag{1}$$

where q is the heat, entering the system, and w is the total work performed on the system by the surroundings. Usually, the total work is given as the sum of two terms: expansion work $(-p\Delta V)$ and so-called useful work (w_{useful})

$$w = -p\Delta V + w_{\text{useful}}. \tag{2}$$

If the process occurs at constant volume, the expansion work is absent and w describes the useful work. In the case of reversible process [3]

$$q = T\Delta S. \tag{3}$$

Thus, eq. (1) is of the form in the case of reversible process

$$w_{\text{useful}} = \Delta U - T\Delta S = \Delta(U - TS) = \Delta A, \tag{4}$$

where ΔA is the change in Helmholtz energy at constant temperature and volume. Since the term "energy" means the capacity of the system to perform work, from eq. (4) it is formally concluded that A is the energy (but in this case eq. (4) can have the second explanation – A is the numerical characteristic of work and not the work itself). Further, from eq. (4) it was concluded that only the part of internal energy U minus TS can be used to produce work. Therefore, the TS quantity was called "bound energy" and (U - TS) – "free energy". The meaning of these notions will be considered in more detail using the Gibbs energy as an example because it is more often used in chemical applications.

2.4. Gibbs energy

According to IUPAC [2], "**Gibbs energy (function),** $G = (H - TS)$, is the enthalpy minus the product of thermodynamic temperature and entropy. It was formerly called free energy or free enthalpy". The reasons for the appearance of the term "Gibbs energy" are similar to those discussed when considering the Helmholtz energy except for the fact that the Gibbs energy describes the reversible useful work performed at constant temperature and pressure. This is readily observed by substituting eq. (3) and eq. (2) into eq. (1) with regard to $V\Delta p = 0$

$$w_{\text{useful}} = \Delta H - T\Delta S = \Delta(H - TS) = \Delta G, \tag{5}$$

where ΔG is the change in Gibbs energy at constant temperature and pressure in the reversible process.

Unfortunately, the word "energy" as defined by IUPAC for a Gibbs energy (and also for a Helmholtz energy), causes great confusion. The Gibbs energy $G = H - TS$ consists of two terms, the enthalpy and the entropy one. The origin of both of the terms is quite different despite the same dimension. The enthalpy, considered above, is not energy.

Consider now the problem of TS nature. In the case of the reversible process $T\Delta S = q$, but in the case of the irreversible process $T\Delta S > q$ and additional contributions to $T\Delta S$ can arise without change in energy. For example, it is well known the increase of the entropy in the process of ideal gas expansion in vacuum without heat consumption ($q = 0$).

Let us consider another example. Let the ideal gas-phase system involves a spontaneous process of the monomolecular transformation of substance A into B. As suggested a change in enthalpy tends to zero in this reaction. Thus, the internal energy, temperature, pressure, and volume will not undergo changes in this process. However, the entropy will increase due to the entropy of the mixing, because the entropy is a function of state. The value of the TS product will increase accordingly. However, the energy and even the bound energy cannot arise from nothing, whereas the entropy, being a function of state, can increase thus reflecting a change in system state without any changes in the internal system energy. Therefore, the TS term is not the energy, which also implies the absence of the term "bound" energy.

Since neither enthalpy nor *TS* are the energy quantities, the difference between them cannot represent energy. Thus, *G* cannot represent energy precisely in terms of this notion. Note that in the irreversible process, occurring at constant temperature and pressure, the Gibbs energy decreases and thus, is not conserved. This is readily demonstrated by e.g., the aforementioned example of a monomolecular transformation of substance A into B. Thus, conservation, as the most important criterion for energy quantity, fails for the Gibbs energy. It is concluded then that the Gibbs energy is not energy [4] but a function of state. In this regard, the Gibbs energy does not differ from heat capacity. The notions of the non-existing in reality quantities of "free energy" and "bound energy" cause only confusion and are, at present, obsolete [5].

Nevertheless, the notions that the Gibbs energy is the energy and thus, obeys conservation laws, prove to be long-lived, which causes erroneous interpretation of a number of processes some of which are of paramount importance.

Let us consider now the reaction of adenosine triphosphate (ATP) hydrolysis in water solution which is of great concern in biochemistry

$$ATP + H_2O = ADP + Pi.$$

Under the standard conditions [6] $\Delta_r G^o = -7$, $\Delta_r H^o = -4$ kcal·mol^{-1}. According to D. Haynie [6, p. 143], "measurement of the enthalpy change of ATP hydrolysis shows that $\Delta H^o = -4$ kcal·mol^{-1}. That is, the hydrolysis of one mole of ATP at 25 ºC results in about 4 kcal·mol^{-1} being transferred to a solution in the form of heat and about 3 kcal·mol^{-1} remaining with ATP and Pi in the form of increased random motion."

The heat of 4 kcal·mol^{-1} is actually released into solution due to hydrolysis. Unfortunately, it is then assumed that the Gibbs energy is conserved which makes his difference in $\Delta_r H^o$ and $\Delta_r G^o$ of 3 кcal·mol^{-1} be located on the degrees of freedom of product molecules. However, in this connection, the product molecules could appear in the non-equilibrium excited states and transfer this energy to solvent molecules which would result in the emission of 7 kcal·mol^{-1} rather than 4 kcal·mol^{-1} which contradicts the experiment. There are no additional 3 кcal·mol^{-1} on the degrees of freedom of ATP and Pi because the Gibbs energy is not conserved.

2.5. Conclusions

Thus, among the four quantities, that claims to be called energy quantity, only the internal energy deserves this name. The other functions, i.e., enthalpy *H*, Helmholtz energy *A*, and Gibbs energy *G* are the parts of a mathematical apparatus for calculating various quantities, such as useful work, equilibrium constants, etc. This also means that the useful work is only calculated by using functions *A* and *G*, but cannot arise from the change in either the Helmholtz or the Gibbs energy. The physical nature of the work performed should be considered separately. Since the Helmholtz energy and the Gibbs energy are not energies, then, to avoid misunderstanding, it is better to exclude the word "energy" from the name of corresponding functions and to use the second variant of the name of these functions according to IUPAC: a Helmholtz function and a Gibbs function [7].

3. Direct conversion of chemical reaction energy to useful work

This Section is devoted to the discussion of the generally accepted theory of the direct conversion of energy [1, 8] produced by chemical reaction to useful work. For simplicity, hereafter exergonic $(\Delta_r G < 0)$ and exothermic $(\Delta_r H < 0)$ reactions will be considered.

The useful work of the chemical reaction occurring at constant temperature and pressure in the reversible conditions can be calculated through the change in the Gibbs function (5). When the interest is the useful work performed by the system in the environment $(-w_{\text{useful}})$, then

$$-w_{\text{useful}} = -\Delta_r H + T\Delta_r S = -\Delta_r G. \tag{6}$$

From eq. (6) it follows formally that the useful work of the reversible system in the environment is the sum of the enthalpy member $(-\Delta_r H)$ and the entropy member $T\Delta_r S$. In this connection it is interesting to discuss the various situations which arise in dependence on the relation between $(-\Delta_r H)$ and $T\Delta_r S$.

From eq. (6) it follows, that for $\Delta_r S > 0$ the useful work in the environment exceeds $(-\Delta_r H)$: $-w_{\text{useful}} > -\Delta_r H$. Therefore, the system must drag the thermal energy from the environment in the volume $T\Delta_r S$ to perform useful work. What is the physical reason for thermal energy consumption? Why does the system consume thermal energy of volume $T\Delta_r S$ neither more or less? How can two different contributions produce the same useful work?

The second case of $\Delta_r S < 0$ is also of interest. In this case $-w_{\text{useful}} < -\Delta_r H$ and the system must evolve the part of reaction heat to the environment. Why cannot the system use the total reaction heat for useful work production if this energy is at its disposal? Why can the system transfer energy of volume $T\Delta_r S$ and neither more or less to the environment?

The third case is $\Delta_r H = 0$. Here the system can use only the thermal energy of the environment to produce useful work.

In the fourth case $\Delta_r S = 0$ and the system performs work formally due to the reaction heat $(-\Delta_r H)$ without exchanging thermal energy with the environment. But it is not the case.

As mentioned in the Introduction, at present, it is generally accepted that the high efficiency of reversible devices is inconsistent with the notions that heat can be used to produce work and that such devices realize the direct conversion of chemical system energy into work. But below it will be shown that in all the cases, the useful work results from the heat of volume $\Delta_r G$ dragged from the environment.

4. A mechanism of useful work production - A Van't Hoff Equilibrium Box

In this Section, it will demonstrate the mechanism of producing useful reversible work which involves no notions of the direct conversion of reaction energy into useful work. To this end, let us consider a Van't Hoff equilibrium box (VHEB) [9 – 11]. A thermodynamic system must provide realization of the reversible process. This means that all changes in the system are infinitely slow at infinitely minor deviation from equilibrium.

It is assumed that in the system the following reaction occurs

$$\sum_i \nu_i A_i = \sum_j \nu_j B_j, \tag{7}$$

where A_i are the reagents and B_j are the products. The reaction takes place in the reactor (Fig. 1) where the reagents and products are in equilibrium. The chemical process is afforded by reservoirs with reagents and products contained in the system. For simplicity, the reagents and products are assumed to be in standard states. The system should have instruments to transport both the reagents from standard vessels to reactor and the products from reactor to standard vessels. In addition, the system should have tools to perform work, because the reversible process must be followed by reversible work production. The instruments and tools for performing work can be used together. The reactor, transporting instruments, and tools for performing work can be placed either separately or together. To provide constant and uniform temperature, it is necessary to locate the system, including reactor, standard vessels, transporting instruments, and tools in a thermostat, which can also imply the environment of constant temperature.

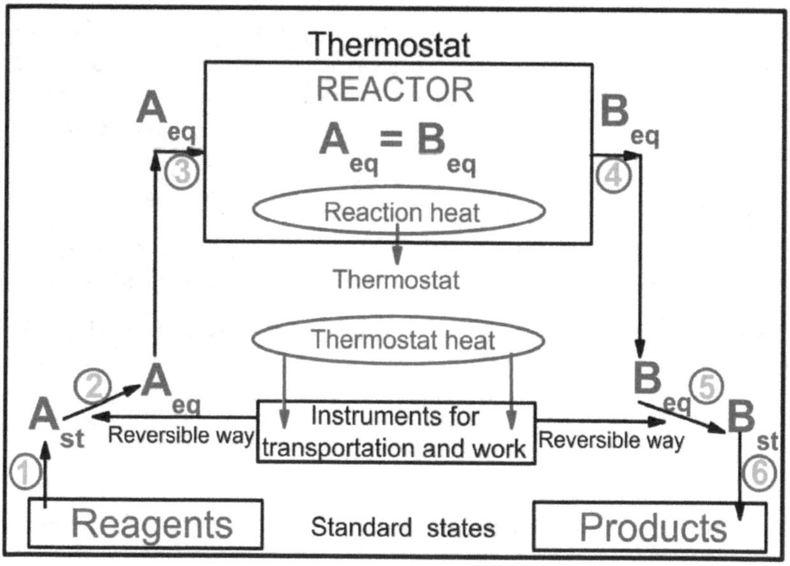

Figure 1. Production of useful work and heat in a closed reversible system. A_{st} (B_{st}) indicate the reagents A_i (products B_j) in the standard states; A_{eq} (B_{eq}) indicate the reagents A_i (products B_j) in the equilibrium states, which correspond to the equilibrium at the reactor; the green figures in circles indicate the step numbers (see text)

Let us consider reversible chemical process in a closed system (Fig. 1). The realization of the reversible chemical process consists in reversible transformation of the reagents to products via chemical reaction. Let us consider the closed system.

4.1. Closed system

The process of reversible work production includes six stages.

Step 1. A small amount of substance A_i is removed from the vessel with reagent A_i in the standard state upon reversible process. Gaseous substances can be removed from a standard vessel and put into a portable cylinder with pistons [11]; solid or liquid substances are placed in lock chambers.

Then, the change in the Gibbs function and the work are zero ($\Delta G_{1i} = 0, w_{1i} = 0$) because a minor portion of substance A_i is in the same standard state as the residual reagent.

Step 2. Reagent A_i is transformed reversibly from the standard into the equilibrium state in the reactor. For example, for ideal gas, the gas pressure will vary from a standard value to the equilibrium partial value in the reactor. In this stage, the reversible work w_{2i} is produced and $\Delta G_{2i} \neq 0$. The work w_{2i} can be done only due to thermostat heat because there are no other energy sources in the system (reaction is in the equilibrium). This work depends on the difference in the physical states of reagent A_i in the initial and equilibrium states. All reagents A_i participate in all stages in quantities proportional to ν_i. For ideal gas, the useful work is

$$w_{2i} = \nu_i RT \ln(p_{i,\text{eq}} / p_{i,\text{st}}), \tag{8}$$

where $p_{i,\text{eq}}$ is the equilibrium pressure of i-th gas in the reactor, and $p_{i,\text{st}}$ is the pressure of i-th gas in a standard vessel. For gaseous components, e.g., the process of reversible gas expansion (compression) in a portable cylinder for producing the maximum useful work, must proceed to the value $p_{i,\text{eq}}$. If expansion (compression) stops at $p_i > p_{i,\text{eq}}$, then the inlet of gas into the reactor causes irreversible gas expansion and thus, the useful work will be less than the maximum one. When due to expansion (compression) the final pressure is less than $p_{i,\text{eq}}$, then the inlet of gas into the reactor causes the irreversible inlet of the i-th gas from an equilibrium mixture in the reactor to the portable cylinder, which also leads to a decrease in useful work. The solid and liquid substances can be transported by lock chambers. The pressure above either solid or liquid substances is varied from 1 bar to the value of the total equilibrium pressure in the reactor. The pressure is created using a minor portion of equilibrium reaction mixture.

The thermostat enthalpy varies as follows: $\Delta H_{2i,\text{thermostat}} = w_{2i}$.

Step 3. Reagent A_i is reversibly introduced into the reactor. Gaseous components are introduced into the reactor through semipermeable membranes using portable cylinders [11]; the solid or liquid ones – by means of lock chambers. Hence, $\Delta G_{3i} = 0,\ w_{3i} = 0$.

The useful work production and the change in thermostat enthalpy ($\Delta H_{2,\text{thermostat}}$) take place only at step 2:

$$w_2 = \Delta H_{2,\text{thermostat}}, \tag{9}$$

where $w_2 = \sum_i w_{2i}$ and $\Delta H_{2,\text{thermostat}} = \sum_i \Delta H_{2i,\text{thermostat}}$.

The same procedure is used to bring products from the standard vessels to reactor.

Step 4. An equivalent amount of product B$_J$ is reversibly removed from the reactor. After this step is $\Delta G_{4j} = 0, w_{4j} = 0$.

Step 5. Product B$_J$, removed from the reactor, is reversibly transformed from the equilibrium state into the standard one to perform work w_{5j}. The change in the Gibbs function is not zero, $\Delta G_{5j} \neq 0$. The change in the thermostat enthalpy is $\Delta H_{5j,\text{thermostat}} = w_{5j}$.

Step 6. Product B$_J$, removed from the reactor, is reversibly introduced into the standard vessel. In this case is $\Delta G_{6j} = 0$ and $w_{6j} = 0$.

The change in the thermostat enthalpy upon thermal energy conversion into useful work at step 5 is

$$w_5 = \Delta H_{5,\text{thermostat}}, \tag{10}$$

where and $w_5 = \sum_j w_{5j}$, $\Delta H_{5,\text{thermostat}} = \sum_j \Delta H_{5j,\text{thermostat}}$

The change in the thermostat enthalpy at the second and fifth steps obeys the equation

$$\Delta H_{2,\text{thermostat}} + \Delta H_{5,\text{thermostat}} = -Q_{\text{dragged}}, \tag{11}$$

where Q_{dragged} is the heat dragged by tools.

For the reversible process, the maximal useful work is numerically equal to $\Delta_r G$, eq. (5) and, hence, the heat dragged by tools from thermostat in the volume

$$w_{\text{useful}} = w_2 + w_5 = \Delta_r G = \Delta_r H - T\Delta_r S = -Q_{\text{dragged}}. \tag{12}$$

The process has resulted in the useful work of the reaction, $\Delta_r G$, but the reaction did not occur yet. To put it otherwise, reaction work was performed without reaction. Only the thermal energy of the thermostat (environment) may be the source of work. This means that the process of useful work production and the reaction itself may be temporally and spatially separated. Thus, eq. (6) numerically connects reaction parameters and the magnitude of the work. However, no reaction energy is needed to produce the work. There is no need to subdivide energy sources into reaction source ($-\Delta_r H$) and thermal $T\Delta_r S$, because there is only one energy source: the thermal energy of thermostat (environment).

For $(\Delta_r S > 0)$, the thermal energy dragged by the tools exceeds $-\Delta_r H$, $(Q_{\text{dragged}} > -\Delta_r H)$; in the case of $(\Delta_r S < 0)$, the dragged thermal energy is less than $-\Delta_r H$ $(Q_{\text{dragged}} < -\Delta_r H)$; in the case of $(\Delta_r S = 0)$ the energy extracted is equal to $-\Delta_r H$ $(Q_{\text{dragged}} = -\Delta_r H)$ and for $(\Delta_r H = 0)$ the dragged thermal energy is of volume $T\Delta_r S$ $(Q_{\text{dragged}} = T\Delta_r S)$. The volume of extracted thermal energy is controlled by chemical equilibrium via $\Delta_r G$.

Thus, the mixture in the reactor is moved off balance to be restored later. As a result, the reaction heat is emitted into the thermostat. Indeed, because of the elementary chemical act in the reactor, the energy released concentrates at the degrees of freedom of the product molecules. As the reactor temperature is constant, this energy is dissipated in the reactor and transferred to the thermostat which causes a $-\Delta_r H$ change in thermostat enthalpy. The total change in thermostat enthalpy is

$$\Delta H_{\text{thermostat}} = w_2 + w_5 - \Delta_r H. \tag{13}$$

The cycle is over. Equations (12, 13) can be used to calculate the total change in thermostat enthalpy

$$\Delta H_{\text{thermostat}} = \Delta_r G - \Delta_r H = -T\Delta_r S. \tag{14}$$

The change in thermostat enthalpy is controlled only by reaction entropy [11].

4.2. The main principles of reversible device functioning in useful work production

This consideration demonstrates the main characteristics of the reversible process of useful work production at constant temperature and pressure in closed systems:

1. The useful work arises from the stage of the reversible transport of reagents from reservoir to reactor and from the stage of the reversible transport of products from the reactor.
2. The only energy source of useful work is the thermal energy of thermostat (or environment).
3. The heat released by chemical reactions is dissipated to the thermostat; the reaction heat is infinitely small in comparison with the volume of the thermostat thermal energy; therefore no reaction heat is really needed to produce work.
4. Although the useful work is produced by the cooling of one body (thermostat), the second law of thermodynamics is not violated, because the process is followed by a change in the amount of reagents and products.
5. The useful work is produced by heat exchange with thermostat (environment) according to the scheme

   reaction heat → thermal thermostat energy → useful work (scheme I)

6. The maximal useful work is equal the heat dragged by tools from thermostat.
7. Useful work depends on the difference in the concentrations of standard and equilibrium states of reagents and products. Therefore, the amount of extracted energy can be calculated via the change in the Gibbs function.

8. There is no direct conversion of the Gibbs energy into useful work. Gibbs energy is equal numerically the thermal energy dragged by systems from the environment for doing work.

4.3. The energy limit of chemical reactions - Open systems

Usually, the total energy which can be produced by chemical system, is $-\Delta_r H$. However, this holds for closed systems only. For open systems, the case is quite different [11].

The open system is depicted in Fig. 2. As compared with the closed one (Fig. 1), the open system consists of two thermostats: the first one contains a reactor and the second one contains standard vessels with substances A and B and tools. The second thermostat can be replaced by the environment. In the process, steps 1, 2, 5, and 6 occur in the second thermostat (environment); steps 3 and 4 take place in the first one. Thus, the heat is released in the first thermostat only and the work is performed by thermal energy of the second thermostat (environment). The processes of heat and work production are spatially separated! The energy potential of the open system obeys the equation

$$q + w_{\text{useful}} = \Delta_r H + \Delta_r G \tag{15}$$

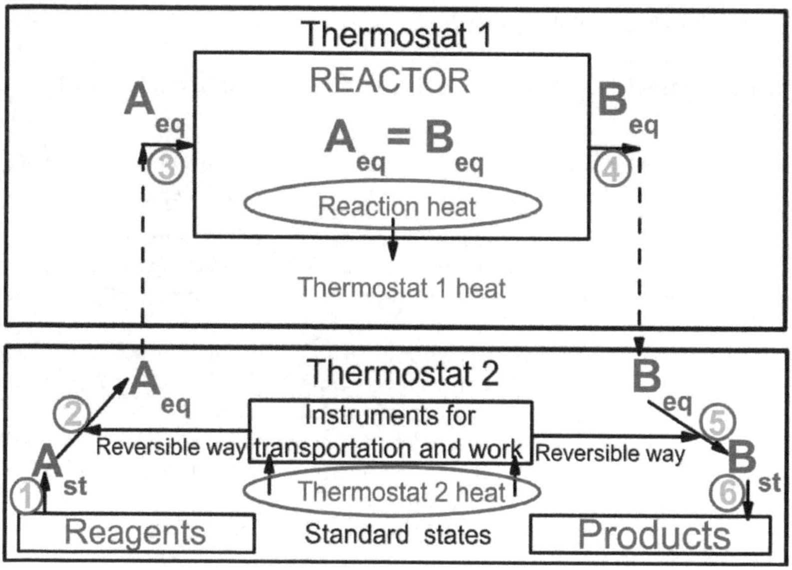

Figure 2. Production of useful work and heat in the open reversible system. The designations see in the subscription to Fig. 1

In the case of coal burning, it is possible to obtain the double total energy [11]. Thus, understanding the mechanism of useful work production in the reversible process allows us to predict an increase in the energy potential of chemical reactions in the open system.

It is worth noting that the open system under study is not a heat pump. The heat pump consumes energy to transfer heat from a cold body to the warm one. The open system studied does not consume external energy and produces heat due to chemical reactions in one thermostat and performs work by extraction of thermal energy from the second one.

4.4. Conclusions

The chemical reaction heat is always released in the reversible chemical processes and passes to the environment independent of the fact whether the system produces work or not, whether it is closed or open. The discussed mechanism of useful work production in the reversible systems did not use such notions as "free energy", "bound energy", "direct Gibbs energy conversion". The useful work arises only due to the heat exchange with a thermal basin in the process, described by the scheme I. The total energy of chemical system can be high and equal to $\Delta_r H + \Delta_r G$.

5. The mechanism of electric work production in a galvanic cell

The current theory of galvanic cells [1, 3, 8] is based on a direct transformation of the energy ($\Delta_r G$) of oxidation-reduction reactions into electric work. However, using VHEB as an example, It is clear that the energy of chemical reactions is first converted into the thermal energy of thermostat (environment) and then the thermal energy is extracted from the thermostat and transformed into work by means of special devices. It is assumed then that in galvanic cells, useful work is produced via the mechanism similar to the VHEB one [12, 13]. The $\Delta_r G$ value is used to calculate electric work which does not, however, mean that the electric work is performed at the expense of the Gibbs energy, all the more it was shown that the Gibbs energy is not energy. The electric work of a galvanic cell results from the electrodes discharged. Electric charging of electrodes is caused by chemical reactions in electrodes.

The mechanism of electric energy production in galvanic cells will be solved by analyzing the behavior of one ion. But it does not denote that thermodynamics will be applied to a real single ion: thermodynamic parameters of one ion imply the averaged parameters of many ions.

5.1. A galvanic cell

For simplicity a Daniell cell will be considered, consisting of zinc (№1) and copper (№2) electrodes (Fig. 3). The activity of salts in solutions is denoted by a_1 and a_2, respectively. Let the cell with an open, external circuit be in equilibrium. Close now the external circuit for the moment and two electrons will transfer from the zinc to the copper electrode. The balance of the cell is distorted. Consider now the establishment of equilibrium on the zinc electrode (Fig 3). To this end, the zinc ion must leave a metallic plate and escape into the bulk. The dissolving of zinc ions is described by the change in a Gibbs function

$$\Delta_r G_1 = \Delta_r G_1^o + RT \ln a_1, \tag{16}$$

where $\Delta_r G_1^o$ is a standard change in the Gibbs function upon dissolving. The ion penetrates further into solution with execution of the work (w_{1g}) in the electric field

$$w_{1g} = nF\Delta_{\mathrm{Met1}}^{\mathrm{sol}}\varphi_1, \tag{17}$$

where n is the number of electrons, participating in the reactions, F is the Faraday constant, and $\Delta_{\mathrm{Met1}}^{\mathrm{sol}}\varphi_1$ is the difference in potentials of solution and metal. The work described in equation (17), is the electric work spent to charge an electrode. It is performed at the extraction of the thermal energy of solution due to the absence of other energy sources in the system. Since in equilibrium, the chemical potential of ions in solution equals the chemical potential of metal, it is possible to derive the equation for electrochemical equilibrium

$$\Delta_r G_1^o + RT\ln a_1 + nF\Delta_{\mathrm{Met1}}^{\mathrm{sol}}\varphi_1 = 0, \tag{18}$$

which readily gives the expression for both the work performed on the first electrode and its galvanic potential [3]

$$w_{1g} = nF\Delta_{\mathrm{Met1}}^{\mathrm{sol}}\varphi_1 = -\Delta_r G_1^o - RT\ln a_1, \tag{19}$$

$$\Delta_{\mathrm{Met1}}^{\mathrm{sol}}\varphi_1 = -\frac{\Delta_r G_1^o}{nF} - \frac{RT\ln a_1}{nF}. \tag{20}$$

The latter is the potential for a half-cell. Thus, the approach, based on the consideration of the behavior of one ion, provides a common expression for electrode potential.

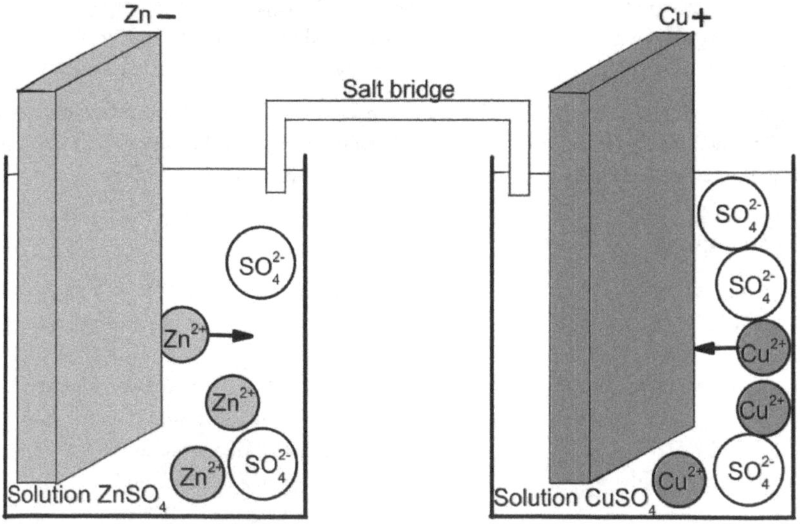

Figure 3. Establishment of equilibrium on the electrodes

The change in electrode enthalpy involves dissolution enthalpy and thermal energy consumption upon ion transport into solution. The equation for dissolution enthalpy is readily obtained from eq. (16)

$$\Delta_r H_1 = \Delta_r H_1^{\circ} - RT^2 \frac{\partial \ln a_1}{\partial T}, \tag{21}$$

where $\Delta_r H_1^{\circ}$ is a standard change in enthalpy during ion dissolving. The total change in the enthalpy of the first electrode ($\Delta H_{1,\text{thermostat}}$) is the sum of the expressions $-\Delta_r H_1$ and $-w_{g1}$

$$\Delta H_{1,\text{thermostat}} = -\Delta_r H_1 - w_{g1} = -T\Delta_r S_1^{\circ} + RT\ln a_1 + RT^2 \frac{\partial \ln a_1}{\partial T} = -T\Delta_r S_1, \tag{22}$$

where $\Delta_r S_1^{\circ}$ is a standard change in entropy upon ion dissolving and $\Delta_r S_1$ is the change in entropy upon ion dissolving on the first electrode which amounts to

$$\Delta_r S_1 = \Delta_r S_1^{\circ} - R\ln a_1 - RT \frac{\partial \ln a_1}{\partial T}. \tag{23}$$

As follows from eq. (22), the change in enthalpy, related to the first electrode, is independent of the processes, occurring on the second one. Therefore, studying either release or absorption of heat on a separate electrode, one may calculate the change in entropy due to the escape of the ions of the same type into the bulk.

A corresponding expression for the second electrode is of the same form but index "1" should be substituted by index "2":

$$\Delta_r G_2 = \Delta_r G_2^{\circ} + RT\ln a_2, \tag{24}$$

$$w_{2g} = nF\Delta_{\text{Met2}}^{\text{sol}}\varphi_2 = -\Delta_r G_2^{\circ} - RT\ln a_2 \tag{25}$$

$$\Delta_{\text{Met2}}^{\text{sol}}\varphi_2 = -\frac{\Delta_r G_2^{\circ}}{nF} - \frac{RT\ln a_2}{nF}. \tag{26}$$

$$\Delta_r H_2 = \Delta_r H_2^{\circ} - RT^2 \frac{\partial \ln a_2}{\partial T}, \tag{27}$$

$$\Delta_r S_2 = \Delta_r S_2^{\circ} - R\ln a_2 - RT \frac{\partial \ln a_2}{\partial T}, \tag{28}$$

$$\Delta H_{2,\text{thermostat}} = -\Delta_r H_2 - w_{2g} = -T\Delta_r S_2^{\circ} + RT\ln a_2 + RT^2 \frac{\partial \ln a_2}{\partial T} = -T\Delta_r S_2. \tag{29}$$

In the operation of the galvanic cell, the processes on the second electrode are oppositely directed which should be taken into account in consideration of the thermodynamic cell parameters.

Equations (20) and (26) allow to get the Nernst equation for the potential of the cell [3]

$$E = -\Delta_{\mathrm{Met2}}^{\mathrm{sol}}\varphi_2 + \Delta_{\mathrm{Met1}}^{\mathrm{sol}}\varphi_1 = -\frac{\Delta_r G^o}{nF} - \frac{RT}{nF}\ln\frac{a_1}{a_2}, \quad (30)$$

where E – the cell potential, $\Delta_r G^o = \Delta_r G_1^o - \Delta_r G_2^o$.

The electric work of the galvanic cell (w_{el}) results from the transformation of the potential energy of the charged electrodes into electric energy. The potential energy arises from the thermal energy of both of the electrodes upon ions transport into solution and equals

$$w_{\mathrm{el}} = w_{1g} - w_{2g} = -\Delta_r G^o - RT\ln\frac{a_1}{a_2} = -w_{\mathrm{useful}}. \quad (31)$$

The change in thermostat enthalpy is of the form

$$\Delta H_{\mathrm{thermostat}} = \Delta H_{1,\mathrm{thermostat}} - \Delta H_{2,\mathrm{thermostat}} = -T\Delta_r S, \quad (32)$$

which is in fair agreement with similar expression, described by eq. (14), for the VHEB. The detailed equation for $\Delta H_{\mathrm{thermostat}}$ can be get after substitution corresponding expressions (22) and (29) into (32)

$$\Delta H_{\mathrm{thermostat}} = -T(\Delta_r S_1^o - \Delta_r S_2^o) - RT\ln(a_2 / a_1) - RT^2\left(\frac{\partial \ln a_2}{\partial T} - \frac{\partial \ln a_1}{\partial T}\right). \quad (33)$$

The sum of eqs. (31) and (33) gives the total energy (electric work + heat), produced by the galvanic cell

$$w_{\mathrm{el}} + \Delta H_{\mathrm{thermostat}} = -\Delta_r H^o - RT^2\left(\frac{\partial \ln a_2}{\partial T} - \frac{\partial \ln a_1}{\partial T}\right). \quad (34)$$

From eq.(34) it follows that the total energy produced by the galvanic cell is equal to the heat emitted by oxidation-reduction reaction.

Thus, the approach, based on the analysis of the behavior of one ion gives the same results as the present-day theory. However, it uses not a mysterious, direct transformation of the chemical energy ($\Delta_r G$) into electric work, but the concept of chemical energy conversion into the thermal one, and then, the thermal energy of thermostat (environment) is converted into the potential energy of charged electrodes [12, 13]. The electric energy of the galvanic cell arises according to the scheme:

reaction heat → thermal thermostat energy → potential energy of charged electrodes → electric energy.

Thus, in various systems with uniform temperature, useful work is produced by the same mechanism through the exchange of thermal energy with thermostat (environment). No direct conversion of chemical energy into useful work is observed. Unfortunately, in the galvanic cell, the processes of heat release and useful work production cannot be spatially separated, because both of them occur in a double layer. Therefore, galvanic cells are unpromising in production of a double amount of energy.

5.2. A concentration cell

Consider now a concentration cell, consisting of two electrodes, e.g., the zinc ones of different solution activity. Standard changes in the Gibbs function, enthalpy, and entropy for the concentration cell tend to zero due to the same chemical nature of both of the electrodes. By definition, it has been considered that $a_2 > a_1$. From eq. (31) it follows

$$w_{\text{el}} = RT\ln(a_2 / a_1) = -w_{\text{useful}}, \tag{35}$$

which is a usual expression for the electric energy of the concentration cell. From eq. (34) it follows

$$w_{\text{el}} + \Delta H_{\text{thermostat}} = -RT^2\left(\frac{\partial \ln a_2}{\partial T} - \frac{\partial \ln a_1}{\partial T}\right). \tag{36}$$

For the system in which the activities are temperature-independent, the electric energy arises from the thermal thermostat energy (environment)

$$w_{\text{el}} = -\Delta H_{\text{thermostat}} = -w_{\text{useful}}, \tag{37}$$

which is in fair agreement with conventional concepts.

6. Useful work of the systems with concentration gradient

An ideal system with concentration gradient has no potential energy because the mixing does not result in heat release and $\Delta H = 0$. Nevertheless, the system with concentration gradient can be used, as any non-equilibrium system, to produce useful work if this system is supplied with special tools for extracting heat from the environment with a simultaneous transformation of the extracted thermal energy into work upon the system approaches to the equilibrium. The volume of useful work is equal to the volume of the heat extracted from the environment

$$w_{\text{useful}} = \Delta H_{\text{thermostat}} \tag{38}$$

The useful work of the system with a concentration gradient w_{useful} obeys expressions (35) and (38). The concentration cell is a good example of such a system.

7. General conclusions

Any non-equilibrium state can serve as an energy source. The thermostat (environment) is an active participant of the process of reversible useful work production in devices operating at constant temperature. The heat released by chemical reactions, always dissipates in the thermostat (environment). The useful work is produced by special tools that provide the extraction of the thermal energy of the thermostat (environment) and the transformation of thermal energy into work at the process of the restoration of the chemical equilibrium. The volume of the useful work is equal, in reversible conditions, to the change in Gibbs function. A spatial separation of reactor and tools can lead to a substantial increase in the energy produced. The direct conversion of the Gibbs energy into useful work does not exist. The concepts of free and bound energy become unnecessary.

Author details

Nikolai Bazhin
Institute of Chemical Kinetics and Combustion, Novosibirsk State University, Institutskaya 3, Novosibirsk, Russia

8. References

[1] Denbigh K (1971) The Principles of Chemical Equilibrium, 3rd ed., The University Press: Cambridge, 494 p.
[2] IUPAC Green Book (1996) Quantities, Units and Symbols in Physical Chemistry 48 p.
[3] Atkins P (2001) The Elements of Physical Chemistry, 3rd ed., Oxford University Press, Oxford,.549 p.
[4] Lower S. (2010) Chemistry. Available http://www.chem1.com/acad/webtext/virtualtextbook.html
[5] Gokcen NA, Reddy RG (1996) Thermodynamics, Second edition, Springer, 416 p.
[6] Haynie DN (2008), Biological Thermodynamics, 2nd ed., Cambridge University Press, 422 p.
[7] Haywood RW (1980) Equilibrium Thermodynamics for Engineers and Scientists, J. Wiley, N.Y., 456 p.
[8] Adam NK (1956) Physical Chemistry, Oxford, Clarendon Press, 658 p.
[9] Physical Chemistry (1980) Gerasimov YaI, Ed., Chimiya, Moscow (in Russian), 1279 p.
[10] Steiner LE (1948) Introduction to Chemical Thermodynamic, 2nd ed.; McGraw-Hill Book Company, New York, 516 p.
[11] Bazhin NM, Parmon VN (2007) Conversion of the Chemical Reaction Energy into Useful Work in the Van't Hoff Equilibrium Box. J. Chem. Ed. 84: 1053 – 1055
[12] Bazhin NM, Parmon VN (2007) Ways of Energy Conversion in Electrochemical Cells. Doklady Physical Chemistry 417: 335-336
[13] Bazhin NM (2011) Mechanism of electric energy production in galvanic and concentration cells. J. Eng. Thermophysics 20: 302-307

Chapter 2

A View from the Conservation of Energy to Chemical Thermodynamics

Ahmet Gürses and Mehtap Ejder-Korucu

Additional information is available at the end of the chapter

1. Introduction

According to the **conservation of energy law**, **energy**, which is the capacity to do work or to supply heat, can be neither created nor destroyed; it can only be converted from one form into another. For example, the water in a reservoir of dam has potential energy because of its height above the outlet stream but has no kinetic energy because it is not moving. As the water starts to fall over the dam, its height and **potential energy**, *(Ep)* is energy due to position or any other form of stored energy, decrease while its velocity and **kinetic energy**, *(E_K)* is the energy related to the motion of an object with mass m and velocity v, increase. The total of potential energy plus kinetic energy always remains constant. When the water reaches the bottom and dashes against the rocks or drives the turbine of a generator, its kinetic energy is converted to other forms of energy-perhaps into heat that raises the temperature of the water or into electrical energy [6]. If any fuel is burned in an open medium, its energy is lost almost entirely as heat, whereas if it is burned in a car engine; a portion of the energy is lost as work to move the car, and less is lost as heat. These are also typical examples of the existence of the law.

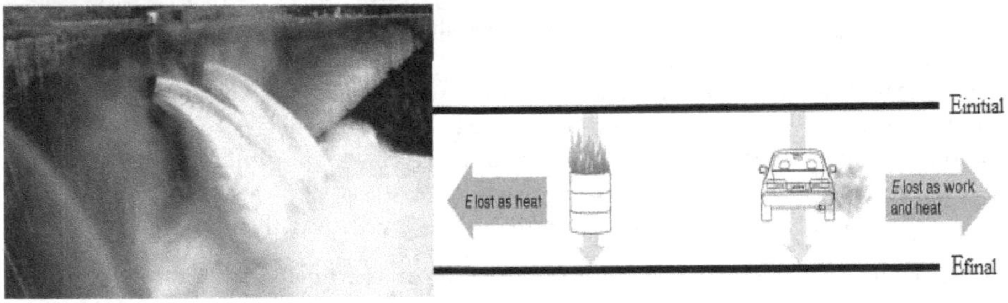

Figure 1. Some examples showing the existence of the conservation of energy law

It is thought that the summation of the introduction as a detailed concept map related with the conservation of energy would be better. This map in Fig.2 presents a concise view for many concepts of thermodynamics and their relations.

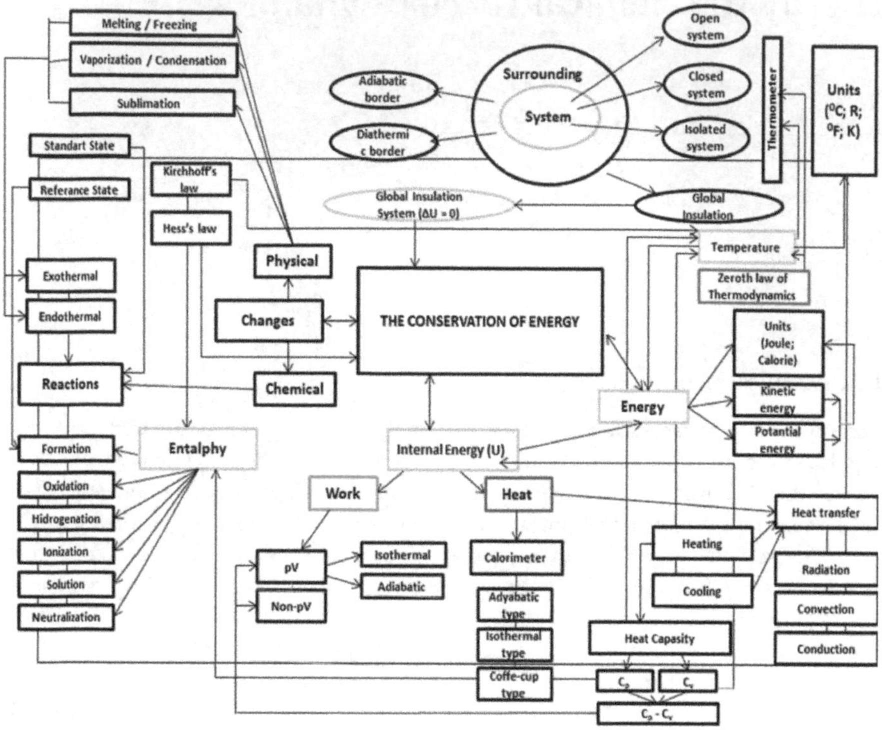

Figure 2. A concept map for the conservation of energy law

The macroscopic part of universe under study in thermodynamics is called the system. The parts of the universe that can interact with the system are called the surroundings [5]. In order to describe the thermodynamic behavior of a physical system, the interaction between the system and its surroundings must be understood. Thermodynamic systems are thus classified into three main types according to the way they interact with the surroundings (Fig.3); **open system:** matter and energy can be exchanged with the surroundings; **closed system:** can exchange energy but not matter with the surroundings and **isolated system:** cannot exchange matter or energy with the surroundings [4, 5]. Addition, a boundary that does permit energy transfer as heat (such as steel and glass) is called **diathermic** and a boundary that does not permit energy transfer as heat is called **adiabatic** [1;4].

The system may be homogeneous or heterogeneous. An exact definition is difficult, but it is convenient to define a **homogeneous system** as one whose properties are the same in all parts, or at least their spatial variation is continuous. A **heterogeneous system** consists of two or more distinct homogeneous regions or **phases**, which are separated from one another by surfaces of discontinuity [4].

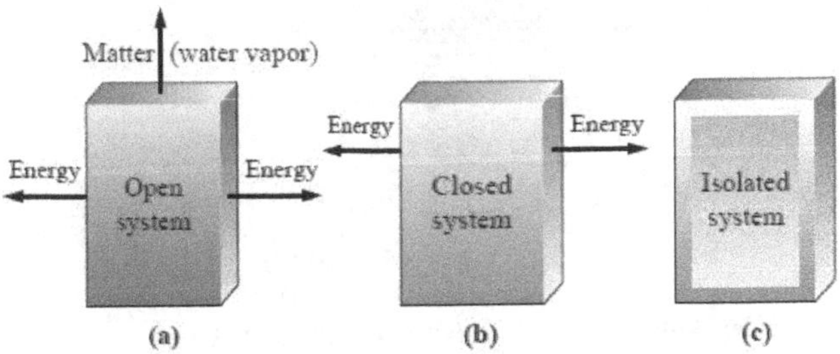

Figure 3. Systems and their surroundings; (a) open system (b) closed system and (c) isolated system.

Heating is a process in which the temperature of system, separated with diathermic border from its surrounding, is increased. This process leads to passing system from a state of lower energy to higher one. Heating process based on the energy difference between system and its surrounding provides identify of an important property which indicates the flow direction of energy. This property is called **temperature.** Temperature is a physical property of matter that quantitatively expresses the common notions of hot and cold. Objects of low temperature are cold, while various degrees of higher temperatures are referred to as warm or hot. Heat spontaneously flows from bodies of a higher temperature to bodies of lower temperature, at a rate that increases with the temperature difference and the thermal conductivity. No heat will be exchanged between bodies at same temperature; such bodies are said to be in "**equilibrium**". On the other hand, kinetic energy associated with the random motion of particles is called thermal energy, and the thermal energy of a given material is proportional to temperature. However, the magnitude of thermal energy in a sample also depends on the number of particles in the sample and so it is an **extensive property**. The water in a swimming pool and a cup of water taken from the pool has the same temperature, so their particles have the same average kinetic energy. The water in the pool has much more thermal energy than the water in the cup, simply because there are a larger number of molecules in the pool. A large number of particles at a given temperature have a higher total energy than a small number of particles at the same temperature [7]. Quantitatively, temperature which is an **intensive property** is measured with thermometers, which may be calibrated to a variety of temperature scales [9].

If two thermodynamic systems, A and B, each of which is in **thermal equilibrium** independently, are brought into thermal contact, one of two things will take place: either a flow of heat from one system to the other or no thermodynamic process will result. In the latter case the two systems are said to be in thermal equilibrium with respect to each other [11]. When same treatment has been repeated other system, C, if there is thermal equilibrium between B and C; the condition of thermodynamic equilibrium between them may be symbolically represented as follows,

$$\text{If } A = B \text{ and } B = C \tag{1}$$

This observation has also been schematically shown in Fig.4

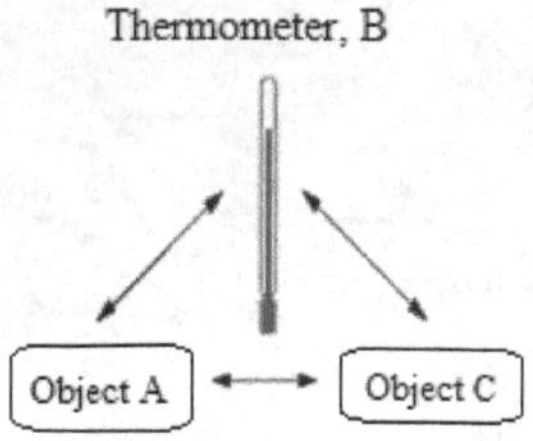

Figure 4. A schematic presentation of the zeroth law of thermodynamics.

Based on preceding observations, some of the physical properties of the system B (Thermometer*)* can be used as a measure of the temperature, such as the volume of a fixed amount of a liquid merqury or any alcohol under standard atmospheric pressure. The **zeroth law of thermodynamics** is the assurance of the existence of a property called the temperatur*e* [11].

The zeroth law allows us to assert the existence of temperature as a state function. Having defined temperature, how do we measure it? Of course, you are familiar with the process of putting a liquid-mercury thermometer in contact with a system, waiting until the volume change of the mercury has ceased, indicating that thermal equilibrium between the thermometer and the system has been reached [5].

It is necessary to know four common different temperature scales, namely Fahrenheit (^{0}F), Celsius (^{0}C), Kelvin (K) and Rankine (^{0}R)**.** To convert these scales one another can be used the following equations [9,10].

$$T / K = \theta / ^0C + 273.15 \tag{2}$$

$$\left[\mathrm{oR}\right] = \left[\mathrm{oF}\right] + 459.67\left[\mathrm{oF}\right] = 1.8\left[\mathrm{oC}\right] + 32\left[\mathrm{oR}\right] = 1.8\left[\mathrm{K}\right] \tag{3}$$

2. The first law of thermodynamic (the conservation of energy)

In thermodynamics, the total energy of a system is called its **internal energy**, *U*. The internal energy is the total kinetic and potential energies of the particles in the system. It is denoted by ΔU the change in internal energy when a system changes from an initial state i with internal energy U_i to a final state of internal energy U_f :

$$\Delta U = U_f - U_i \tag{4}$$

The internal energy is a state function in the sense that its value depends only on the current state of the system and is independent of how that state has been prepared. In other words, internal energy is a function of the properties as variables that determine the current state of the system. Changing any one of the state variables, such as the pressure and temperature, results in a change in internal energy. The internal energy is a state function that has consequences of the greatest importance. The internal energy is an extensive property of a system and is measured in joules (1 J = 1 kg m^2 s^{-2}). The molar internal energy, U_m, is the internal energy divided by the amount of substance in a system, $U_m = U/n$; it is an intensive property and commonly reported in kilojoules per mole (kJ mol^{-1}) [12].

A particle has a certain number of motional degrees of freedom, such as the ability to translate (the motion of its centre of mass through space), rotate around its centre of mass, or vibrate (as its bond lengths and angles change, leaving its centre of mass unmoved). Many physical and chemical properties depend on the energy associated with each of these modes of motion. For example, a chemical bond might break if a lot of energy becomes concentrated in it, for instance as vigorous vibration. According to it, the average energy of each quadratic contribution to the energy is 1/2 kT. The mean energy of the atoms free to move in three dimensions is kT and the total energy of a monatomic perfect gas is NkT, or nRT (because $N = nNA$ and $R = NAk$, NA: Avogadro's number and k: Boltzman's constant). I t can therefore be written;

$$U_m(T) = U_m(0) + RT(\textit{monatomic gas; translation only}) \tag{5}$$

where $U_m(0)$ is the molar internal energy at $T = 0$, when all translational motions have ceased and the sole contribution to the internal energy arises from the internal structure of the atoms. This equation shows that the internal energy of a perfect gas increases linearly with temperature. At 25°C, 3/2 RT = 3.7 kJ mol^{-1}, so translational motion contributes about 4 kJ mol^{-1} to the molar internal energy of a gaseous sample of atoms or molecules.

When the gas consists of molecules, we need to take into account the effect of rotation and vibration. A linear molecule, such as N_2 and CO_2, can rotate around two axes perpendicular to the line of the atoms, so it has two rotational modes of motion, each contributing a term 1/2 kT to the internal energy. Therefore, the mean rotational energy is kT and the rotational contribution to the molar internal energy is RT.

$$U_m(T) = U_m(0) + 5/2RT(\textit{linear molecule; translation and rotation only}) \tag{6}$$

A nonlinear molecule, such as CH_4 or H_2O, can rotate around three axes and, again, each mode of motion contributes a term 1/2 kT to the internal energy. Therefore, the mean rotational energy is 3/2 kT and there is a rotational contribution of 3/2 RT to the molar internal energy. That is,

$$U_m(T) = U_m(0) + 3RT(\textit{nonlinear molecule; translation and rotation only}) \tag{7}$$

The internal energy now increases twice as rapidly with temperature compared with the monatomic gas. Another way: for a gas consisting of 1 mol of nonlinear molecules to undergo the same rise in temperature as 1 mol of monatomic gas, twice as more energy must be supplied. Molecules do not vibrate significantly at room temperature and, as a first approximation; the contribution of molecular vibrations to the internal energy is negligible except for very large molecules such as polymers and biological macromolecules. None of the expressions we have been derived depends on the volume occupied by the molecules: there are no intermolecular interactions in a perfect gas, so the distance between the molecules has no effect on the energy. That is, the internal energy of a perfect gas is independent of the volume it occupies. The internal energy of interacting molecules in condensed phases also has a contribution from the potential energy of their interaction. However, no simple expressions can be written down in general. Nevertheless, the crucial molecular point is that, as the temperature of a system is raised, the internal energy increases as the various modes of motion become more highly excited [12].

By considering how the internal energy varies with temperature when the pressure of the system is kept constant; many useful results and also some unfamiliar quantities can be obtained. If it is divided both sides of eqn (($dU = (\partial U/\partial V)_T\, dV + (\partial U/\partial T)_v\, dT$) (($(\partial U/\partial V)_T = \pi_T$ and π_T called as **the internal pressure** ; $(\partial U/\partial T)_v = C_v$ and it is called as **heat capacity at constant volume**)) by dT and impose the condition of constant pressure on the resulting differentials, so that dU/dT on the left becomes $(\partial U/\partial T)_p$, So;

$$(\partial U / \partial T)_p = \pi_T (\partial V / \partial T)_p + C_v \tag{8}$$

It is usually sensible in thermodynamics to inspect the output of a manipulation like this to see if it contains any recognizable physical quantity. The differential coefficient on the right in this expression is the slope of the plot of volume against temperature (at constant pressure). This property is normally identified as thermal expansion coefficient, α, of a substance, which is defined as

$$\alpha = 1/V \, (\partial V / \partial T)_p \tag{9}$$

When it is introduced the definition of α into the equation for $(\partial U/\partial T)_p = \alpha \pi_T V + C_v$,this equation is entirely general (provided the system is closed and its composition is constant). It expresses the dependence of the internal energy on the temperature at constant pressure in terms of C_v, which can be measured in one experiment, in terms of α, which can be measured in another, and in terms of the quantity π_T, for a perfect gas, $\pi_T = 0$, so

$$(\partial U / \partial T)_p = C_v \tag{10}$$

That is, the constant-volume heat capacity of a perfect gas is equal to the slope of a plot of internal energy against temperature at constant pressure as well as (by definition) to the slope at constant volume. It can also be predicted that the change of internal energy with temperature at constant pressure means a total energy change raised from increase in both energy of kinetics and potential energy of particles in higher temperature. The translational motion of particles against constant external pressure will lead to expansion, but thermal expansion characteristics of substance control its magnitude. Thus, we can see that heating in constant volume only changes internal energy as q_v ($\Delta U = q_v$), whereas its change in constant pressure additionally includes changing of potential energy of particles due to translation motion. That is, changes in constant pressure require a different definition of the transferred energy [12].

3. From conservation of energy to heat and work

It has been found experimentally that the internal energy of a system may be changed either by doing work on the system or by heating it. Whereas we may know how the energy transfer has occurred (because we can see if a weight has been raised or lowered in the surroundings, indicating transfer of energy by doing work, or if ice has melted in the surroundings, indicating transfer of energy as heat), the system is blind to the mode employed. Heat and work are equivalent ways of changing a system's internal energy. It is also found experimentally that, if a system is isolated from its surroundings, then no change in internal energy takes place. This summary of observations is now known as the **First Law of thermodynamics** or the **Conservation of Energy** and is expressed as follows [1;12].

The internal energy of an isolated system is constant.

$$\Delta U = 0 \tag{11}$$

A system cannot be used to do work, leave it isolated, and then come back expecting to find it restored to its original state with the same capacity for doing work. The experimental evidence for this observation is that no 'perpetual motion machine', a machine that does work without consuming fuel or using some other source of energy, has ever been built. These remarks may be summarized as follows. If we write w for the work done on a system, q for the energy transferred as heat to a system, and ΔU for the resulting change in internal energy, and then it follows that

$$\Delta U = q + w \tag{12}$$

Equation summarizes the equivalence of heat and work and the fact that the internal energy is constant in an isolated system (for which $q = 0$ and $w = 0$). The equation states that the change in internal energy of a closed system is equal to the energy that passes through its boundary as heat or work. It employs the 'acquisitive convention', in which w and q are positive if energy is transferred to the system as work or heat and are negative if energy is lost from the system. In other words, we view the flow of energy as work or heat from the system's perspective [1;12].

Heat (q)

Heat flows by virtue of a temperature difference. Heat will flow until temperature gradients disappear [8]. When a heater is immersed in a beaker of water (the system), the capacity of the system to do work increases because hot water can be used to do more work than cold water [1].

An **exothermic process** is a process that releases energy as heat. All combustion reactions are exothermic. An **endothermic process** is a process that absorbs energy as heat. An example of an endothermic process is the vaporization of water. An endothermic process in a diathermic container results in energy flowing into the system as heat. An exothermic process in a similar diathermic container results in a release of energy as heat into the surroundings. If an endothermic process in nature was taken place by a system divided an adiabatic boundary from its surroundings, a lowering of temperature of the system results; conversely if the process is exothermic, temperature rises [1]. These are the expected results of conversation of energy.

Work (*w*)

Energy is the essence of our existence as individuals and as a society. Just as energy is important to our society on a macroscopic scale, it is critically important to each living organism on a microscopic scale. The living cell is a miniature chemical factory powered by energy from chemical reactions. The process of cellular respiration extracts the energy stored in sugars and other nutrients to drive the various tasks of the cell. Although the extraction process is more complex and more subtle, the energy obtained from "fuel" molecules by the cell is the same as would be obtained from burning the fuel to power an internal combustion engine [3].

The fundamental physical property in thermodynamics is work is done when an object is moved against an opposing force. Doing work is equivalent into raising a weight somewhere in the surrounding. An example of doing work is the expansion of a gas that pushes out a piston and raises a weight. A chemical reaction that derives an electric current through a resistance also does work, because the same current could be driven through a motor and used to raise a weight [1].

Work is the transferred energy by virtue of a difference in mechanical properties from a boundary between the system and the surroundings. There are many types of work; such as mechanical work, electrical work, magnetic work, and surface tension [8].

The SI unit of both heat and work (kg m^2/s^2) is given the name joule (J), after the English physicist James Prescott Joule (1818-1889) [6].

$$1\ \mathrm{J} = 1\ \mathrm{kg\ m^2/s^2}$$

In addition to the SI unit joule, some chemist's still use the unit calorie (cal). Originally defined as the amount of energy necessary to raise the temperature of 1 g of water by 1°C (specially, from 14.5 °C to 15.5 °C), one calorie is now defined as exactly 4,184 J [6].

The work–energy theorem

In mechanics, the work–energy theorem demonstrates that the total work done on a system is transformed into kinetic energy. This is represented in a very simple and meaningful equation as follows:

$$W_{total} = \Delta E_k \tag{13}$$

in which, W_{total} is the total work done on the system, including the work carried out by all the external forces (W_{ex}), as well as the work developed by the internal forces within the system (W_{in}). Thus;

$$W_{total} = W_{ex} + W_{in} \tag{14}$$

Now, if the external work is separated into two terms, namely the work done by the external conservative forces ($W_{ex,c}$), which are associated with an external potential energy ($Ep_{,ex}$), and the non-conservative external work ($W_{non-¢ex}$), the whole external work can be written as:

$$W_{ex} = W_{ex,c} + W_{non-¢ex} = -\Delta E_{p,ex} + W_{non-¢ex} \tag{15}$$

Similarly, the work developed by the internal forces within the system can be also expressed as the sum of a conservative work term plus the non-conservative internal work. Thus;

$$W_{in} = W_{in,c} + W_{non-¢'in} = -\Delta E_{p,in} + W_{non-¢in} \tag{16}$$

Where, $\Delta E_{p,in}$ is the internal potential energy of the system.

As for the kinetic energy of a system, mechanics shows that it can be considered as consisting of two terms, as follows:

$$E_k = 1/2Mv^2{}_{CM} + E_{k,CM} \tag{17}$$

M being the total mass of the system,v_{CM} the velocity of its center of mass, and $E_{k,CM}$ the kinetic energy of the system with respect to its center of mass. The first term on the right-hand side of equation above represents the kinetic energy of the center of mass of the system, as if it had the mass of the whole system. Thus, as the velocity v_{CM} is taken with respect to an external reference frame, this first term can be called the external kinetic energy of the system ($E_{k,ex}$), whereas the second term would be its internal kinetic energy ($E_{k,in}$). Accordingly, the increase of the kinetic energy of a system can be written in the following way:

$$\Delta E_k = \Delta E_{k,ex} + \Delta E_{k,in} \tag{18}$$

Then, the substitution of equations 14, 15, 16 and 18 into equation 13 allows us to order terms as follows:

$$W_{ex} = \Delta E_{p,ex} + \Delta E_{k,ex} + \Delta E_{p,in} + \Delta E_{k,in} - W_{in} \tag{19}$$

Equation 19 is a general developed expression of the work–energy theorem derived from mechanics. It should be noticed that, though it does not describe the details of the energetic terms, each of them is explicitly stated, which will be of great help both to define and to understand the contribution of thermodynamics when establishing the first law [17].

Reversibility and Reversible changes

A reversible change in thermodynamics is a change that can be reversed by an infinitesimal modification of a variable. One example of reversibility that we have encountered already is the thermal equilibrium of two systems with the same temperature. The transfer of energy as heat between the two is reversible because, if the temperature of either system is lowered infinitesimally, then energy flows into the system with the lower temperature. If the temperature of either system at thermal equilibrium is raised infinitesimally, then energy flows out of the hotter system. There is obviously a very close relationship between reversibility and equilibrium: systems at equilibrium are poised to undergo reversible change. Suppose a gas is confined by a piston and that the external pressure, p_{ex}, is set equal to the pressure, p, of the confined gas. Such a system is in mechanical equilibrium with its surroundings because an infinitesimal change in the external pressure in either direction causes changes in volume in opposite directions. If the external pressure is reduced infinitesimally, the gas expands slightly. If the external pressure is increased infinitesimally, the gas contracts slightly. In either case the change is reversible in the thermodynamic sense. If, on the other hand, the external pressure differs measurably from the internal pressure, then changing p_{ex} infinitesimally will not decrease it below the pressure of the gas, so will not change the direction of the process. Such a system is not in mechanical equilibrium with its surroundings and the expansion is thermodynamically irreversible [12].

To achieve reversible expansion we set p_{ex} equal to p at each stage of the expansion. In practice, this equalization could be achieved by gradually removing weights from the piston so that the downward force due to the weights always matches the changing upward force due to the pressure of the gas. When we set $p_{ex} = p$, eqn ($dw = -p_{ex}dV$) becomes

$$dw = -p_{ex}dV = -pdV \qquad \textit{reversible expansion work} \tag{20}$$

(Equations valid only for reversible processes are labeled with a subscript rev.) Although the pressure inside the system appears in this expression for the work, it does so only because p_{ex} has been set equal to p to ensure reversibility. The total work of reversible expansion from an initial volume V_i to a final volume V_f is therefore

$$w = -\int_{Vi}^{Vf} pdV \tag{21}$$

The integral can be evaluated once it is known how the pressure of the confined gas depends on its volume [12].

At the isothermal, reversible expansion of a perfect gas, the work made by keeping the system in thermal contact with its surroundings can be stated as follows;

$$w = -nRT\int_{Vi}^{Vf} dV / V = -nRT\ln V_f / V_i \tag{22}$$

When the final volume is greater than the initial volume, as in an expansion, the logarithm in Eqn. 22 is positive and hence $w < 0$. In this case, the system has done work on the surroundings and there is a corresponding reduction in its internal energy, but due to there is a compensating influx of energy as heat, overall the internal energy is constant for the isothermal expansion of a perfect gas, which clearly indicates **the Conservation of Energy Law.** The equations also show that more work is done for a given change of volume when the temperature is increased. at a higher temperature the greater pressure of the confined gas needs a higher opposing pressure to ensure reversibility and the work done is correspondingly greater (Fig. 5) [12].

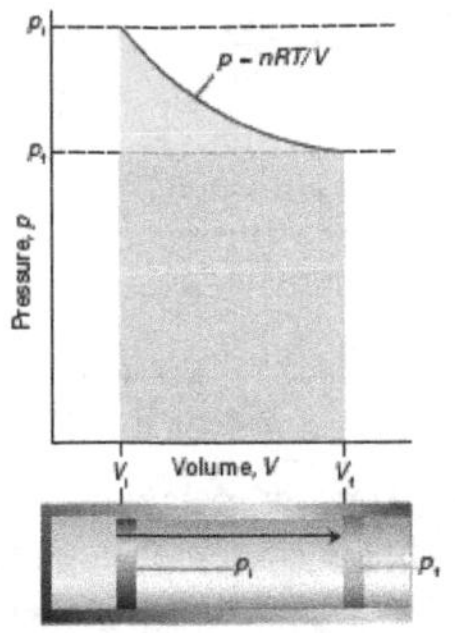

Figure 5. The work done by a perfect gas when it expands reversibly and isothermally is equal to the area under the isotherm $p = nRT/V$. The work done during the irreversible expansion against the same final pressure is equal to the rectangular area shown slightly darker. It can be seen that the reversible work is greater than the irreversible work [12].

Adiabatic changes

We are now equipped to deal with the changes that occur when a perfect gas expands adiabatically. A decrease in temperature should be expected: because work is done but no heat enters the system, the internal energy decrease, and therefore the temperature of the working gas also decrease. In molecular terms, the kinetic energy of the particles decrease as work is done, so their average speed decreases, and hence the temperature decrease. This means that in the case of perfect gas, change in the distance between particles cannot be responsible for the changing of internal energy but **the Conservation of Energy Law** requires a measurable reduction in kinetic energy of particles, i.e. a reduction in their velocities. The change in internal energy of a perfect gas when the temperature is changed from T_i to T_f and the volume is changed from V_i to V_f can be expressed as the sum of two steps (Fig. 6).

In the first step, only the volume changes and the temperature is held constant at its initial value. However, because the internal energy of a perfect gas is independent of the volume the molecules occupy, the overall change in internal energy arises solely from the second step, the change in temperature at constant volume. Provided the heat capacity is independent of temperature, this change is

$$\Delta U = C_V\left(T_f - T_i\right) = C_V \Delta T \quad (23)$$

Because the expansion is adiabatic, we know that $q = 0$; because $\Delta U = q + w$, it then follows that ΔU = ad. The subscript 'ad' denotes an adiabatic process. Therefore, by equating the two expressions we have obtained for ΔU, we obtain

$$W_{ad} = C_V \Delta T \quad (24)$$

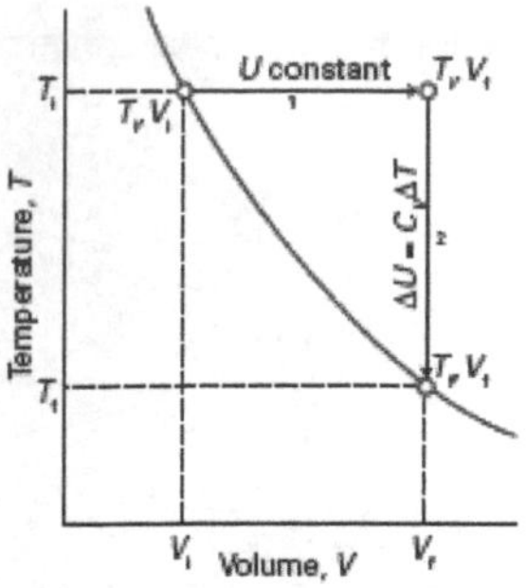

Figure 6. To achieve a change of state from one temperature and volume to another temperature and volume, we may consider the overall change as composed of two steps. In the first step, the system expands at constant temperature; there is no change in internal energy if the system consists of a perfect gas. In the second step, the temperature of the system is reduced at constant volume. The overall change in internal energy is the sum of the changes for the two steps [12].

That is, the work done during an adiabatic expansion of a perfect gas is proportional to the temperature difference between the initial and final states. That is exactly what it can be expected on molecular grounds and according to **Conversation of Energy**, because the mean kinetic energy is proportional to *T*, so a change in internal energy arising from temperature alone is also expected to be proportional to ΔT [12].

$$T_f = Ti\left(V_i / V_f\right)^{1/c} \tag{25}$$

where $c = \mathbf{Cv}\ /R$ *(Cv :molar heat capacity in constant volume and R: universal gas constant)*. By raising each side of this expression to the power *c*, an equivalent expression is

$$V_i T_i^c = V_f T_f^{\,c} \tag{26}$$

This result is often summarized in the form VT^c = constant

4. From the conservation of energy to heat transfer, heat capacity and the enthalpy

The process of heat moving from one object into another is called heat transfer. The difference in temperature defines the direction in which the heat flows when two objects come into contact; heat always flows from a hotter object at a higher temperature into a colder object at a lower temperature. Heat transfer which can mainly occur in three ways, namely conduction, convection and radiation changes the temperature of matter; it can also cause changes in phase or state [7].

When energy is added to a substance and no work is done, the temperature of the substance usually rises i.e. substance is heated; exception to the case in which a substance undergoes a change of state-also called a phase transition, such as vaporization/condensation, melting/freezing, and sublimation. The quantity of energy required to raise the temperature of given mass of a substance by some amount varies from one substance to another. The amount of increase also depends on the conditions in which heating takes place. Heat capacity (C) of a substance is the quantity of energy required to raise its temperature by 1°C or 1 K. The heat capacity differs from substance to substance. In the case of a monatomic gas such as helium under constant volume, if it is assumed that no electronic or nuclear quantum excitations occur, each atom in the gas has only 3 degrees of freedom, all of a translational type. No energy dependence is associated with the degrees of freedom which define the position of the atoms. While, in fact, the degrees of freedom corresponding to the momenta of the atoms are quadratic, and thus contribute to the heat capacity. In the somewhat more complex case of a perfect gas of diatomic molecules, the presence of internal degrees of freedom is apparent. Table1 shows C_v values for some mono and diatomic gases at 1 atm and 25°C.

Monoatomic Gases	C_v (J/Kmol)	Diatomic Gases	C_v (J/Kmol)
He	12.5	H_2	20.18
Ne	12.5	CO	20.2
Ar	12.5	N_2	19.9
Kr	12.5	Cl_2	24.1
Xe	12.5	Br_2 (vapor)	28.2

Table 1. C_v values for some mono and diatomic gases at 1 atm and 25°C.

From Table 1, it can be seen clearly that the heat capacities of all monoatomic gasses have exactly same values, but they are lower than those of diatomic gasses which include the contributions of translation vibration, and rotation.

The heat capacity is directly proportional to the amount of substance. Heat capacity is an extensive property, meaning it is a physical property that scales with the size of a physical system .That means by doubling the mass of substance, heat capacity can be doubled. The heat required to increase the temperature from T_1 to T_2 of a substance can be calculated using the following equation.

$$Q = C \, x \, \Delta T \quad (27)$$

The unit of heat capacity is J^oC^{-1} or JK^{-1}. For many experimental and theoretical purposes it is more convenient to report heat capacity as an **intensive property**, as an intrinsically characteristic property of a particular substance. This is most often accomplished by the specification of the property per a unit of mass. In science and engineering, such properties are often prefixed with the term *specific.* International standards now recommend that specific heat capacity always refer to division by mass. The units for the specific heat capacity are [C] = J/kg K in chemistry, the heat capacity is also often specified relative one mole, the unit for amount of substance, and is called the molar heat capacity, having the unit, J/mol K. For some considerations it is useful to specify the volume-specific heat capacity, commonly called volumetric heat capacity, which is the heat capacity per unit volume and has SI units [S] = J/m^3 K. This is used almost exclusively for liquids and solids, since for gasses it may be confused with specific heat capacity at constant volume.

In thermodynamics, two types of heat capacities are defined; C_p, the heat capacity at constant pressure and C_v, heat capacity at constant volume. The total energy of a system in thermodynamics is called internal energy which specifies the total kinetic and potential energy of particles in the system. Internal energy of a system can be changed either by doing work on the system or heating it as a result of the conversation of energy law. The internal energy of a substance increases when its temperature is increased. By considering the total change in internal energy of a substance which is heated at constant pressure, the difference between heat capacities at two different conditions can be meaningfully interpreted. Heat capacity in terms of derivative at constant volume is expressed as follows:

$$C_v = (\partial U / \partial T)_v \tag{28}$$

The first law of thermodynamics argues that the internal energy of a system which is heated at constant-pressure differs from that at the constant-volume by the work needed to change the volume of the system to maintain constant pressure. This work arises in two ways: One is the work of driving back the atmosphere (external work); the other is the work of stretching the bonds in the material, including any weak intermolecular interactions (internal work). In the case of a perfect gas, the second makes no contribution.

In order to find out how the internal energy varies with temperature when the pressure rather than the volume of the system is kept constant; it can be divided both sides of ($dU = (\partial U/\partial V)_T\, dV + (\partial U/\partial T)_v\, dT$) by dT and thus;

$$(\partial U / \partial T)p = (\partial U / \partial V)_T (\partial V / \partial T)p + (\partial U / \partial T)_v \tag{29}$$

It is usually sensible in thermodynamics to inspect the output of a manipulation like this to see if it contains any recognizable physical quantity. The partial derivatives on the right in this expression are the slope of the plot of volume against temperature at constant pressure, the slope of the plot of internal energy against volume at constant temperature and the slope of the plot of internal energy against temperature at constant volume, respectively. These properties are normally tabulated as the expansion coefficient, α, of a substance, which is defined as $\alpha = 1/V\ (\partial U/\partial T)_p$, internal pressure, π_T, and constant volume heat capacity, C_v, respectively. Thus;

$$(\partial U / \partial T)_p = \alpha \pi_T V + C_v \tag{30}$$

Equation (30) is entirely general for a closed system, which may be in solid, liquid, or gas states, with constant composition. It expresses the changing of internal energy with the temperature at constant pressure depends on two terms on the right in this expression The first term is related to the potential energy of particles and it comprises internal work made against internal pressure due to thermal expansion which can be considered for all substances in solid, liquid, or gas states, but weakness of inter-particles interactions in gas state requires to take account of an additional contribution to identify the real change in internal energy of any system in gas state heated at constant pressure. The fact that for a perfect gas, $\pi_T = 0$ and so $(\partial U/\partial T)p = C_v$,supports to this remark. The second term also is interested in the kinetic energies of particles. That is, the energy at the constant-pressure of any substance must defined by an another thermodynamic property or function and this function must include external work made to external pressure due to volumetric expansion of any system in gas state. This thermodynamic function which takes account of external work is the enthalpy.

The enthalpy

The change in internal energy is not equal to the heat supplied when the system is free to change its volume. Under these circumstances some of the energy supplied as heat to the system is returned to the surroundings as expansion work, so ΔU is less than q [1].

To determine ΔE, it must be measured both heat and work which done by expanding of a gas. It can be found the quantity of pV work done by multiplying the external pressure (P) by the change in volume of the gas (ΔV, or $V_{final} - V_{initial}$). In an open flask (or a cylinder with a weightless, frictionless piston), a gas does work by pushing back the atmosphere (Figure 7) [18].

$$w = -p\Delta V \tag{31}$$

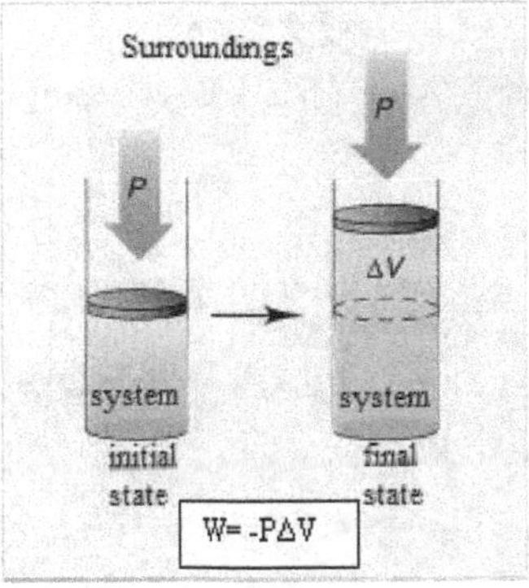

Figure 7. Pressure-volume work. When the volume (V) of a system increases by an amount of ΔV against an external pressure (p), the system pushes back, and thus does pV work on the surroundings (w = -p ΔV) [18].

For changes at constant pressure, a thermodynamic variable called the enthalpy, H, is mathematically defined as follows [1],

$$H = U + pV \tag{32}$$

where U is the internal energy of the system, p is the pressure of the system, and V is the volume of the system [3].

Since internal energy, pressure, and volume are all state functions, enthalpy is also a state function. But what exactly is enthalpy? To help answer this question, consider a process carried out at constant pressure and where the only work allowed is pressure-volume work ($w= - p\Delta V$). Under this conditions, the expression

$$\Delta U = q_p + w \tag{33}$$

becomes

$$\Delta U = q_p - p\Delta V \tag{34}$$

or

$$q_p = \Delta U + p\Delta V \tag{35}$$

where q_p, ΔU and $p\Delta V$ are the transferred energy as heat to the system heated at constant pressure, change in internal energy and change in pV, respectively. Since p is constant; the change in pV is due only to a change in volume. Thus

$$(pV) = p\,\Delta V \tag{36}$$

$$\Delta H = q_p \tag{37}$$

Heat capacity in terms of derivative at constant pressure, i.e. changes in the energy of a system heated at constant pressure is also expressed as follows:

$$C_p = (\partial H / \partial T)_p \tag{38}$$

The slope of a plot of internal energy against temperature at constant volume, for a perfect gas *Cv* is also the slope at constant pressure. In order to obtain an easy way to derive the relation between *Cp* and *Cv* for a perfect gas, both heat capacities can be used in terms of derivatives at constant pressure:

$$Cp - Cv = (\partial H / \partial T)p - (\partial U / \partial T)v = (\partial H / \partial T)p - (\partial U / \partial T)p \tag{39}$$

Then, if $(H = U + pV = U + nRT)$ is introduced into the first term, which results in

$$Cp - Cv = (\partial U / \partial T)p + nR - (\partial U / \partial T)p \quad \text{and}(C_p - C_v = nR) \tag{40}$$

This means that in the case of a perfect gas, R, universal gas constant may be considered as the work done to push back the atmosphere per unit increase in temperature. However, the general relation between the two heat capacities for any pure substance is demonstrated as follows;

$$C_p - C_v = \alpha^2 TV / \kappa_T \tag{41}$$

This formula is a thermodynamic expression, which means that it applies to any substance (that is, it is universally true). It reduces to previous equation for a perfect gas when it was set as $\alpha = 1/T$ and $\kappa_T = 1/p$

Because thermal expansion coefficients, α, of liquids and solids are small, it is tempting to deduce from last equation that for them $Cp \approx Cv$. But this is not always so, because the compressibility κ_T might also be small, so α^2/κ_T might be large. That is, although only a little work need be done to push back the atmosphere, a great deal of work may have to be done to pull atoms apart from one another as the solid expands. As an illustration, for water at 25 ^{0}C, Eqn (41) gives C_p = 75.3 J/Kmol compared with C_v = 74.8 J/Kmol. In some cases, the two heat capacities differ by as much as 30 per cent [21]. The constant-pressure heat capacity C_p differs from the constant-volume heat capacity C_v by the work needed to change the volume of the system to maintain constant pressure. This work arises in two ways. One is the work of driving back the atmosphere (external work); the other is the work of stretching the bonds in the material, including any weak intermolecular interactions (internal work). In the case of a perfect gas, the second makes no contribution. This suggests that the difference between two heat capacities is related to both internal work and external work done by the particles of a substance as an expended result of the conservation of energy.

By considering the variation of H with temperature at constant volume, the validity of the enthalpy function can be differently verified. Firstly; for a closed system of constant composition, H is expressed in the total differential of T and p;

$$dH = ((\partial H / \partial p)_T \, dp + (\partial H / \partial T)_p \, dT \tag{42}$$

and then, divided this equation though by dT;

$$(\partial H / \partial T)_v = (\partial H / \partial p)_T (\partial p / \partial T)_v + C_p \tag{43}$$

The manipulation of this expression provides more involved equation which can be applied to any substance. Because all the quantities that appear in it can be measured in suitable experiments.

$$(\partial H / \partial T)_v = (1 - \alpha\mu / \kappa_T)\, C_p \tag{44}$$

where the isothermal compressibility, κ_T, is defined as

$$\kappa_T = -1/V \, (\partial V / \partial p)_T \tag{45}$$

and the Joule – Thomson coefficient, μ, is defined as

$$\mu = (\partial T / \partial p)_H \tag{46}$$

This expression derived for the changing of the enthalpy with temperature at constant volume suggests that change in H with increased temperature at constant V is lower than that at constant p and the difference between them depend on some characteristic properties of particles, such as κ_T, μ and α,, indicating its relation with the absence of external work.

The measurement of an enthalpy change

An enthalpy change can be measured calorimetrically by monitoring the temperature change that accompanies a physical or chemical change occurring at constant pressure. A calorimeter for studying processes at constant pressure is called an isobaric calorimeter. A simple example is a thermally insulated vessel open to the atmosphere: the heat released in the reaction is monitored by measuring the change in temperature of the contents. For a combustion reaction an adiabatic flame calorimeter may be used to measure ΔT when a given amount of substance burns in a supply of oxygen. Another route to ΔH is to measure the internal energy change using a bomb calorimeter, and then to convert ΔU to ΔH. Because solids and liquids have small molar volumes, for them both pV_m , external work and internal work is so small that the molar enthalpy and molar internal energy are almost identical ($H_m = U_m + pV_m \approx U_m$). Consequently, if a process involves only solids or liquids, the values of ΔH and ΔU are almost identical. Physically, such processes are accompanied by a very small change in volume; the system does negligible work on the surroundings when the process occurs, so the energy supplied as heat stays entirely within the system, as a expended result of the conversation of energy.

Calorimeters

The heat that is given out or taken in, when a chemical reaction occurs can be measured using a calorimeter. A simple, constant-pressure calorimeter (Coffee-cup calorimeter) for measuring heat for reactions in solution is shown in Fig. 8. This figure also shows a bomb calorimeter. The container is an expanded polystyrene cup with a lid. This material provides insulation which ensures that heat loss to, or gains from the surroundings is minimized; the outer cup in Fig. 8 provides additional insulation. As the reaction takes place, the thermometer records any change in temperature.

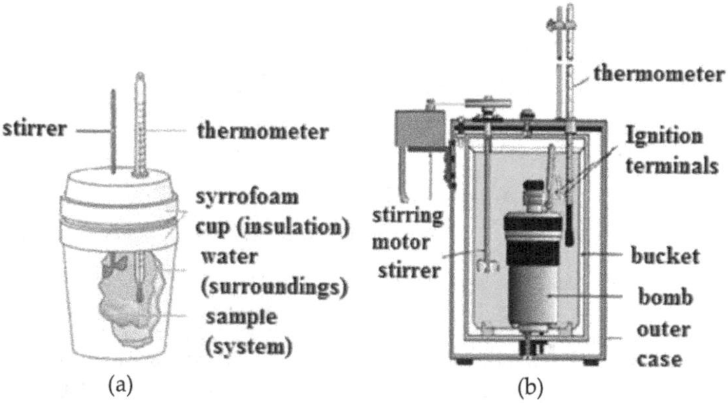

Figure 8. Calorimeter types; (a) Coffee-cup calorimeter [18] and (b) Bomb calorimeter [22].

The relationship between the temperature change and the energy is given as follows;

$$Energy\ change\ in\ J = m\ (g).C\ (J\ /\ gK).\ \Delta T\ (K) \tag{47}$$

where C is the specific heat capacity of the solution. Since the reaction is carried out at constant pressure, the energy is equal to the enthalpy change. For dilute aqueous solutions, it is usually sufficient to assume that the specific heat capacity of the solution is the same as for water: C_{water} 4.18 J/gK. It is assumed that no heat is used to change the temperature of the calorimeter itself. Where a calorimeter is made from expanded polystyrene cups, this is a reasonable assumption because the specific heat capacity of the calorimeter material is so small. However, the approximation is not valid for many types of calorimeter and such pieces of apparatus must be calibrated before use. Measurements made in the crude apparatus shown together with in Figure 8 are not accurate, and more specialized calorimeters, such as **bomb calorimeter** must be used if accurate results are required [18;19]. A bomb calorimeter is ideally suited for measuring the heat evolved in a combustion reaction. The system is everything within the double-walled outer jacket of the calorimeter. This includes the bomb and its contents, the water in which the bomb is immersed, the thermometer, the stirrer, and so on. Before using above equation, it must be emphasized that a rise in temperature of a system insulated with its surrounding does not occur by any heat transferred from the surroundings to the system because of the temperature difference between them. The difference in temperature observed during the measuring may be arisen from the change in the composition of system and considered as if the involved reaction occurs as exothermic or endothermic in diathermic condition. At constant pressure (where only pV work is allowed), the change in enthalpy ΔH of the system is equal to the energy flow as heat. This means that for a reaction studied at constant pressure, the flow of heat is a measure of the change in enthalpy for the system. For this reason, the terms heat of reaction and change in enthalpy are used interchangeably for reactions studied at constant pressure. For a chemical reaction, the enthalpy change is given by the equation

$$\Delta H = H_{products} - H_{reactants} \tag{48}$$

In a case in which the products of a reaction have a greater enthalpy than the reactants, ΔH will be positive. Thus heat will be absorbed by the system, and the reaction is **endothermic**. On the other hand, if the enthalpy of the products is less than that of the reactants, ΔH will be negative. In this case the overall decrease in enthalpy is achieved by the generation of heat, and the reaction is **exothermic** [3]. Energy diagrams for exothermic and endothermic reactions are shown in Figure 9.

Standard enthalpy change

The standard enthalpy change of a reaction refers to the enthalpy change when all the reactants and products are in their **standard states**. The notation for this thermochemical quantity is $\Delta_r H^0$ (T) where the subscript 'r' stands for 'reaction', the superscript 'o' means 'standard state conditions', and (T) means 'at temperature T'. This type of notation is found for other thermodynamic functions that we meet later on.

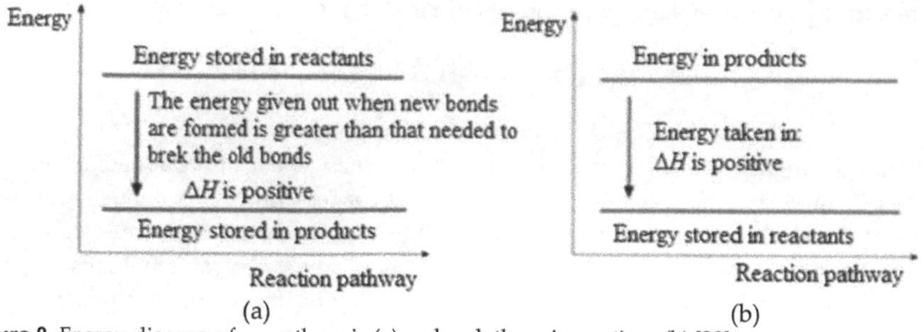

Figure 9. Energy diagrams for exothermic (a) and endothermic reactions (b) [20].

The standard state of a substance is its most thermodynamically stable state under a pressure of 1 bar (1.00 x 10^5 Pa) and at a specified temperature, T. Most commonly, T = 298.15 K, and the notation for the standard enthalpy change of a reaction at 298.15K is then $\Delta_r H^0$ (298.15 K). It is usually sufficient to write $\Delta_r H^0$ (298 K) [19].

Standard enthalpies of formation

The standard enthalpy of formation (ΔH_f^0) of a compound is defined as the change in enthalpy that accompanies the formation of one mole of a compound from its elements with all substances in their reference states. The reference state of a element is its most thermodynamically stable state under a pressure of 1 bar (1.00 x 10^5 Pa) and at a specified temperature, T.

A degree symbol on a thermodynamic function, for example, ΔH, indicates that the corresponding process has been carried out under standard conditions. The standard state for a substance is a precisely defined reference state. Because thermodynamic functions often depend on the concentrations (or pressures) of the substances involved, it must be used a common reference state to properly compare the thermodynamic properties of two substances. This is especially important because, for most thermodynamic properties, it can be measured only *changes* in the property

Enthalpy is a state function, so it can be chosen *any* convenient pathway from reactants to products and then sums the enthalpy changes along the chosen pathway. A convenient pathway, shown in Fig. 10, involves taking the reactants apart to the respective elements in their reference states in reactions (a) and (b) and then forming the products from these elements in reactions (c) and (d). This general pathway will work for any reaction, since atoms are conserved in a chemical reaction [3].

From Fig. 10, it can be seen that reaction (a), where methane is taken apart into its elements,

$$CH_4(g) \rightarrow C(s) + 2H_2(g) \tag{49}$$

is just the reverse of the formation reaction for methane:

$$C(s) + 2H_2(g) \rightarrow CH_4(g)\ \Delta H^0{}_f = -75\ kJ/mol \tag{50}$$

Since reversing a reaction means changing the sign of ΔH, but keeping the magnitude the same, ΔH for reaction (a) is *$-\Delta H^0{}_f$, or 75 kJ. Thus $\Delta H^0\,(a) = 75$ kJ.*

It can be secondarily considered reaction (b). Here oxygen is already an element in its reference state, so no change is needed. Thus $\Delta H^0\,(\mathrm{b}) = 0$.

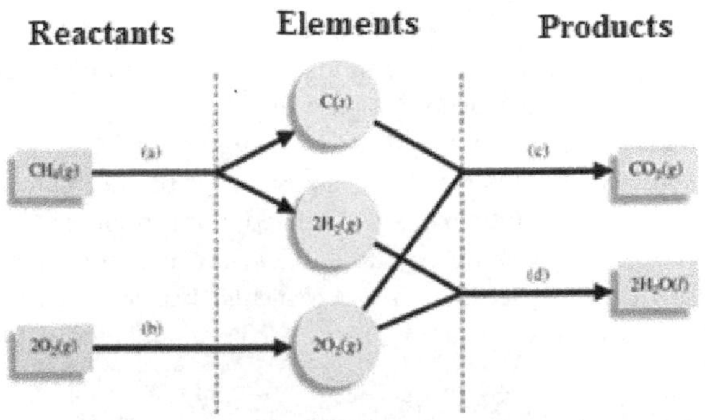

Figure 10. In this pathway for the combustion of methane, the reactants are first taken apart in reactions (a) and (b) to form the constituent elements in their reference states, which are then used to assemble the products in reactions (c) and (d) [3].

The next steps, reactions (c) and (d), use the elements formed in reactions (a) and (b) to form the products. That is, reaction (c) is simply the formation reaction for carbon dioxide:

$$C(s) + O_2(g) \rightarrow CO_2(g) \quad \Delta H^0{}_f = -394\ kJ\,/\,mol \tag{51}$$

and

$$\Delta H^0{}_{(c)} = \Delta H^0{}_f \ for\ CO_2(g) = -394\ kJ\,/\,mol \tag{52}$$

Reaction (d) is the formation reaction for water:

$$H_2(g) + 1/2O_2(g) \rightarrow H_2O\ (l) \quad \Delta H^0{}_f = -286\ kJ\,/\,mol \tag{53}$$

However, since 2 moles of water are required in the balanced equation, it must be formed 2 moles of water from the elements:

$$2H_2(g) + O_2(g) \rightarrow 2H_2O \tag{54}$$

Thus

$$\Delta H^0{}_{(d)} = 2\ x\ \Delta H^0{}_f \ for\ H_2O\ (l) = 2(-286\ kJ\,/) = -572\ kJ\,/\,mol \tag{55}$$

It has now been completed the pathway from the reactants to the products. The change in enthalpy for the reaction is the sum of the ΔH values (including their signs) for the steps:

$$\Delta H^0{}_{reaction} = \Delta H^0{}_{(a)} + \Delta H^0{}_{(b)} + \Delta H^0{}_{(c)} + \Delta H^0{}_{(d)} \tag{56}$$

$$\begin{aligned} &= \left[-\Delta H^0{}_f \text{ for } CH_4(g) + 0 + \left[\Delta H^0{}_f \text{ for } CO_2(g)\right] + \left[2\, x\, \Delta H^0{}_f \text{ for } H_2O\,(l)\right]\right. \\ &= -\,(-75\, kJ) + 0 + (-394kJ) + (-572\, kJ) = -891\, kJ \end{aligned} \tag{57}$$

This process is diagramed in Fig. 11. It can be seen that the reactants are taken apart and converted to elements [not necessary for $O_{2\,(g)}$] that are then used to form products. It can be seen that this is a very exothermic reaction because very little energy is required to convert the reactants to the respective elements but a great deal of energy is released when these elements form the products. This is why this reaction is so useful for producing heat to warm homes and offices. If it is examined carefully the pathway used in this example, it can be understood that first, the reactants were broken down into the elements in their reference states and then the products were then constructed from these elements. This involved formation reactions and thus enthalpies of formation. The entire process can be summarized as follows:

$$\Delta H^0{}_{reaction} = \sum v_p \Delta H^0{}_{f\ (products)} - \sum v_p \Delta H^0{}_{f\ (reactants)} \tag{58}$$

where, v is stoichiometric coefficients for both reactants and products. The enthalpy change for a given reaction can be calculated by subtracting the enthalpies of formation of the reactants from the enthalpies of formation of the products.

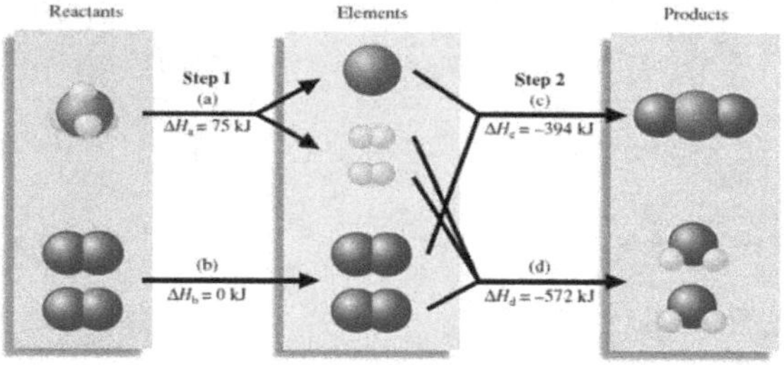

Figure 11. A schematic diagram of the energy changes for the reaction $CH_4\,(g) + 2O_2\,(g) \rightarrow CO_2\,(g) + 2H_2O\,(l)$ [3].

Hess's law

Another way to calculate values of ΔH for reactions involves manipulating equations for other reactions with known ΔH values. When chemical equations are added to yield a different chemical equation, the corresponding ΔH values are added to get the ΔH for the desired equation. This principle is called **Hess's law** and it is an application of the **first law of thermodynamic** or the **conservation of energy**. For example, it can be calculated the ΔH for the reaction of carbon with oxygen gas to yield carbon dioxide from the values for the reaction of carbon with oxygen to yield carbon monoxide and that of carbon monoxide plus oxygen to yield carbon dioxide [9].

Desired

$$C(s) + O_2(g) \rightarrow CO_2(g) \tag{59}$$

Given

$$C(s) + 1/2\ O_2(g) \rightarrow CO(g) \quad \Delta H = -110 kJ \tag{60}$$

$$CO(g) + 1/2\ O_2(g) \rightarrow CO_2(g) \quad \Delta H = -283 kJ \tag{61}$$

Adding the two chemical equations given:

$$C_{(s)} + CO_{(g)} + O_{2(g)} \rightarrow CO_{2(g)} + CO_{(g)} \tag{62}$$

Eliminating CO from both sides results in the desired equation:

$$C(s) + O_2(g) \rightarrow CO_2(g) \tag{63}$$

Therefore, adding these two ΔH values will give the ΔH desired:

$$\Delta H = (-110\ kJ) + (-283\ kJ) = -393\ kJ \tag{64}$$

It must be noticed that enthalpies of formation have not been used explicitly in this process [9]. The Hess's law is shown schematically in Fig 12 [3].

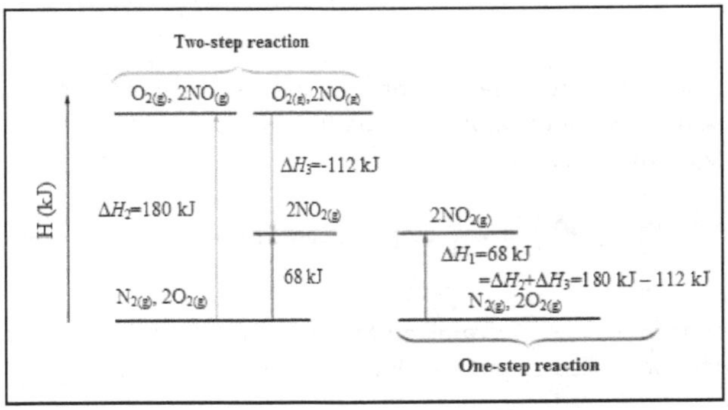

Figure 12. A schematic diagram of Hess's law. The same change in enthalpy occurs when nitrogen and oxygen react to form nitrogen dioxide, regardless of whether the reaction occurs in one (red) or two (blue) steps [3].

The overall reaction can be written in one step, where the enthalpy change is represented by ΔH_1 [3].

$$N_2(g) + 2O_2(g) \rightarrow 2NO_2(g) \quad \Delta H_1 = 68\ kJ \tag{65}$$

This reaction also can be carried out in two distinct steps, with enthalpy changes designated by ΔH_2 and ΔH_3:

$$N_2(g) + O_2(g) \rightarrow 2NO\ (g) \quad \Delta H_2 = 180\ kJ \tag{66}$$

$$2NO\ (g) + O_2(g) \rightarrow 2NO_2(g) \quad \Delta H_3 = -112\ kJ \tag{67}$$

Net reaction:

$$N_2(g) + 2O_2(g) \rightarrow 2NO_2(g) \quad \Delta H_2 + \Delta H_3 = 68\ kJ \tag{68}$$

The sum of the two steps gives the net, or overall, reaction and that

$$\Delta H_1 = \Delta H_2 + \Delta H_3 = 68\ kJ \tag{69}$$

The temperature dependence of reaction enthalpies

The standard enthalpies of many important reactions have been measured at different temperatures. However, in the absence of this information, standard reaction enthalpies at different temperatures may be calculated from heat capacities and the reaction enthalpy at some other temperature (Fig. 13). In many cases heat capacity data are more accurate than reaction enthalpies. It follows from eqn ($dH = C_p dT$ at constant pressure) that, when a substance is heated from T_1 to T_2, its enthalpy changes from

$H\ (T_1)$ to

$$H(T_2) = H(T_1) + \int_{T1}^{T2} CpdT \tag{70}$$

(It has been assumed that no phase transition takes place in the temperature range of interest.) Because this equation applies to each substance in the reaction, the Standard reaction enthalpy changes from $\Delta H^0_r\ (T_1)$ to

$$\Delta H_r^{\ 0}(T_2) = \Delta H_r^{\ 0}(T_1) + \int_{T1}^{T2} \Delta CpdT \tag{71}$$

where ΔCp is the difference of the molar heat capacities of products and reactants under standard conditions weighted by the stoichiometric coefficients that appear in the chemical equation [12];

$$\Delta C_p = \sum \nu C_{p(Products)} - \sum \nu C_{p(Reactants)} \tag{72}$$

Equation 71 is known as **Kirchhoff's law**. It is normally a good approximation to assume that ΔCp_r is independent of the temperature, at least over reasonably limited ranges. Although the individual heat capacities may vary, their difference varies less significantly. In some cases the temperature dependence of heat capacities is taken into account by using equation below.

$$C_p = a + bT + c / T^2 \quad (73)$$

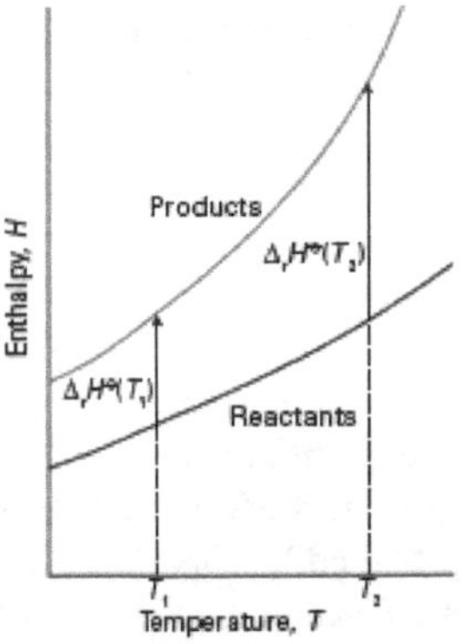

Figure 13. An illustration of the content of Kirchhoff's law. When the temperature is increased, the enthalpy of both the products and the reactants increases. In each case, the change in enthalpy depends on the heat capacities of the substances. The change in reaction enthalpy reflects the difference in the changes of the enthalpies [12].

Author details

Ahmet Gürses
Atatürk University, K.K. Education Faculty, Department of Chemistry, Erzurum, Turkey

Mehtap Ejder-Korucu
Kafkas University, Faculty of Science and Literature, Department of Chemistry, Kars, Turkey

5. References

[1] Atkins. P., and de Paula. J., Atkıns' Physical Chemistry. Oxford Universty Press. Seventh Edition. 2002.

[2] Petrucci. R. H., Harwood. W. S., Genaral Chemistry Principles and Modern Applications. Sıxth Edition. MacMillan Publishing Company. New York. 1993

[3] Zumdahl. S. S and Zumdahl S. A., Chemistry.Houghton Mifflin Company. Seventg Edition. 2007.

[4] Svein Stolen and Tor Grande. Chemical Thermodynamics of Materials. 2004. John Wiley & Sons, Ltd ISBN 0 471 492320 2.

[5] Levine. I. N., Physical Chemistry. McGRAW-HILL International Edition. Third Edition 1988.

[6] McMURRAY. F., Chemistry. Fourth edition.

[7] Gilbert. T. R., Kirss. R. V., Foster. N., and Davies. G., Chemistry the science in context. Second edition. 2009. W. W. Norton & Company, Inc.

[8] Lee. H-G., Chemical Thermodynamics for Metals and Materials. Imperial College Press. 1999.

[9] Goldberg. D. E., Fundamentals of Chemistry. Fifth edition. The McGraw-Hill. 2007

[10] Powers. J. M., Lecture Notes on Thermodynamics. Notre Dame, Indiana; USA. 2012

[11] Tanaka. T., The Laws of Thermodynamics. Cambridge Universty Press.

[12] Atkins. P., and de Paula. J., Atkıns' Physical Chemistry. Oxford Universty Press 2009.

[13] Laider, Keith, J. (1993). *The World of Physical Chemistry* (http:/ / books. google. com) Oxford University Press. ISBN 0-19-855919-4.

[14] International Union of Pure and Applied Chemistry, Physical Chemistry Division. "Quantities, Units and Symbols in Physical Chemistry" (http:/ / old. iupac. org/ publications/ books/ gbook/ gren book). Blackwell Sciences. p. 7. "The adjective specific before the name of an extensive quantity is often used to mean divided by mass."

[15] International Bureau of Weights and Measures (2006), *The International System of Units (SI)* (http:/ / www. bipm. org) (8th ed.), ISBN 92-822-2213-6,

[16] Gilbert. T. R., 7.chemistry-the science in context-second edition. 2008.

[17] Minguez. J. M., The work-energy theorem and the first law of thermodynamics. International Journal of Mechanical engineering Education 33/1.

[18] Silberg. M. S., Principles of General Chemistry. McGraw-Hill Higher Education. 2007.

[19] Housecroft. C. E and Constable. E. C., Chemistry: An Introduction to Organic, Inorganic and Physical Chemistry. 3rd Edition. Pearson Prentice Hall. 2006.

[20] Jones. A., Clement. M., Higton. A., and Golding. E., Access to Chemistry. Royal Society of Chemistry. 1999.

[21] Atkins. P. W., Physical Chemistry. Oxford University Press. Sixth Edition. 1998.

[22] Mortimer. R. G., Physical Chemistry. Third edition. Academic Press in an imprint of Elsevier.2008.

Section 2

Statistical Thermodynamics

Chapter 3

Gibbs Free Energy Formula for Protein Folding

Yi Fang

Additional information is available at the end of the chapter

1. Introduction

Proteins are life's working horses and nature's robots. They participate in every life process. They form supporting structures of cell, fibre, tissue, and organs; they are catalysts, speed up various life critical chemical reactions; they transfer signals so that we can see, hear, and smell; they protect us against intruders such as bacteria and virus; they regulate life cycles to keep that everything is in order; etc., just mention only a few of their functions.

The first thing drawing our attention of proteins are their size. Proteins are macromolecules, that is, large molecules. Non-organic molecules usually are small, consisting of from a couple of atoms to a couple of dozen atoms. A small protein will have thousands of atoms, large ones have over ten thousand atoms. With their huge number of atoms, one can imaging that how complicated should be of a protein molecule. Fortunately, there are some regularities in these huge molecules, i.e., proteins are polymers building up by monomers or smaller building blocks. The monomers of proteins are amino acids, life employs 20 different amino acids to form proteins. In cell, a series of amino acids joined one by one into amino acids sequences. The order and length of this amino acid sequence is translated from DNA sequences by the universal genetic code. The bond joining one amino acid to the next one in sequence is peptide bond (a covalent bond) with quite regular specific geometric pattern. Thus amino acids sequences are also called peptide chains. But the easy translation and geometric regularities stop here. The peptide chain has everything required to a molecule, all covalent bonds are correctly formed. But to perform a protein's biological function, the peptide chain has to form a specific shape, called the protein's **native structure**. Only in this native structure a protein performs its biological function. Proteins fall to wrong shapes not only will not perform its function, but also will cause disasters. Many disease are known to be caused by some proteins taking wrong structure.

How the peptide chain take its native structure? Is there another genetic code to guide the process of taking to the native structure? In fact, at this stage, life's most remarkable drama takes stage. Once synthesized, the peptide chain of a protein spontaneously (some need the help of other proteins and molecules) fold to its native structure. This process is called **protein folding**. At this stage, everything is governed by simple but fundamental physical laws.

The **protein folding problem** then can be roughly divided into three aspects: 1. folding process: such that how fast a peptide chain folds, what are the intermediate structures between the initial shape and the native structure. 2. the mechanics of the folding, such as what is the deriving force. 3. the most direct application to biological study is the prediction of the native structure of a protein from its peptide chain. All three parts of the protein folding problem can have a unified treatment: writing down the Gibbs free energy formula $G(\mathbf{X})$ for any conformation $\mathbf{X} = (\mathbf{x}_1, \cdots, \mathbf{x}_i, \cdots, \mathbf{x}_M) \in \mathbb{R}^{3M}$ of protein, where $\mathbf{x}_i \in \mathbb{R}^3$ is the atom $\mathbf{a}_i$'s atomic center.

The fundamental law for protein folding is the **Thermodynamic Principle**: the amino acid sequence of a protein determines its native structure and the native structure of the protein has the minimum Gibbs free energy among all possible conformations as stated in Anfinsen (1973). Let $\mathbf{X}$ be a conformation of a protein, is there a natural Gibbs free energy function $G(\mathbf{X})$? The answer must be positive, as G. N. Lewis said in 1933: "There can be no doubt but that in quantum mechanics one has the complete solution to the problems of chemistry." (quoted from Bader (1990), page 130.) Protein folding is a problem in biochemistry, why such a formula $G(\mathbf{X})$ has not been found and what is the formula? This chapter is trying to give the answers.

First, the Gibbs free energy formula is given, it has two versions, the chemical balance version (1) and the geometric version (2).

1.1. The formula

Atoms in a protein are classified into classes H_i, $1 \leq i \leq H$, according to their levels of hydrophobicity. The formula has two versions, the chemical balance version is:

$$G(\mathbf{X}) = \mu_e N_e(\mathbf{X}) + \sum_{i=1}^{H} \mu_i N_i(\mathbf{X}), \tag{1}$$

where $N_e(\mathbf{X})$ is the mean number of electrons in the space included by the first hydration shell of $\mathbf{X}$, μ_e is its chemical potential. $N_i(\mathbf{X})$ is the mean number of water molecules in the first hydration layer that directly contact to the atoms in H_i, μ_i is the chemical potential.

Let $M_{\mathbf{X}}$ (see FIGURE 3) be the molecular surface for the conformation $\mathbf{X}$, defining $M_{\mathbf{X}i} \subset M_{\mathbf{X}}$ as the set of points in $M_{\mathbf{X}}$ that are closer to atoms in H_i than to any atoms in H_j, $j \neq i$. Then the geometric version of $G(\mathbf{X})$ is:

$$G(\mathbf{X}) = \nu_e \mu_e V(\Omega_{\mathbf{X}}) + d_w \nu_e \mu_e A(M_{\mathbf{X}}) + \sum_{i=1}^{H} \nu_i \mu_i A(M_{\mathbf{X}_i}), \quad \nu_e, \nu_i > 0, \tag{2}$$

where $V(\Omega_{\mathbf{X}})$ is the volume of the domain $\Omega_{\mathbf{X}}$ enclosed by $M_{\mathbf{X}}$, d_w the diameter of a water molecule, and $A(M_{\mathbf{X}})$ and $A(M_{\mathbf{X}_i})$ the areas of $M_{\mathbf{X}}$ and $M_{\mathbf{X}_i}$, $\nu_e[V(\Omega_{\mathbf{X}}) + d_w A(M_{\mathbf{X}})] = N_e$, $\nu_i A(M_{\mathbf{X}_i}) = N_i(\mathbf{X})$, $1 \leq i \leq H$. The ν_e and ν_i are independent of $\mathbf{X}$, they are the average numbers of particles per unit volume and area.

Before the actual derivation is given, some basic facts should be stated, such as hydrophobicity, protein structures, and the environment in which the protein folds. Brief description of the methods in the experimental measurements and theoretical derivation of the Gibbs free energy of the protein folding is introduced to give the motivation and idea of the derivation. By making critics on the previous derivation, the necessary concepts would be clarified, what are important in the derivation would be identified, and would set the thermodynamic system that most fit the reality currently known about the protein folding process. Then both classical and quantum statistical derivations were given, the only difference is that in the classical statistically derived formula, the volume and the whole surface area terms in formula (2) are missing. Thus it is that only quantum statistical method gives us the volume and whole surface terms in formula (2). After the derivations, some remarks are made. A direct application of the Gibbs free energy formula (2) is the *ab initio* prediction of proteins' natives structures. Gradient formulas of $G(\mathbf{X})$ are given to be able to apply the Newton's fastest descending method. Finally, it should be emphasized that the gradient $\nabla G(\mathbf{X})$ not only can be used to predict the native structure, it is actually the force that forces the proteins to fold as stated in Ben-Naim (2012). In Appendix, integrated gradient formulas of $G(\mathbf{X})$ on the molecular surface are given.

2. Proteins

2.1. Amino acids

There are 20 different amino acids that appear in natural proteins. All amino acids have a common part, or the **back bone** consisting of 9 atoms in FIGURE 1 (except the R).

NH_2 is the amino group and COOH is the carboxyl group of the back bone. Single amino acid is in polar state, so the amino group gains one more hydrogen from the carboxyl group, or perhaps the amino group losses one electron to the carboxyl group. Geometrically it is irrelevant since after forming peptide bonds the amino group will loss one H to become NH and the carboxyl group will loss one OH to become CO. Thus an amino acid in the sequence is also called a **residue**.

$$
\begin{array}{ccccc}
\mathrm{H} & & \mathrm{H}_\alpha & & \mathrm{O} \\
\backslash & & | & & \| \\
\mathrm{N} & \!\!\!-\!\!\!- & \mathrm{C}_\alpha & \!\!\!-\!\!\!- & \mathrm{C} \\
/ & & | & & | \\
\mathrm{H} & & \mathrm{R} & & \mathrm{O} \!-\!\!\!- \mathrm{H}
\end{array}
$$

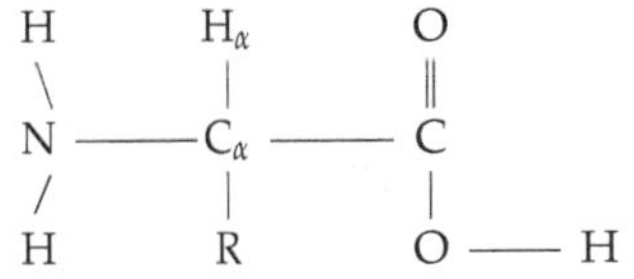

Figure 1. An generic amino acid.

The group R in FIGURE 1 is called **side chain**, it distinguishes the 20 different amino acids. A side chain can be as small as a single hydrogen atom as in Glycine, or as large as consisting of 18 atoms including two rings as in Tryptophan. 15 amino acids have side chains that contain more than 7 atoms, i.e., more atoms than that of the back bone in an amino acid sequence. Except Glycine, a C_β carbon in a side chain forms a covalent bond with the **central carbon** C_α of the back bone.

2.2. Hydrogen bonds

A hydrogen bond is the attractive interaction of a hydrogen atom with an electronegative atom (the **accepter**), like nitrogen, oxygen or fluorine (thus the name "hydrogen bond", which must not be confused with a covalent bond to hydrogen). The hydrogen must be covalently bonded to another electronegative atom (forming a **donor group**) to create the hydrogen bond. These bonds can occur between molecules (intermolecular), or within different parts of a single molecule (intramolecular). The hydrogen bond is stronger than the van der Waals interaction, but weaker than covalent or ionic bond. Hydrogen bond occurs in both inorganic molecules such as water and organic molecules such as DNA, RNA, and proteins.

Some amino acids' side chains contain hydrogen bond donors or acceptors that can form hydrogen bond with either other side chains in the same protein (intramolecular hydrogen bond) or with surrounding water molecules (inter-molecular hydrogen bond). Those amino acids whose side chains do not contain either donors or acceptors of hydrogen bond are classified as hydrophobic.

2.3. Hydrophobicity levels

Every atom in a protein belongs to a moiety or atom group, according to the moiety's level of ability to form hydrogen bond, the atom is assigned a hydrophobicity level. All the hydrophobicity scales are tested or theorized in some aspects of individual amino acid, either as a independent molecule or as a residue in a protein, in various environments such as solvent, PH value, temperature, pressure, etc. That is just like taking a snap shot of an object with complicated shape. All snap shots are different if taking from different angles of view. Therefore, there are many different classifications of hydrophobicity, for example, in Eisenberg and McLachlan (1986) there are five classes, C, O/N, O^-, N^+, S. Let a protein have M atoms $\{\mathbf{a}_1, \cdots, \mathbf{a}_i, \cdots, \mathbf{a}_M\}$. One can assume that there are H hydrophobic classes, such that $\{\mathbf{a}_1, \cdots, \mathbf{a}_i, \cdots, \mathbf{a}_M\} = \cup_{i=1}^{H} H_i$.

2.4. Protein structures

Let a molecule have M atoms, listed as $(\mathbf{a}_1, \cdots, \mathbf{a}_i, \cdots, \mathbf{a}_M)$. A presentation of a structure $\mathbf{X}$ of this molecule is a series atomic centers (nuclear centers) of the atoms $\mathbf{a}_i$, $\mathbf{x}_i \in \mathbb{R}^3$. Hence it can be written as a point in $\mathbb{R}^{3M}$, $\mathbf{X} = (\mathbf{x}_1, \cdots, \mathbf{x}_i, \cdots, \mathbf{x}_M)$. The space $\mathbb{R}^{3M}$ then is called the *control space*. The real shape of the structure $\mathbf{X}$ is realized in $\mathbb{R}^3$, called the *behavior space* as defined in Bader (1990), it is a bunch of overlapping balls (spheres), $P_{\mathbf{X}} = \cup_{i=1}^{M} B(\mathbf{x}_i, r_i)$, where r_i is the van der Waals radius of the atom $\mathbf{a}_i$ and $B(\mathbf{x}, r)$ is the closed ball $\{\mathbf{y} : |\mathbf{y} - \mathbf{x}| \leq r\} \subset \mathbb{R}^3$, of center x and radius r.

Protein native structures are complicated. Unlike the famous double-helix structure of DNA structure, the only general pattern for protein structure is no pattern at all. To study the native structures of proteins people divide the structures in different levels and make structure classifications.

The amino acid sequence of a protein is called its **primary structure**. Regular patterns of local (along the sequences) structures such as helix, strand, and turn are called the **secondary structure** which contain many intramolecular hydrogen bonds in regular patterns. The global assembly of these secondary structures, connected by turns and irregular loops, is called the **tertiary structure**. For proteins having multiple amino acid sequences or structurally associated with other molecules there are also **quaternary structures**, see Branden and Tooze (1999) and Finkelstein and Ptitsyn (2002).

The secondary structures are local structures, they are usually in helix, strand, and turn. A common feature of them is that they have regular geometric arrangement of their main chain atoms, such that there are good opportunities to form hydrogen bonds between different residues. Several strands may form sheet, stabilized by regular pattern of hydrogen bonds. Turns and loops are necessary for the extended long chain to transfer to a sphere like shape. Turns are short, 3 or 4 residues long. Loops involves many residues, but without any regular pattern of hydrogen bonds. Loops often form the working place of the protein, therefore appear on the out surface of the native structure.

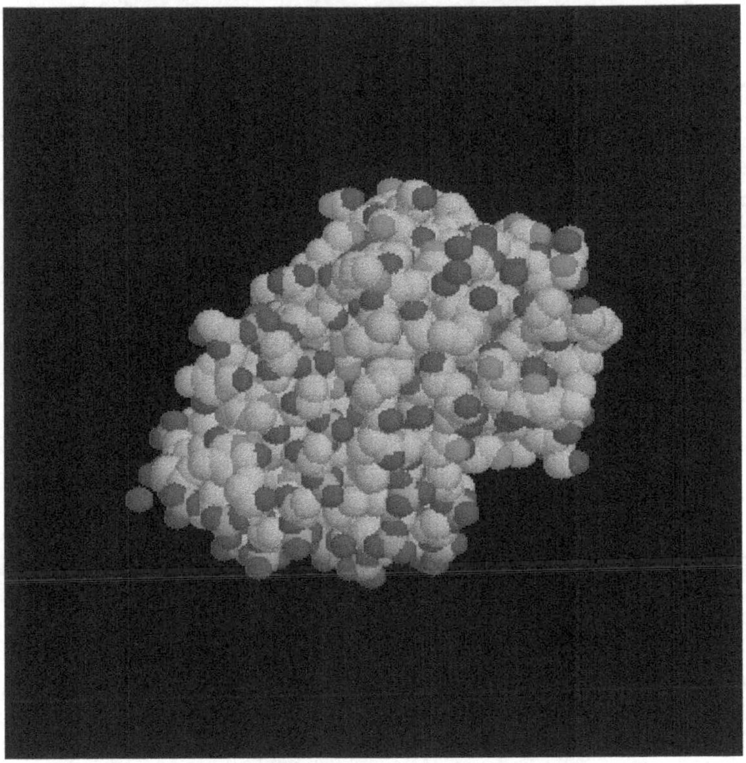

Figure 2. P_X is a bunch of overlapping balls, called the space-filling model, or CPK model.

3. Some functions in thermodynamics

A thermodynamic system consists of particles in a region $\Gamma \subset \mathbb{R}^3$ and a bath or environment surrounding it. A wall, usually the boundary $\partial\Gamma$ separating the system with its surrounding.

If no energy and matter can be exchanged through the wall, the system is an **isolated system**. If only energy can be exchanged, the system is a **closed system**. If both energy and matter can be exchanged with the surrounding, the system is an **open system**.

For an open system Γ of variable particles contacting with surrounding thermal and particle bath, let U, T, S, P, V, μ and N be the inner energy, temperature, entropy, pressure, volume, chemical potential, and the number of particles of the system Γ respectively, then

$$dU = TdS - PdV + \mu dN, \tag{3}$$

By Legendre transformations various extensive quantities can be derived,

$$F = U - TS, \quad G = U - TS + PV, \quad \phi = F - \mu N = U - TS - \mu N \tag{4}$$

where F, G, and ϕ are Helmholtz, Gibbs free energies, and thermodynamic potential respectively. Then

$$dF = -SdT - PdV + \mu dN, \quad dG = -SdT + VdP + \mu dN, \quad d\phi = -SdT - PdV - Nd\mu. \tag{5}$$

Which shows that $U = U(S,V,N)$, $F = F(T,V,N)$, $G = G(T,P,N)$, $\phi = \phi(T,V,\mu)$. All **extensive** quantities satisfy a linear homogeneous relation, i.e., consider a scaling transformation which enlarges the actual amount of matter by a factor λ, then all extensive quantities are multiplied by a factor λ. U, S, V, N, F, G, ϕ are extensive, while T, P, μ are **intensive**. Thus

$$\lambda U = U(\lambda S, \lambda V, \lambda N), \quad \lambda F = F(T, \lambda V, \lambda N), \quad \lambda G = G(T, P, \lambda N), \quad \lambda\phi(T,V,\mu) = \phi(T, \lambda V, \mu). \tag{6}$$

From equations in (5) $(\frac{\partial \phi}{\partial V})_{T,\mu} = -P$. By equations in (6)

$$\phi = \frac{d(\lambda\phi)}{d\lambda} = V\left(\frac{\partial\phi}{\partial V}\right)_{T,\mu} = -PV \tag{7}$$

and

$$\phi(T,V,\mu) = -PV. \tag{8}$$

Equation (8) is true for any open thermodynamics system.

4. Statistical mechanics

Thermodynamics is a phenomenological theory of macroscopic phenomena that neglects the individual properties of particles in a system. Statistical mechanics is the bridge between the macroscopic and microscopic behavior. In statistical mechanics, the particles in a system obey either classical or quantum dynamic laws, and the macroscopic quantities are statistical averages of the corresponding microscopic quantities. If the particles obey classical dynamical law, it is the classical statistical mechanics. If the particles obey quantum dynamical law, it is the quantum statistical mechanics. But the averaging to get macroscopic quantities from microscopic ones are in the same principle and formality.

Protein folding studies the structure of the protein molecule, what is the native structure and why and how the protein folds to it. All these aspects are specific properties of a particle, the protein molecule. To get the Gibbs free energy formula $G(\mathbf{X})$ for each conformation $\mathbf{X}$, statistical mechanics is needed with careful specification of the thermodynamic system.

4.1. The canonic ensemble

Statistical mechanics uses ensembles of all microscopic states under the same macroscopic character, for example, all microscopic states corresponding to the same energy E. The probability of this ensemble then is proportional to

$$p_E \propto \exp(-\beta E), \tag{9}$$

where $\beta = 1/kT$, k the Boltzmann constant and T the temperature. If there are only a series energy levels $E_1, E_2, \cdots$, then the probability distribution for canonic ensemble is

$$p_i = \frac{\exp(-\beta E_i)}{\sum_{n=1}^{\infty} \exp(-\beta E_n)}. \tag{10}$$

Various of thermodynamic quantities, such as the inner energy of the system, can be put as the means:

$$U = \langle E_i \rangle = \frac{\sum_{i=1}^{\infty} E_i \exp(-\beta E_i)}{\sum_{n=1}^{\infty} \exp(-\beta E_n)}. \tag{11}$$

If only the Halminltonian $H(\mathbf{q}, \mathbf{p})$ is known, where $\mathbf{q} = (\mathbf{q}_1, \cdots, \mathbf{q}_i, \cdots, \mathbf{q}_N) \in \Gamma^N$ is the position of the N particles in the thermodynamic system $\Gamma \subset \mathbb{R}^3$ under study, and $\mathbf{p} = (\mathbf{p}_1, \cdots, \mathbf{p}_i, \cdots, \mathbf{p}_N)$ momentums of these particles, the *canonical phase-space density* of the system then is

$$p_c(\mathbf{q}, \mathbf{p}) = \frac{\exp[-\beta H(\mathbf{q}, \mathbf{p})]}{\frac{1}{N!h^{3N}} \int_{\Gamma^N} d\mathbf{q}^N \int_{\mathbb{R}^{3N}} \exp[-\beta H(\mathbf{q}, \mathbf{p})] d\mathbf{p}^N} = \frac{\exp[-\beta H(\mathbf{q}, \mathbf{p})]}{\mathcal{Z}(T, V, N)}. \tag{12}$$

where $N!$ is the Gibbs corrector because that the particles in the system is indistinguishable. $\mathcal{Z}(T, V, N)$ is called the *canonic partition function*, it depends on the system's temperature T, volume V, and particle number N. Note that under the assumption of the canonic ensemble, they are all fixed for the fixed thermodynamical system Γ. Especially, $V = V(\Gamma) = \int_\Gamma d\mathbf{q}$ implicitly set that $\Gamma \subset \mathbb{R}^3$ has a volume.

Then the entropy S is

$$S = S(\Gamma) = \langle -k \ln p_c \rangle = \frac{k}{N!h^{3N}} \int_{\Gamma^N} d\mathbf{q}^N \int_{\mathbb{R}^{3N}} [\beta H(\mathbf{q}, \mathbf{p}) + \ln \mathcal{Z}(T, V, N)] p_c(\mathbf{q}, \mathbf{p}) d\mathbf{p}^N$$
$$= \frac{1}{T} [\langle H \rangle + kT \ln \mathcal{Z}(T, V, N)]. \tag{13}$$

From which the Helmholtz free energy $F = F(\Gamma)$ and the Gibbs free energy $G = G(\Gamma)$ are obtained,

$$F = U - TS = -kT \ln \mathcal{Z}(T, V, N), \quad G = PV + F = PV - kT \ln \mathcal{Z}(T, V, N). \tag{14}$$

Therefore, to obtain the Gibbs free energy one has to really calculate $\ln \mathcal{Z}(T,V,N)$, a task that often cannot be done.

4.2. The grand canonic ensemble

The grand canonic ensemble or macroscopic ensemble deals with an open thermodynamic system Γ, i.e., not only energy can be exchanged, matter particles can also be exchanged between Γ and environment. Therefore, the particle number N in Γ is variable.

In classical mechanics, suppose that the phase space is $(\mathbf{q},\mathbf{p}) \in \Gamma^N \times \mathbb{R}^{3N}$. Let H be the Hamiltonian, the grand canonic phase-space density is

$$p_{gc}(\mathbf{q},\mathbf{p},N) = \frac{\exp[-\beta(H-\mu N)]}{\sum_{N=0}^{\infty} \frac{1}{N!h^{3N}} \int_{\Gamma^N} d\mathbf{q}^N \int_{\mathbb{R}^{3N}} \exp[-\beta(H(\mathbf{q},\mathbf{p})-\mu N] d\mathbf{p}^N} = \frac{\exp[-\beta(H-\mu N)]}{\mathcal{Z}(T,V,\mu)}, \tag{15}$$

where $V = V(\Gamma)$ is the volume of the system. By definition the entropy is

$$\begin{aligned} S(\Gamma) &= \langle -k \ln p_{gc} \rangle = k \sum_{N=0}^{\infty} \int_{\Gamma^N} d\mathbf{q}^N \int_{\mathbb{R}^{3N}} \{\beta[H(\mathbf{q},\mathbf{p}) - \mu N] + \ln \mathcal{Z}\} p_{gc}(\mathbf{q},\mathbf{p}) d\mathbf{p}^N \\ &= \frac{1}{T} \left[\langle H \rangle - \mu \langle N \rangle + kT \ln \mathcal{Z} \right]. \end{aligned} \tag{16}$$

Here $\langle H \rangle = U$ is the inner energy of the system Γ, $\langle N \rangle = N(\Gamma)$ is the mean number of particles in Γ. More importantly, the function $-kT \ln \mathcal{Z}(T,V,\mu)$ is nothing but the grand canonic potential ϕ, from equation (8) it is just $-PV$. Thus

$$G = U + PV - TS = \mu \langle N \rangle. \tag{17}$$

5. Experimental measuring and theoretical derivation of the Gibbs free energy of protein folding

The newly synthesized peptide chain of a protein automatically folds to its native structure in the physiological environment. Change of environment will make a protein denatured, i.e., the protein no longer performs its biological function. The facts that denaturation does not change the protein molecule, that the only thing changed is its structure, was first theorized by Hisen Wu based on his own extensive experiments, Hisen Wu (1931). It was found that after removing the agents that caused the change of environment, some protein can automatically retake its native structure, this is called renaturation or refolding. After many experiments in denaturation and renaturation, Anfinsen summarized the Thermodynamic Principle as the fundamental law of the protein folding, Anfinsen (1973). Anfinsen's work actually show that protein refolds spontaneously after removing denaturation agents. Therefore, in the physiological or similar environment, the native structure has the minimum Gibbs free energy; and in a changed environment, the denatured structure(s) will have the smaller Gibbs free energy. The Thermodynamic Principle of protein folding then is the general thermodynamics law, if a change happens spontaneously, then the end state will have smaller Gibbs free energy than the initial state.

To apply the Thermodynamic Principle in the research of protein folding, it is necessary to know the Gibbs free energy formula $G(\mathbf{X})$ for each conformation $\mathbf{X}$. Until now, theoretical derivation of $G(\mathbf{X})$ is unsuccessful and rarely being tried. Most knowledge of the Gibbs free energy of protein folding comes from experiment observations.

5.1. Experimental measuring of $\triangle G$

The basic principle of experimentally measuring $\triangle G$, the difference in Gibbs free energy between the native and the denatured structures of a protein is as follows. For protein molecules in a solution, the criterion of the protein is in the native structure is that it performes its biological function, otherwise the protein is denatured or not in the native structure. The level of biological function indicates the degree of the denaturation. Let B be the native structure, denote its molar concentration as $[B]$. Denote A as an non-native structure of the same protein in the solution and $[A]$ its molar concentration.

Three things to be borne in mind: 1. the environment is the physiological environment or similar one such that the protein can spontaneously fold; 2: individual molecule cannot be directly measured, so the measuring is in per mole term, $R = N_A k$ instead of k should be used, where N_A is the Avogadro's number; 3: the environment in reality has constant pressure P, hence the enthalpy $H = U + PV$ can replace the inner energy U, where V is the volume of the system (it is a subset of the whole $\mathbb{R}^3$).

As expressed in (9), the probabilities of the protein takes the conformations A and B are

$$p_A \propto W_A \exp\left(-\frac{H_A}{RT}\right), \quad p_B \propto W_B \exp\left(-\frac{H_B}{RT}\right), \tag{18}$$

where $H_A = U_A + PV$ and $H_B = U_B + PV$ are the enthalpy per mole for A and B, W_A (W_B) is the number of ways of the enthalpy H_A (H_B) can be achieved by microscopic states. The quantities $[A]$ and $[B]$ are assumed to be measurable in experiment. Therefore their ration $K = [A]/[B]$ is also measurable. Then

$$\triangle G^o = -RT \ln K. \tag{19}$$

To see that equation (19) is true, note that the ratio K is equal to the ratio p_A/p_B and the entropies per mole are $S_A = R \ln W_A$, $S_B = R \ln W_B$, therefore

$$\begin{aligned} -RT\ln K &= -RT\ln\frac{p_A}{p_B} = -RT\left(\frac{H_B}{RT} - \frac{H_A}{RT}\right) - RT(\ln W_A - \ln W_B) \\ &= H_A - H_B - T(R\ln W_A - R\ln W_B) = H_A - H_B - T(S_A - S_B) \\ &= H_A - TS_A - (H_B - TS_B) = G_A - G_B = \triangle G^o. \end{aligned} \tag{20}$$

But in reality, the ratio K is measurable in experiment is only theoretical, since in physiological environment $K \cong 0$, i.e., almost all protein molecules take the native structure B.

There is no way to change the native structure B to A while keeping the environment unchanged. In experiments, one has to change the environment to get the protein denatured, that is, to change its shape from the native structure B to another conformation A. Heating the solution is a simple way to change the environment, during the heating, the system absorbs an amount of heat H, the system's temperature increased from T_0 to T_1. Then

$$G(A,T_1) - G(B,T_0) = f(H), \tag{21}$$

where $f(H)$ is a function depending on H and its value is obtained from experiment. What really needed is

$$\triangle G = G(A,T_0) - G(B,T_0). \tag{22}$$

To get $\triangle G$, interpolation to equation (21) is used to estimate the value in T_0. Other methods of changing environment face the same problem, i.e., interpolation has neither theoretical nor observation basis.

Equation (19) may give the reason why $\triangle G$ is used whenever referring the Gibbs free energy. For experiment, only $\triangle G$ can be got. In theoretical derivation, this rule no longer to be followed and moreover, without a base structure to compare to, the notation $\triangle G$ will look strange.

More importantly, it should be emphasized again, that the Thermodynamic Principle really says that in the physiological environment the native structure has the minimum Gibbs free energy; and in other environment, the native structure no longer has the minimum Gibbs free energy. Summarizing, it is

$$G(B,T_0) < G(A,T_0), \quad G(A,T_1) < G(B,T_1). \tag{23}$$

It should always keep in mind that before comparison, first clarify the environment.

When deriving the Gibbs free energy formula, the first thing is also to make clear what is the environment. Another reality that should be borne in mind is that during the protein folding process, the environment does not change.

Remember that after removing the denaturation agent some proteins will spontaneously refold to their native structure, this is called the refolding or renaturation. Distinguish the original protein folding problem and protein refolding problem is another important issue. Only in the refolding case, a theoretical derivation can make the environment change, for example, lower the temperature to the room temperature (around 300K). Some discussions on protein folding are really talking about refolding, because they start from changing the environment from nonphysological to physiological.

While experiment has no way to change the native structure without disturbing the environment, theory can play a role instead. Formulas (1) and (2) give us the chance to compare $\triangle G$, as long as the accurate chemical potentials' values are known.

5.2. Theoretical consideration of the protein folding problem

Protein folding is a highly practical field. Very few attention was paid to its theoretical part. For example, almost nobody has seriously considered the Gibbs free energy formula. Instead, all kinds of empirical models are tried in computer simulation, without any justification in fundamental principle.

One attempt to theoretically get the Gibbs free energy formula from canonic ensemble is summarized by Lazaridis and Karplus (2003), the theoretical part of it is reported below and why it is not successful will be briefly pointed out. Their notations such as $\mathbf{R} = \mathbf{X}$ as conformation, $A = F$ as the Helmholtz free energy, $Q = \mathcal{Z}$ as the partition function, $\Lambda = h$, etc., will be kept in this section.

Treating the protein folding system as the set of all conformations plus surrounding water molecules with a phase point $(\mathbf{R}, \mathbf{r})$, where $\mathbf{r}$ are coordinates of N water molecules plus their orientations. The Hamiltonian H can be decomposed as

$$H = H_{mm} + H_{mw} + H_{ww}, \tag{24}$$

where mm means interactions inside the protein, mw between protein and water molecules, and ww water to water, all in the atomic level. Triplet interactions mmm, mmw, etc., can also be considered, but for simplicity only take the pairwise atomic interactions.

Applying the canonic ensemble, the canonic partition function is

$$Q = \frac{\int \exp(-\beta H) d\mathbf{r}^N d\mathbf{R}^M}{N! \Lambda^{3M} \Lambda^{3N}} = \frac{Z}{N! \Lambda^{3M} \Lambda^{3N}},$$

and the Helmholtz free energy is given by

$$A = -kT \ln Q = -kT \ln \left[\int \exp(-\beta H) d\mathbf{r}^N d\mathbf{R}^M \right] + kT \ln(N! \Lambda^{3M} \Lambda^{3N}). \tag{25}$$

To separate the contributions made by water molecules and the conformations, the *effective energy* W is defined,

$$\exp(-\beta W) = \exp(-\beta H_{mm}) \frac{\int \exp(-\beta H_{mw} - \beta H_{ww}) d\mathbf{r}^N}{\int \exp(-\beta H_{ww}) d\mathbf{r}^N} = \exp(-\beta H_{mm}) \exp(-\beta X), \tag{26}$$

Define

$$\langle \exp(-\beta H_{mw}) \rangle_o = \frac{\int \exp(-\beta H_{mw}) \exp(-\beta H_{ww}) d\mathbf{r}^N}{\int \exp(-\beta H_{ww}) d\mathbf{r}^N}. \tag{27}$$

The effective energy $W(\mathbf{R})$ is:

$$W(\mathbf{R}) = H_{mm}(\mathbf{R}) + X(\mathbf{R}) = H_{mm}(\mathbf{R}) - kT \ln \langle \exp(-\beta H_{mw}) \rangle_o \equiv H_{mm}(\mathbf{R}) + \triangle G^{\text{slv}}(\mathbf{R}). \tag{28}$$

The term $\triangle G^{\text{slv}}(\mathbf{R})$ is called the *solvation free energy* while H_{mm} is the *intra-macromolecular energy*.

After changing $\mathbf{R}$ to interior coordinates $\mathbf{q}$, it is stated that

$$Z = V 8\pi^2 \int \exp(-\beta H_{ww}) d\mathbf{r}^N \int \exp(-\beta W) d\mathbf{q}, \tag{29}$$

because the interior coordinates has only $3M - 6$ dimension, the integration of the remaining 6 dimension over the system getting the value $V 8\pi^2$, implying that each $\mathbf{x}_i$ in $\mathbf{R}$ can be any point in the system that has volume V. As usual, the probability of finding the system at the configuration $(\mathbf{q})$ is:

$$p(\mathbf{q}) = \frac{\exp[-\beta W(\mathbf{q})]}{\int \exp[-\beta W(\mathbf{q})]d\mathbf{q}}. \tag{30}$$

Consequently,

$$\int p(\mathbf{q}) \ln p(\mathbf{q}) d\mathbf{q} = -\ln Z + \ln \int \exp(-\beta H_{ww}) d\mathbf{r}^N + \ln(V8\pi^2) - \beta \int p(\mathbf{q}) W(\mathbf{q}) d\mathbf{q}, \tag{31}$$

From equation (25),

$$\begin{aligned} A &= -kT \int \exp(-\beta H_{ww}) d\mathbf{r}^N + kT \ln\left(\frac{\Lambda^{3M}}{V8\pi^2}\right) + \int p(\mathbf{q}) W(\mathbf{q}) d\mathbf{q} + kT \int p(\mathbf{q}) \ln p(\mathbf{q}) d\mathbf{q} \\ &= A^o + kT \ln\left(\frac{\Lambda^{3M}}{V8\pi^2}\right) + \langle W \rangle - TS^{\text{conf}}, \end{aligned} \tag{32}$$

where $A^o = -kT \int \exp(-\beta H_{ww}) d\mathbf{r}^N$ is the pure Helmholtz free energy of pure solvent; the term $-TS^{\text{conf}} = kT \int p(\mathbf{q}) \ln p(\mathbf{q}) d\mathbf{q}$ is the contribution of the configurational entropy of the macromolecule to the free energy.

The Gibbs free energy is $G = A + PV$. Since the volume is thought negligible under ambient conditions so Gibbs and Helmholtz free energies are considered identical.

Now for any subset of $A \subset \Gamma$, integrals restricted on A gives the Helmholtz energy A_A, i.e.,

$$A_A = A^o + kT \ln\left(\frac{\Lambda^{3M}}{V8\pi^2}\right) + \langle W \rangle_A - TS_A^{\text{conf}}. \tag{33}$$

Thus for two different subsets A and B, the difference in the Helmholtz free energy is

$$\begin{aligned} \triangle A = A_B - A_A &= \langle W \rangle_B - \langle W \rangle_A - T(S_B^{\text{conf}} - S_A^{\text{conf}}) \\ &= \triangle \langle H_{mm} \rangle + \triangle \langle \triangle G^{\text{slv}} \rangle - T \triangle S^{\text{conf}}. \end{aligned} \tag{34}$$

Especially, "If A is the denatured state and B the native state, both of which have to be defined in some way and both of which include many configurations, Eq. (34) gives the free energy of folding."

5.3. Critics of the derivation in Lazaridis and Karplus (2003)

Protein folding is considered a very practical research field, dominating activities are computer simulations with empirical models. There are very few theoretical discussions about protein folding. This derivation in Lazaridis and Karplus (2003) is a rare example deserving an analysis to see why for decades there has been no theoretic progress in this field. Many lessons can be learned from this example.

One important lesson from the derivation Lazaridis and Karplus (2003) is that when dealing with thermodynamics and statistical mechanics, the thermodynamic system must be clearly defined. The system will occupy a space in $\mathbb{R}^3$, what is it? How to delimit it?

More importantly, it is not just one conformation $\mathbf{R}$, but all conformations of a single protein are considered in the derivation. As a single point $\mathbf{R} \in \mathbb{R}^{3M}$, no structural features of the conformation $\mathbf{R}$ are considered, i.e., this particle is structureless. Remember that the research object is the conformation of the protein, we cannot treat them as structureless particles. Yes, classical derivations such as the ideal gas system are defined this way, that is because that the interest is not in the individual particle's structure but the macroscopic properties of the idea gas. The lesson then is that instead of considering all conformations together in a system, specific thermodynamic system has to be tailored for each individual conformation $\mathbf{R}$. And such a system contains only one conformation $\mathbf{R}$, with its structure geometry, and other particles such as water molecules, thus the Gibbs free energy of such a system will be indexed by $\mathbf{R}$, $G = G(\mathbf{R})$.

Perhaps the biggest lesson to be learned is that when solving a problem, one should concentrate on the specific features of the problem to design the ways to attack it, not just imitate successful classical examples.

The derivation of Lazaridis and Karplus (2003) gives the effective energy $W(\mathbf{R})$ as some substitute of the Gibbs free energy without theoretic basis for its relation to the Thermodynamic Principle. Moreover, the formula $W(\mathbf{R})$ tells us nothing of how to calculate it, all are buried in multiple-integrations without clear delimitation. Being the only function for individual conformation $\mathbf{R}$, it was pointed out in Lazaridis and Karplus (2003) that "The function W defines a hyper-surface in the conformation space of the macromolecule in the presence of equilibrated solvent and, therefore, includes the solvation entropy. This hyper-surface is now often called an 'energy landscape'. It determines the thermodynamics and kinetics of macromolecular conformational transitions." From this comment it can be seen that the authors are not against individual quantities such as $W(\mathbf{R})$ and think they are important to the study of protein folding. Changing the "effective energy" $W(\mathbf{R})$ to the Gibbs free energy $G(\mathbf{R})$, the comment really makes sense. The lesson should be learned is that never invent theoretical concepts without firm theoretical basis. Another one is that always keep in mind that useful Gibbs free energy formula should be calculable.

From now on, the notation $\mathbf{X} = \mathbf{R}$ will be used to represent a conformation. To put the Thermodynamic Principle in practice, not merely as a talking show, what really needed is $G(\mathbf{X})$, the Gibbs free energy of each individual conformation $\mathbf{X}$, not the effective energy $W(\mathbf{R})$. One hopes that the formula $G(\mathbf{X})$ should be calculable, not buried in multiple integrations. To get such a formula, the grand canonic ensemble and eventually the quantum statistics have to be applied.

6. Necessary preparations for the derivation of the Gibbs free energy formula

Summarizing what have learned from the critics of the derivation in Lazaridis and Karplus (2003), in any attempt of derivation of the Gibbs free energy formula one has to: 1. clearly state all assumptions used in the derivation; and 2. for each conformation $\mathbf{X}$, set a thermodynamic system $\mathcal{T}_{\mathbf{X}}$ associated with $\mathbf{X}$; 3. use the grand canonic ensemble.

6.1. The assumptions

All assumptions here are based on well-known facts of consensus among protein folding students. Let $\mathfrak{U}$ be a protein with M atoms $(\mathbf{a}_1, \cdots, \mathbf{a}_i, \cdots, \mathbf{a}_M)$. A structure (conformation) of $\mathfrak{U}$ is a point $\mathbf{X} = (\mathbf{x}_1, \cdots, \mathbf{x}_i, \cdots, \mathbf{x}_M) \in \mathbb{R}^{3M}$, $\mathbf{x}_i \in \mathbb{R}^3$ is the atomic center (nuclear) position of $\mathbf{a}_i$. Alternatively, the conformation $\mathbf{X}$ corresponds to a subset in $\mathbb{R}^3$, $P_{\mathbf{X}} = \cup_{i=1}^{M} B(\mathbf{x}_i, r_i) \subset \mathbb{R}^3$ where r_i's are van der Waals radii.

1. The proteins discussed here are monomeric, single domain, self folding globular proteins.
2. Therefore, in the case of our selected proteins, the environment of the protein folding, the physiological environment, is pure water, there are no other elements in the environment, no chaperonins, no co-factors, etc. This is a rational simplification, at least when one considers the environment as only the first hydration shell of a conformation, as in our derivation of the $G(\mathbf{X})$.
3. During the folding, the environment does not change.
4. Anfinsen (1973) showed that before folding, the polypeptide chain already has its main chain's and each residue's covalent bonds correctly formed. Hence, our conformations should satisfy the following steric conditions set in Fang (2005) and Fang and Jing (2010): there are $\epsilon_{ij} > 0$, $1 \leq i < j \leq M$ such that for any two atoms $\mathbf{a}_i$ and $\mathbf{a}_j$ in $P_{\mathbf{X}} = \cup_{k=1}^{M} B(\mathbf{x}_k, r_k)$,

$$\begin{array}{ll} \epsilon_{ij} \leq |\mathbf{x}_i - \mathbf{x}_j|, & \text{no covalent bond between } \mathbf{a}_i \text{ and } \mathbf{a}_j; \\ d_{ij} - \epsilon_{ij} \leq |\mathbf{x}_i - \mathbf{x}_j| \leq d_{ij} + \epsilon_{ij}, & d_{ij} \text{ is the standard bond length between } \mathbf{a}_i \text{ and } \mathbf{a}_j. \end{array} \tag{35}$$

   All conformations satisfying the steric conditions (35) will be denoted as $\mathfrak{X}$ and in this chapter only $\mathbf{X} \in \mathfrak{X}$ will be considered.
5. A water molecule is treated as a single particle centered at the oxygen nuclear position $\mathbf{w} \in \mathbb{R}^3$, and the covalent bonds in it are fixed. In the Born-Oppenheimer approximation, only the conformation $\mathbf{X}$ is fixed, all particles, water molecules or electrons in the first hydration shell of $P_{\mathbf{X}}$, are moving.
6. As in section 2.3, there are H hydrophobic levels H_i, $i = 1, \cdots, H$, such that $\cup_{i=1}^{H} H_i = (\mathbf{a}_1, \cdots, \mathbf{a}_i, \cdots, \mathbf{a}_M)$.

6.2. The thermodynamic system $\mathcal{T}_{\mathbf{X}}$

Let d_w be the diameter of a water molecule and $M_{\mathbf{X}}$ be the molecular surface of $P_{\mathbf{X}}$ as defined in Richards (1977) with the probe radius $d_w/2$, see FIGURE 3. Define

$$\mathcal{R}_{\mathbf{X}} = \{\mathbf{x} \in \mathbb{R}^3 : \text{ dist}(\mathbf{x}, M_{\mathbf{X}}) \leq d_w\} \setminus P_{\mathbf{X}} \tag{36}$$

as the first hydration shell surrounding $P_{\mathbf{X}}$, where $\text{dist}(\mathbf{x}, S) = \inf_{\mathbf{y} \in S} |\mathbf{x} - \mathbf{y}|$. Then $\mathcal{T}_{\mathbf{X}} = P_{\mathbf{X}} \cup \mathcal{R}_{\mathbf{X}}$ will be our thermodynamic system of protein folding at the conformation $\mathbf{X}$.

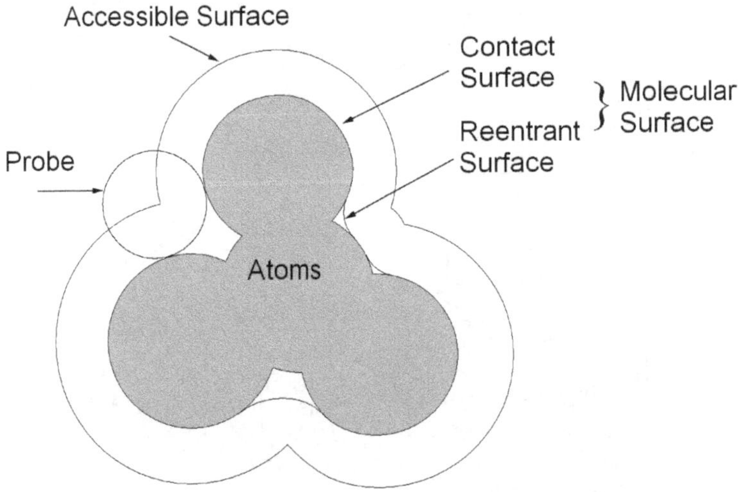

Figure 3. Two dimensional presenting of molecular surface Richards (1977) and solvent accessible surface Lee and Richards (1971). This figure was originally in Fang and Jing (2010).

Let $I_i \subset \{1,2,\cdots,M\}$ be the subset such that $\mathbf{a}_j \in H_i$ if and only if $j \in I_i$. Define $P_{\mathbf{X}i} = \cup_{j\in I_i} B(\mathbf{x}_j, r_j) \subset P_{\mathbf{X}}$ and as shown in FIGURE 4,

$$\mathcal{R}_{\mathbf{X}i} = \{\mathbf{x} \in \mathcal{R}_{\mathbf{X}} : \ \mathrm{dist}(\mathbf{x}, P_{\mathbf{X}_i}) \le \mathrm{dist}(\mathbf{x}, P_{\mathbf{X}} \backslash P_{\mathbf{X}i})\}, \quad 1 \le i \le H. \tag{37}$$

Let $V(\Omega)$ be the volume of $\Omega \subset \mathbb{R}^3$, then

$$\mathcal{R}_{\mathbf{X}} = \cup_{i=1}^{H} \mathcal{R}_{\mathbf{X}i}, \quad V(\mathcal{R}_{\mathbf{X}}) = \sum_{i=1}^{H} V(\mathcal{R}_{\mathbf{X}i}), \text{ and for } i \ne j, \quad V(\mathcal{R}_{\mathbf{X}i} \cap \mathcal{R}_{\mathbf{X}j}) = 0. \tag{38}$$

Since $M_{\mathbf{X}}$ is a closed surface, it divides $\mathbb{R}^3$ into two regions $\Omega_{\mathbf{X}}$ and $\Omega'_{\mathbf{X}}$ such that $\partial\Omega_{\mathbf{X}} = \partial\Omega'_{\mathbf{X}} = M_{\mathbf{X}}$ and $\mathbb{R}^3 = \Omega_{\mathbf{X}} \cup M_{\mathbf{X}} \cup \Omega'_{\mathbf{X}}$. Note that $P_{\mathbf{X}} \subset \Omega_{\mathbf{X}}$ and all nuclear centers of atoms in the water molecules in $\mathcal{R}_{\mathbf{X}}$ are contained in $\Omega'_{\mathbf{X}}$. Moreover, $\Omega_{\mathbf{X}}$ is bounded, therefore, has a volume $V(\Omega_{\mathbf{X}})$. For $S \subset \mathbb{R}^3$, denote $\overline{S}$ as the closure of S. Define the hydrophobicity subsurface $M_{\mathbf{X}i}, 1 \le i \le H$, as

$$M_{\mathbf{X}i} = M_{\mathbf{X}} \cap \overline{\mathcal{R}_{\mathbf{X}i}}. \tag{39}$$

Let $A(S)$ be the area of a surface $S \subset \mathbb{R}^3$, then

$$M_{\mathbf{X}} = \cup_{i=1}^{H} M_{\mathbf{X}i}, \quad A(M_{\mathbf{X}}) = \sum_{i=1}^{H} A(M_{\mathbf{X}i}), \text{ and if } i \ne j, \text{ then } A(M_{\mathbf{X}i} \cap M_{\mathbf{X}j}) = 0. \tag{40}$$

Although the shape of each atom in a molecule is well defined by the theory of atoms in molecules as in Bader (1990) and Popelier (2000), what concerning us here is the overall

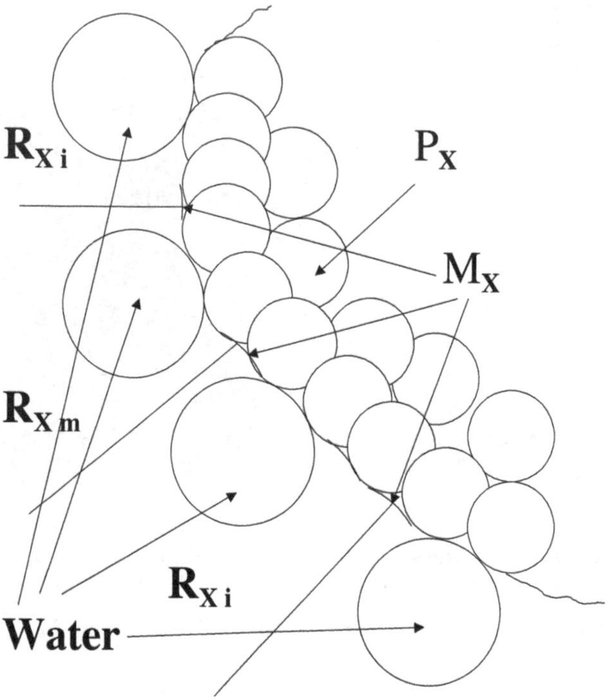

Figure 4. Note that $\mathcal{R}_{\mathbf{X}i}$ generally are not connected, i.e., having more than one block.

shape of the structure $P_{\mathbf{X}}$. The cutoff of electron density $\rho \geq 0.001$au in Bader (1990) and Popelier (2000) gives the overall shape of a molecular structure that is just like $P_{\mathbf{X}}$, a bunch of overlapping balls. Moreover, the boundary of the $\rho \geq 0.001$au cutoff is very similar to the molecular surface $M_{\mathbf{X}}$ which was defined by Richards (1977) and was shown has more physical meaning as the boundary surface of the conformation $P_{\mathbf{X}}$ in Tuñón *et. al.* (1992) and Jackson and Sternberg (1993).

7. Gibbs free energy formula: Classical statistical mechanics derivation

The grand canonic ensemble or macroscopic ensemble will be applied to derive the desired Gibbs free energy formula $G(\mathbf{X})$. In addition to let the number of water molecules vary, the assumptions is that the chemical potential μ will be different for water molecules contacting to different hydrophobicity levels H_i (or falling in $\mathcal{R}_{\mathbf{X}i}$). Counting the numbers N_i of water molecules that contact to atoms in H_i, the N and μ in equation (15) should be modified to $N = (N_1, \cdots, N_i, \cdots, N_H)$, $\mu = (\mu_1, \cdots, \mu_i, \cdots, \mu_H)$. Let $(\mathbf{q}, \mathbf{p}) \subset \mathcal{R}_{\mathbf{X}}^M \times \mathbb{R}^{3M}$ be the water molecules' phase space for a fixed N, where $M = \sum_{i=1}^H N_i$. Let $H_{\mathbf{X}} = H_{\mathbf{X}}(\mathbf{q}, \mathbf{p})$ be the Hamiltonian. The grand canonic phase density function will be

$$p_{\mathbf{X}}(\mathbf{q},\mathbf{p},N) = \frac{\exp\{-\beta[H_{\mathbf{X}}(\mathbf{q},\mathbf{p}) - \sum_{i=1}^{H}\mu_i N_i]\}}{\sum_{M=0}^{\infty}\frac{1}{M!h^{3M}}\sum_{\sum N_i=M}\prod_{i=1}^{H}\int_{\mathcal{R}_{\mathbf{X}i}^{N_i}}\mathbf{dq}^{N_i}\int_{\mathbb{R}^{3M}}\exp\{-\beta[H_{\mathbf{X}} - \sum_{i=1}^{H}\mu_i N_i]\}\mathbf{dp}^{3M}}$$

$$= \frac{\exp\{-\beta[H_{\mathbf{X}}(\mathbf{q},\mathbf{p}) - \sum_{i=1}^{H}\mu_i N_i]\}}{\mathcal{Z}(T,V,\mu)}. \quad (41)$$

The entropy $S(\mathbf{X}) = S(\mathcal{T}_{\mathbf{X}})$ is

$$S = \langle -k\ln p_{\mathbf{X}}\rangle = -k\sum_{M=0}^{\infty}\sum_{N_1+\cdots+N_H=M}\prod_{i=1}^{H}\int_{\mathcal{R}_{\mathbf{X}i}^{N_i}}\mathbf{dq}^{N_i}\int_{\mathbb{R}^{3M}}\ln p_{\mathbf{X}}p_{\mathbf{X}}\mathbf{dp}^{M}$$

$$= -k\sum_{M=0}^{\infty}\sum_{N_1+\cdots+N_H=M}\prod_{i=1}^{H}\int_{\mathcal{R}_{\mathbf{X}i}^{N_i}}\mathbf{dq}^{N_i}\left[\beta\sum_{i=1}^{H}\mu_i N_i - \beta H_{\mathbf{X}}(\mathbf{q},\mathbf{p}) - \ln\mathcal{Z}\right]p_{\mathbf{X}}\mathbf{dp}^{M} \quad (42)$$

$$= \frac{1}{T}\left[\langle H\rangle - \sum_{i=1}^{H}\mu_i\langle N_i\rangle + kT\ln\mathcal{Z}(T,V,\mu)\right] \quad (43)$$

$$= \frac{1}{T}\left[U(\mathcal{T}_{\mathbf{X}}) - \sum_{i=1}^{H}\mu_i N_i(\mathcal{T}_{\mathbf{X}}) + kT\ln\mathcal{Z}(T,V,\mu)\right]$$

where $U(\mathbf{X}) = U(\mathcal{T}_{\mathbf{X}}) = \langle H\rangle$ is the inner energy, $N_i(\mathbf{X}) = \langle N_i\rangle$ the mean number of water molecules in $\mathcal{R}_{\mathbf{X}i}$. By equation (8), $kT\ln\mathcal{Z}(T,V,\mu) = -\phi(T,V,\mu) = PV(\mathcal{T}_{\mathbf{X}})$. Therefore, from $G = U + PV - TS$,

$$G(\mathbf{X}) = G(\mathcal{T}_{\mathbf{X}}) = U(\mathbf{X}) + PV(\mathcal{T}_{\mathbf{X}}) - TS(\mathcal{T}_{\mathbf{X}}) = \sum_{i=1}^{H}\mu N_i(\mathbf{X}). \quad (44)$$

The Gibbs free energy given in formula (44) does not involve any integration at all, just counting the number of water molecules contacting atoms in H_i. Furthermore, against the effective energy, potential function H_{mm} plays no role at all, a surprise indeed. But formula (44) also is not easy to calculate, counting the number of water molecules actually need more knowledge of the conformation's boundary, the molecular surface $M_{\mathbf{X}}$. Formula (44) can be directly transfered into a geometric version.

7.1. Converting formula (44) to a geometric version

Since every water molecule in $\mathcal{R}_{\mathbf{X}i}$ has contact with the surface $M_{\mathbf{X}i}$, $N_i(\mathbf{X})$ is proportional to the area $A(M_{\mathbf{X}i})$. Therefore, there are $\nu_i > 0$, such that

$$\nu_i A(M_{\mathbf{X}i}) = N_i(\mathbf{X}), \quad 1 \le i \le H. \quad (45)$$

Substitute in (44),

$$G(\mathbf{X}) = G(\mathcal{T}_{\mathbf{X}}) = \sum_{i=1}^{H}\mu_i\nu_i A(M_{\mathbf{X}i}). \quad (46)$$

For each conformation $\mathbf{X}$, the molecular surface $M_{\mathbf{X}}$ is calculable, see Connolly (1983). The areas $A(M_{\mathbf{X}})$ and $A(M_{\mathbf{X}\,i})$ are also calculable. Therefore, unlike the formula given in (34), this formula is calculable. Moreover, our derivation theoretically justified the surface area models that will be discussed later, only difference is that the molecular surface area is used here instead of the solvent accessible surface area.

But still something is missing. That is, the volume $V(\mathcal{T}_{\mathbf{X}})$, an important thermodynamic quantity, does not show here at all. It seems that no way to put the $V(\mathcal{T}_{\mathbf{X}})$ here in the classical statistical mechanics. To resolve this, the quantum statistical mechanics is necessary.

8. A quantum statistical theory of protein folding

In 1929 Dirac wrote: "The underlying physical laws necessary for the mathematical theory of ... the whole of chemistry are thus completely known, and the difficulty is only that the exact application of these laws leads to equations much too complicated to be soluble." (quoted from Bader (1990), page 132). Yes, the multidimensional Shrödinger equation for protein folding is beyond our ability to solve, no matter how fast and how powerful our computers are. But mathematical theory guarantees that there is a complete set of eigenvalues (energy levels) and eigenfunctions to the Shrödinger equation in the Born-Oppenheimer approximation. Then consider that in the statistical mechanics, ensembles collect all (energy) states of the same system. Although one cannot have exact solutions to the Shrödinger equation, the eigenvalues of it are theoretically known. Thus one can apply the grand canonical ensemble to obtain the desired Gibbs free energy formula $G(\mathbf{X})$. This is the main idea of the derivation.

8.1. The Shrödinger equation

For any conformation $\mathbf{X} \in \mathfrak{X}$, let $\mathbf{W} = (\mathbf{w}_1, \cdots, \mathbf{w}_i, \cdots, \mathbf{w}_N) \in \mathbb{R}^{3N}$ be the nuclear centers of water molecules in $\mathcal{R}_{\mathbf{X}}$ and $\mathbf{E} = (\mathbf{e}_1, \cdots, \mathbf{e}_i, \cdots, \mathbf{e}_L) \in \mathbb{R}^{3L}$ be electronic positions of all electrons in $\mathcal{T}_{\mathbf{X}}$. Then the Hamiltonian for the system $\mathcal{T}_{\mathbf{X}}$ is

$$\hat{H} = \hat{T} + \hat{V} = -\sum_{i=1}^{M} \frac{\hbar^2}{2m_i} \nabla_i^2 - \frac{\hbar^2}{2m_w} \sum_{i=1}^{N} \nabla_i^2 - \frac{\hbar^2}{2m_e} \sum_{i=1}^{L} \nabla_i^2 + \hat{V}(\mathbf{X}, \mathbf{W}, \mathbf{E}), \tag{47}$$

where m_i is the nuclear mass of atom $\mathbf{a}_i$, m_w and m_e are the masses of water molecule and electron; ∇_i^2 is Laplacian in corresponding $\mathbb{R}^3$; and V the potential.

8.2. The first step of the Born-Oppenheimer approximation

Depending on the shape of $P_{\mathbf{X}}$, for each i, $1 \le i \le H$, the maximum numbers $N_{\mathbf{X}\,i}$ of water molecules contained in $\mathcal{R}_{\mathbf{X}\,i}$ vary. Theoretically all cases are considered, i.e., there are $0 \le N_i \le N_{\mathbf{X}\,i}$ water molecules in $\mathcal{R}_{\mathbf{X}\,i}$, $1 \le i \le H$. Let $M_0 = 0$ and $M_i = \sum_{j \le i} N_j$ and $\mathbf{W}_i = (\mathbf{w}_{M_{i-1}+1}, \cdots, \mathbf{w}_{M_{i-1}+j}, \cdots, \mathbf{w}_{M_i}) \in \mathcal{R}_{\mathbf{X}\,i}^{N_i}$, $1 \le i \le H$, and $\mathbf{W} = (\mathbf{W}_1, \mathbf{W}_2, \cdots, \mathbf{W}_{M_H}) \in \prod_{i=1}^{H} \mathcal{R}_{\mathbf{X}\,i}^{N_i}$ denote the nuclear positions of water molecules in $\mathcal{R}_{\mathbf{X}}$. As well, there will be all possible numbers $0 \le N_e < \infty$ of electrons in $\mathcal{T}_{\mathbf{X}}$. Let $\mathbf{E} = (\mathbf{e}_1, \mathbf{e}_2, \cdots, \mathbf{e}_{N_e}) \in \mathbb{R}^{3N_e}$ denote their nuclear positions. For each fixed $\mathbf{X} \in \mathfrak{X}$ and $N = (N_1, \cdots, N_H, N_e)$, the Born-Oppenheimer approximation has the Hamiltonian

$$\hat{H}_{\mathbf{X}} = -\frac{\hbar^2}{2}\left\{\frac{1}{m_w}\sum_{j=1}^{M_H}\nabla_j^2 + \frac{1}{m_e}\sum_{v=1}^{N_e}\nabla_v^2\right\} + \hat{V}(\mathbf{X},\mathbf{W},\mathbf{E}). \tag{48}$$

The eigenfunctions $\psi_i^{\mathbf{X},N}(\mathbf{W},\mathbf{E}) \in L_0^2(\prod_{i=1}^H \mathcal{R}_{\mathbf{X}i}^{N_i} \times \mathcal{T}_{\mathbf{X}}^{N_e}) = \mathcal{H}_{\mathbf{X},N}$, $1 \leq i < \infty$, comprise an orthonormal basis of $\mathcal{H}_{\mathbf{X},N}$. Denote their eigenvalues (energy levels) as $E_{\mathbf{X},N}^i$, then $\hat{H}_{\mathbf{X}}\psi_i^{\mathbf{X},N} = E_{\mathbf{X},N}^i\psi_i^{\mathbf{X},N}$.

8.3. Grand partition function and grand canonic density operator

Since the numbers N_i and N_e vary, the grand canonic ensemble should be adopted. Let μ_i be the chemical potentials, that is, the Gibbs free energy per water molecule in $\mathcal{R}_{\mathbf{X}i}$. Let μ_e be electron chemical potential. The grand canonic density operator is like in equation (15), or see Greiner *et. al.* (1994) and Dai (2007)

$$\hat{\rho}_{\mathbf{X}} = \exp\left\{-\beta\left[\hat{H}_{\mathbf{X}} - \sum_{i=1}^{H}\mu_i\hat{N}_i - \mu_e\hat{N}_e - \phi(\mathbf{X})\right]\right\}. \tag{49}$$

where $\phi(\mathbf{X})$ is the grand canonic potential ϕ in equation (8) with the index $\mathbf{X}$ and the grand partition function is

$$\begin{aligned}\exp[-\beta\phi(\mathbf{X})] &= \mathrm{Trace}\left\{\exp\left[-\beta\left(\hat{H}_{\mathbf{X}} - \sum_{i=1}^{H}\mu_i\hat{N}_i - \mu_e\hat{N}_e\right)\right]\right\} \\ &= \sum_{i,N}\exp\left\{-\beta\left[E_{\mathbf{X},N}^i - \sum_{i=1}^{H}\mu_i N_i - \mu_e N_e\right]\right\}.\end{aligned} \tag{50}$$

8.4. The Gibbs free energy $G(\mathbf{X})$

As in equation (16), under the grand canonic ensemble the entropy $S(\mathbf{X}) = S(\mathcal{T}_{\mathbf{X}})$ of the system $\mathcal{T}_{\mathbf{X}}$ is

$$\begin{aligned}S(\mathbf{X}) &= -k\mathrm{Trace}(\hat{\rho}_{\mathbf{X}}\ln\hat{\rho}_{\mathbf{X}}) = -k\langle\ln\hat{\rho}_{\mathbf{X}}\rangle = k\beta\left\langle\hat{H}_{\mathbf{X}} - \phi(\mathbf{X}) - \sum_{i=1}^{H}\mu_i\hat{N}_i - \mu_e\hat{N}_e\right\rangle \\ &= \frac{1}{T}\left[\langle\hat{H}_{\mathbf{X}}\rangle - \langle\phi(\mathbf{X})\rangle - \sum_{i=1}^{H}\mu_i\langle\hat{N}_i\rangle - \mu_e\langle\hat{N}_e\rangle\right] \\ &= \frac{1}{T}\left[U(\mathbf{X}) - \phi(\mathbf{X}) - \sum_{i=1}^{H}\mu_i N_i(\mathbf{X}) - \mu_e N_e(\mathbf{X})\right].\end{aligned} \tag{51}$$

Denote $\langle \hat{N}_i \rangle = N_i(\mathbf{X})$ as the mean number of water molecules in $\mathcal{R}_{\mathbf{X}i}$, $1 \le i \le H$, and $\langle \hat{N}_e \rangle = N_e(\mathbf{X})$ the mean number of electrons in $\mathcal{T}_{\mathbf{X}}$. The inner energy $\langle \hat{H}_{\mathbf{X}} \rangle$ of the system $\mathcal{T}_{\mathbf{X}}$ is denoted as $U(\mathbf{X}) = U(\mathcal{T}_{\mathbf{X}})$. By equation (8) and the remark after it $\phi(\mathbf{X})(T, V, \mu_1, \cdots, \mu_H, \mu_e) = -PV(\mathbf{X})$, where $V(\mathbf{X}) = V(\mathcal{T}_{\mathbf{X}})$ is the volume of the thermodynamic system $\mathcal{T}_{\mathbf{X}}$. Thus by equation (51) the Gibbs free energy $G(\mathbf{X}) = G(\mathcal{T}_{\mathbf{X}})$ in formula (1) is obtained:

$$G(\mathbf{X}) = G(\mathcal{T}_{\mathbf{X}}) = PV(\mathbf{X}) + U(\mathbf{X}) - TS(\mathbf{X}) = \sum_{i=1}^{H} \mu_i N_i(\mathbf{X}) + \mu_e N_e(\mathbf{X}). \tag{52}$$

8.5. Converting formula (1) to geometric form (2)

As in the classical statistical mechanics case,

$$\nu_i A(M_{\mathbf{X}i}) = N_i(\mathbf{X}), \quad 1 \le i \le H. \tag{53}$$

Similarly, there will be a $\nu_e > 0$ such that $\nu_e V(\mathcal{T}_{\mathbf{X}}) = N_e(\mathbf{X})$. By the definition of $\mathcal{T}_{\mathbf{X}}$ and $\Omega_{\mathbf{X}}$, it is roughly $V(\mathcal{T}_{\mathbf{X}} \backslash \Omega_{\mathbf{X}}) = d_w A(M_{\mathbf{X}})$. Thus

$$N_e(\mathbf{X}) = \nu_e V(\mathcal{T}_{\mathbf{X}}) = \nu_e [V(\Omega_{\mathbf{X}}) + V(\mathcal{T}_{\mathbf{X}} \backslash \Omega_{\mathbf{X}})] = \nu_e V(\Omega_{\mathbf{X}}) + d_w \nu_e A(M_{\mathbf{X}}). \tag{54}$$

Substitute equations (45) and (54) into formula (1), formula (2) is obtained.

9. Some remarks

The question to applying fundamental physical laws directly to the protein folding problem is, can it be done? It should be checked that how rigorous is the derivation and be asked that are there any fundamental errors? Possible ways to modify the formula or the derivation will also be discussed.

By applying quantum statistics the protein folding problem is theoretically treated. A theory is useful only if it can make explanations to the observed facts and if it can simplify and improve research methods as well as clarify concepts. It will be shown that $G(\mathbf{X})$ can do exactly these.

If the same theoretical result can be derived from two different disciplines, it is often not just by chance. An early phenomenological mathematical model Fang (2005), starting from purely geometric reasoning, has achieved formula (2), with just two hydrophobic levels, hydrophobic and hydrophilic.

A theory also has to be falsifiable, that is making a prediction to be checked. The fundamental prediction is that minimizing formula (1) or (2) the native structures will be obtained for the amino acid sequences of proteins considered in the assumptions of the formulas. That can only be done after the actual values of the chemical potentials appear in the formulas, for the physiological environment, are determined.

9.1. How rigorous is the derivation?

Two common tools in physics, the first step of the Born-Oppenheimer approximation in quantum mechanics and the grand canonic ensemble in statistical physics, are applied to obtain formula (1).

9.1.1. The Born-Oppenheimer approximation

The Born-Oppenheimer approximation "treats the electrons as if they are moving in the field of fixed nuclei. This is a good approximation because, loosely speaking, electrons move much faster than nuclei and will almost instantly adjust themselves to a change in nuclear position." Popelier (2000). Since the mass of a water molecule is much less than the mass of a protein, this approximation can be extended to the case of when $\mathbf{X}$ changes the other particles, electrons and water molecules, will quickly adjust themselves to the change as well.

9.1.2. The statistical physics in general and the grand canonic ensemble in particular

"Up to now there is no evidence to show that statistical physics itself is responsible for any mistakes," the Preface of Dai (2007). Via the ensemble theory of statistical mechanics only one protein molecule and particles in its immediate environment are considered, it is justified since as pointed out in Dai (2007) page 10, "When the duration of measurement is short, or the number of particles is not large enough, the concept of ensemble theory is still valid." And among different ensembles, "Generally speaking, the grand canonic ensemble, with the least restrictions, is the most convenient in the mathematical treatment." Dai (2007) page 16. In fact, the canonic ensemble has been tried and ended with a result that the eigenvalues of the quantum mechanics system have to be really calculated, to do it accurately is impossible.

The derivations in this chapter only puts together the two very common and sound practices: the Born-Oppenheimer approximation (only the first step) and the grand canonic ensemble, and apply them to the protein folding problem. As long as protein folding obeys the fundamental physical laws, there should not be any serious error with the derivation.

9.2. Equilibrium and quasi-equilibrium

A protein's structure will never be in equilibrium, in fact, even the native structure is only a snapshot of the constant vibration state of the structure. The best description of conformation $\mathbf{X}$ is given in Chapter 3 of Bader (1990). Simply speaking, a conformation $\mathbf{X}$ actually is any point $\mathbf{Y}$ such that all $\mathbf{y}_i$ are contained in a union of tiny balls centered at $\mathbf{x}_i$, $i = 1, \cdots, M$. In this sense, it can only be anticipated that a quasi-equilibrium description (such as the heat engine, Bailyn (1994) page 94) of the thermodynamic states of the protein folding. This has been built-in in the Thermodynamical Principle of Protein Folding. So the quantities such as $S(\mathbf{X})$, $\phi(\mathbf{X})$, and $G(\mathbf{X})$ can only be understood in this sense. That is, observing a concrete folding process one will see a series of conformations $\mathbf{X}_i$, $i = 1, 2, 3, \cdots$. The Thermodynamic Principle then says that measuring the Gibbs free energy $G(\mathbf{X}_i)$ one will observe that eventually $G(\mathbf{X}_i)$ will converge to a minimum value and the $\mathbf{X}_i$ will eventually approach to the native structure. While all the time, no conformation $\mathbf{X}_i$ and thermodynamic system $\mathcal{T}_{\mathbf{X}_i}$ are really in equilibrium state.

9.3. Potential energy plays no role in protein folding

Formulas (1) and (2) theoretically show that hydrophobic effect is the driving force of protein folding, it is not just solvent free energy besides the pairwise interactions such as the Coulombs, etc., as all force fields assumed. Only in the physiological environment the hydrophobic effect works towards to native structure, otherwise it will push denaturation as discussed in explanation of folding and unfolding. Formulas (1) and (2) show that the Gibbs free energy is actually independent of the potential energy, against one's intuition and a bit of surprising. The explanation is that during the folding process, all covalent bonds in the main chain and each side chain are kept invariant, the potential energy has already played its role in the synthesis process of forming the peptide chain, which of course can also be described by quantum mechanics. According to Anfinsen (1973), protein folding is after the synthesis of the whole peptide chain, so the synthesis process can be skipped and the concentration can be focused on the folding process.

The steric conditions (35) will just keep this early synthesis result, not any $\mathbf{X} = (\mathbf{x}_1, \cdots, \mathbf{x}_i, \cdots, \mathbf{x}_M)$ is eligible to be a conformation, it has to satisfy the steric conditions (35). The steric conditions not only pay respect to the bond length, it also reflect a lot of physic-chemical properties of a conformation: They are defined via the allowed minimal atomic distances, such that for non-bonding atoms, the allowed minimal distances are: shorter between differently charged or polarized atoms; a little longer between non-polar ones; and much longer (generally greater than the sum of their radii) between the same charged ones, etc. For example, minimal distance between sulfur atoms in Cysteine residues to form disulfide bonds is allowed. And for any newly found intramolecular covalent bond between side chains, such as the isopeptide bonds in Kang and Baker (2011), the steric conditions can be easily modified to allow the newly found phenomenon.

The drawback of the steric conditions is that the minimization in equation (57) becomes a constrained minimization.

9.4. Unified explanation of folding and denaturation

Protein denaturation is easy to happen, even if the environment is slightly changed, as described by Hsien Wu (1931). (Hsien Wu (1931) is the 13th article that theorizes the results of a series experiments, and a preliminary report was read before the Xlllth International Congress of Physiology at Boston, August 19-24, 1929, and published in the *Am. J. Physiol.* for October 1929. In which Hsien Wu first suggested that the denatured protein is still the same molecule, only structure has been changed.) Anfinsen in various experiments showed that after denaturation by changed environment, if removing the denature agent, certain globular proteins can spontaneously refold to its native structure, Anfinsen (1973). The spontaneous renaturation suggests that protein folding does not need outside help, at least to the class of proteins in this chapter. Therefore, the fundamental law of thermodynamics asserts that in the environments in which a protein can fold, the native structure must have the minimum Gibbs free energy. The same is true for denaturation, under the denatured environment, the native structure no longer has the minimum Gibbs free energy, some other structure(s), will have the minimum Gibbs free energy. Thus let En present environment, any formula of Gibbs free energy should be stated as $G(\mathbf{X}, En)$ instead of just $G(\mathbf{X})$, unless the environment is specified like in this chapter. Let En_N be the physiological environment and En_U be some denatured environment, $\mathbf{X}_N$ be the native structure and $\mathbf{X}_U$ be one of the denatured stable structure in En_U, then the thermodynamic principle for both of protein folding and unfolding should be that

$$G(\mathbf{X}_N, En_N) < G(\mathbf{X}_U, En_N), \quad G(\mathbf{X}_N, En_U) > G(\mathbf{X}_U, En_U). \tag{55}$$

To check this, an experiment should be designed that can suddenly put proteins in a different environment. Formulas (1) and (2) should be written as $G(\mathbf{X}, En_N)$. Indeed, the chemical potentials μ_e and μ_i's are Gibbs free energies per corresponding particles, $\mu = u + Pv - Ts$. Two environment parameters, temperature T and pressure P, explicitly appear in μ, the inner energy u and entropy s may also implicitly depend on the environment. According to formulas (1) and (2), if $\mu_i < 0$, then make more H_i atoms to expose to water (make larger $A(M_{\mathbf{X}i})$) will reduce the Gibbs free energy. If $\mu_i > 0$, then the reverse will happen. Increase or reduce the H_i atoms' exposure to water $(A(M_{\mathbf{X}i}))$, the conformation has to change. The conformation changes to adjust until a conformation $\mathbf{X}_N$ is obtained, such that the net effect of any change of the conformation will either increase some H_i atoms' exposure to water while $\mu_i > 0$ or reduce H_i atoms' exposure to water while $\mu_i < 0$. In other words, the $G(\mathbf{X}, En_N)$ achieves its minimum at $G(\mathbf{X}_N, En_N)$. Protein folding, at least for the proteins considered in the assumptions, is explained very well by formulas (1) and (2).

In changed environment, the chemical potentials μ_e and μ_i's in formulas (1) and (2) changed their values. With the changed chemical potentials, $G(\mathbf{X}, En_U)$ has the same form as $G(\mathbf{X}, En_N)$ but different chemical potentials. Therefore, the structure $\mathbf{X}_U$ will be stable, according to the second inequality in (55), the process is exactly the same as described for the protein folding if the changing environment method does not include introducing new kinds (non-water) of particles, for example, if only temperature or pressure is changed.

Even in the new environment including new kinds of particles, formulas (1) and (2) can still partially explain the denaturation, only that more obstructs prevent the protein to denature to $\mathbf{X}_U$, but any way it will end in some structure other than the $\mathbf{X}_N$, the protein is denatured. Actually, this is a hint of how to modify the current formulas to extend to general proteins.

9.5. Explain hydrophobic effect and the role played by hydrogen bonding

In 1959, by reviewing the literature Kauzmann concluded that the hydrophobic effect is the main driving force in protein folding, Kauzmann (1959). Empirical correlation between hydrophobic free energy and aqueous cavity surface area was noted as early as by Reynolds *et.al.* (1974), giving justification of the hydrophobic effect. Various justifications of hydrophobic effect were published, based on empirical models of protein folding, for example, Dill (1990). But the debate continues to present, some still insist that it is the hydrogen bond instead of hydrophobic effect plays the main role of driving force in protein folding, for example, Rose *et. al* (2006). The theoretically derived formulas (1) and (2) can explain why the hydrophobic effect is indeed the driving force. A simulation of reducing hydrophobic area alone by Fang and Jing (2010) shows that the result is the appearance of regularly patterned intramolecular hydrogen bonds associated to the secondary structures.

In fact, according to formulas (1) and (2), if $\mu_i < 0$, then make more H_i atoms to appear in the boundary of $P_{\mathbf{X}}$ will reduce the Gibbs free energy. If $\mu_i > 0$, then the reverse will happen, reducing the exposure of H_i atoms to water will reduce the Gibbs free energy. This gives a theoretical explanation of the hydrophobic effect. The kinetic formulas $\mathbf{F}_i = - \nabla_{\mathbf{x}_i} G(\mathbf{X})$ (will be discussed later) is the force that push the conformation to change to the native structure.

The mechanics stated above works through the chemical potentials μ_i for various levels of hydrophobicity. In physiological environment, all hydrophobic H_i's will have positive μ_i, all hydrophilic H_i's will have negative μ_i. Thus changing conformation P_X such that the most hydrophilic H_i ($\mu_i = \min(\mu_1, \cdots, \mu_H)$) gets the first priority to appear on the boundary, and the most hydrophobic H_i ($\mu_i = \max(\mu_1, \cdots, \mu_H)$) gets the first priority to hide in the hydrophobic core to avoid contacting with water molecules, etc. One should keep in mind that all the time, the steric conditions (35) have to be obeyed.

But the hydrophobic effect is actually partially working through hydrogen bond formation. This is well presented in the chemical potentials in formulas (1) and (2). In fact, the values of the chemical potentials reflect the ability of the atoms or atom groups to form hydrogen bond, either with another atom group in the protein or with water molecules. This gives a way to theoretically or experimentally determine the values of hydrophilic chemical potentials: checking the actual energy value of the hydrogen bond.

According to Fikelstein and Ptitsyn (2002), energies of hydrogen bonds appearing in protein (intermolecular or intramolecular) are (the positive sign means that to break it energy is needed) and their energies are:

O–H : : : O (21 kJ mol^{-1} or 5.0 kcal mol^{-1}); O–H : : : N (29 kJ mol^{-1} or 6.9 kcal mol^{-1});
N–H : : : N (13 kJ mol^{-1} or 3.1 kcal mol^{-1}); N–H : : : O (8 kJ mol^{-1} or 1.9 kcal mol^{-1}).

For hydrophobic ones, it will be more complicated, common sense is that it reduces the entropy that certainly comes from the inability of forming hydrogen bonds with water molecules. Hence although hydrophobic effect is the driving force of protein folding, it works through the atom's ability or inability to form hydrogen bonds with water molecules.

How to explain the intramolecular hydrogen bonds? It seems that formulas (1) and (2) do not address this issue. The possible theory is that the amino acid sequence of a protein is highly selectable in evolution, in fact only a tiny number of amino acid sequences can really become a protein.

Indeed, suppose in average each species (or "kind" of prokaryote) has 10^5 proteins (*Homo sapiens* has around 3×10^5), and assume that per protein has 100 variants (versions with tiny difference in the peptide sequence of the protein), then there are at most 10^{47} peptide sequences that can really produce a natural protein. Now further suppose that only one in 10^{13} theoretically protein producing peptide sequences on the earth get a chance to be realized, then there will be at most 10^{60} possible protein producing peptide sequence. A huge number! The number of peptide sequences of length less than or equal to n is

$$N(n) = \sum_{i=1}^{n} 20^i = \frac{20^{n+1} - 20}{19} = \frac{20}{19}(20^n - 1) \cong 20^{n+0.0171} \cong 10^{1.301(n+0.0171)}. \quad (56)$$

The longest amino acid sequence in the record of ExPASy Proteomics Server has 35,213 residues. Then $N(35,213) > 10^{1.3\times 35,213} > 10^{45060}$ and the ratio of the number of potentially protein producing peptide sequences to the number of all possible sequence of length up to 35,213 is less than $10^{60}/10^{45060} = 10^{-45000}$, so tiny a number that it is undistinguishable from zero. Even assuming that the longest peptide sequence is only 400, the ratio is still less than 10^{-460}. How small a chance that a random peptide sequence happens to be a protein's peptide sequence!

With these highly specially selected peptide sequences, one can assume that while shrinking the various hydrophobic surfaces to form a hydrophobic core, residues are put in positions to form secondary structures and their associated hydrogen bonds. This sounds a little bit too arbitrary. But the huge number of candidate peptide sequences makes the evolutional selection not only possible but also probable. Moreover, a simulation of shrinking hydrophobic surface area alone indeed produced secondary structures and hydrogen bonds. The simulation was reported by Fang and Jing (2010). Without calculating any dihedral angles or electronic charges, without any arbitrary parameter, paying no attention to any particular atom's position, by just reducing hydrophobic surface area (there it was assumed that there are only two kinds of atoms, hydrophobic and hydrophilic), secondary structures and hydrogen bonds duly appeared. The proteins used in the simulation are 2i9c, 2hng, and 2ib0, with 123, 127, and 162 residues. No simulation of any kind of empirical or theoretical models had achieved such a success. More than anything, this simulation should prove that hydrophobic effect alone will give more chance of forming intramolecular hydrogen bonds. Indeed, pushing hydrophilic atoms to make hydrogen bonds with water molecules will give other non-boundary hydrophilic groups more chance to form intramolecular hydrogen bonds.

Again formula (2) can partly explain the success of this simulation, when there are only two hydrophobic classes in formula (2), the hydrophobic area presents the main positive part of the Gibbs free energy, reducing it is reducing the Gibbs free energy, no matter what is the chemical potential's real value.

9.6. Explanation of the successes of surface area models

In 1995, Wang *et al* (1995) compared 8 empirical energy models by testing their ability to distinguish native structures and their close neighboring compact non-native structures. Their models WZS are accessible surface area models with 14 hydrophobicity classes of atoms, $\sum_{i=1}^{14} \sigma_i A_i$. Each two combination of three targeting proteins were used to train WZS to get σ_i, hence there are three models WZS1, WZS2, and WZS3. Among the 8 models, all WZS's performed the best, distinguishing all 6 targeting proteins. The worst performer is the force field AMBER 4.0, it failed in distinguishing any of the 6 targets.

These testing and the successes of various surface area models such as Eisenberg and MacLachlan (1986), showed that instead of watching numerous pairwise atomic interactions, the surface area models, though looking too simple, have surprising powers. Now the formula (2) gives them a theoretic justification. On the other hand, the successes of these models also reenforce the theoretical results.

There is a gap between the accessible surface area model in Eisenberg and Maclanchlan (1986) and the experiment results (surface tension), as pointed out in Tuñón *et. al.* (1992). The gap disappeared when one uses the molecular surface area to replace the accessible surface area, in Tuñón *et. al.* (1992) it was shown that molecular surface area assigned of 72-73 cal/mol/Å^2 perfectly fits with the macroscopic experiment data. Later it was asserted that the molecular surface is the real boundary of protein in its native structure by Jackson and Sternberg (1993).

By the definition of $\Omega'_{\mathbf{X}}$, as shown in FIGURE 3 and FIGURE 4, water molecules contact to $P_{\mathbf{X}}$ must be outside the molecular surface $M_{\mathbf{X}}$. Since the assessable surface is in the middle of the first hydration shell, it is better to use the molecular surface $M_{\mathbf{X}}$ as the boundary of the conformation $P_{\mathbf{X}}$. Moreover, the conversion of the mean numbers $N_i(\mathbf{X})$ to surface area, $N_i(\mathbf{X}) = \nu_i A(M_{\mathbf{X}.i})$, only works for the molecular surface, not for the accessible surface. This can explain the conclusions that molecular surface is a much better boundary than accessible surface as stated in Tuñón *et. al.* (1992) and Jackson and Sternberg (1993).

In fact, the advantage of the solvent accessible surface is that by definition of it one knows exactly each atom occupies which part of the surface, therefore, one can calculate its share in surface area. This fact may partly account why there are so many models based on the solvent accessible surface, even people knew the afore mentioned gap. For other surfaces, one has to define the part of surface that belongs to a specific hydrophobicity class. This was resolved in Fang (2005) via the distance function definition as is used here.

All surface area models neglected one element, the volume of the structure. As early as in the 1970's, Richards and his colleagues already pointed out that the native structure of globular proteins is very dense, or compact, (density = 0.75, Richards (1977)). To make a conformation denser, obviously we should shrink the volume $V(\Omega_{\mathbf{X}})$. The model in Fang (2005) introduced volume term but kept the oversimplification of all atoms are either hydrophobic or hydrophilic. The derivation of formulas (1) and (2) shows that volume term should be counted, but it may be that $\nu_e \mu_e$ is very small, in that case, volume maybe really is irrelevant.

9.7. Coincidence with phenomenological mathematical model

If a theoretical result can be derived from two different disciplines, its possibility of correctness will be dramatically increased. Indeed, from a pure geometric consideration, a phenomenological mathematical model, $G(\mathbf{X}) = aV(\Omega_{\mathbf{X}}) + bA(M_{\mathbf{X}}) + cA(M_{\mathbf{X}1}), a, b, c > 0$ (it was assumed that there are only two hydrophobicity levels, hydrophobic and hydrophilic, the hydrophilic surface area $A(M_{\mathbf{X}2})$ is absorbed in $A(M_{\mathbf{X}})$ by $A(M_{\mathbf{X}2}) = A(M_{\mathbf{X}}) - A(M_{\mathbf{X}1})$), was created in Fang (2005). It was based on the well-known global geometric characteristics of the native structure of globular proteins: 1. high density; 2. smaller surface area; 3. hydrophobic core, as demonstrated and summarized in Richards (1977) and Novotny *et.al* (1984). So that to obtain the native structure, one should shrink the volume (increasing the density) and surface area, and form better hydrophobic core (reducing the hydrophobic surface area $A(M_{\mathbf{X}1})$) simultaneously and cohesively.

The coincidence of formula (2) and the phenomenological mathematical model of Fang (2005) cannot be just a coincidence. Most likely, it is the same natural law reflected in different disciplines. The advantage of formula (2) is that everything there has its physical meaning.

10. Applications

After the derivation it is suitable to point out some immediate applications of the formula $G(\mathbf{X})$.

10.1. Energy surface or landscape

An obvious application is the construction of Gibbs free energy surface or landscape. Empirical estimate is no longer needed, the Gibbs free energy formula $G : \mathfrak{X} \to \mathbb{R}$ gives a graph $(\mathbf{X}, G(\mathbf{X}))$ over the space $\mathfrak{X}$ (all eligible conformations for a given protein), and this is nothing but the Gibbs free energy surface. Mathematically it is a $3M$ dimensional hyper-surface. Its characteristics concerned by students of energy surface theory, such as how rugged it is? how many local minimums are there? is there a funnel? etc., can be answered by simple calculations of the formula.

Since the function G is actually defined on the whole $\mathbb{R}^{3M}$ (on an domain of $\mathbb{R}^{3M}$ containing all $\mathfrak{X}$ is enough), mathematical tools can be explored to study its graph, and compare the results with the restricted conformations. One important question is: Does the absolute minimum structure belongs to $\mathfrak{X}$?

10.2. Structure prediction

Prediction of protein structures is the most important method to reveal proteins' functions and working mechanics, it becomes a bottle neck in the rapidly developing life science. With more and more powerful computers, this problem is attacked in full front. Various models, homologous or *ab initio*, full atom model or coarse grained, with numerous parameters of which many are quite arbitrary, are used to achieve the goal. Although our computer power growths exponentially, prediction power does not follow that way. At this moment, one should take a deep breath and remind what the great physicist Fermi said: "There are two ways of doing calculations in theoretical physics. One way, and this is the way I prefer, is to have a clear physical picture of the process that you are calculating. The other way is to have a precise and self consistent mathematical formalism." And "I remember my friend Johnny von Neumann used to say, with four parameters I can fit an elephant, and with five I can make him wiggle his trunk." Quoted from Dyson (2004).

These remarks should also apply to any scientific calculation, not just theoretical physics. Look at the current situation, all *ab initio* prediction models are actually just empirical with many parameters to ensure some success. Fermi's comments remind us that a theory should be based on fundamental physical laws, and contain no arbitrary parameters. Look at formulas (1) and (2), one sees immediately that they are neat, precise and self consistent mathematical formulas. Furthermore, they including no arbitrary parameter, all terms in them have clear physical meanings. Chemical potentials μ_e and μ_i's, geometric constants ν_e and ν_i's, can be evalued by theory or experiments, they are not arbitrary at all.

But a theory has to be developed, tested, until justified or falsified. For interested researchers, the tasks are to determine the correct values of the chemical potentials in formula (1) and the geometric ratios ν_e and ν_i in formula (2). There are many estimates to them, but they are either for the solvent accessible surface area such as in Eisenberg and MaLachlan (1986) hence not suit to the experiment data as pointed out in Tuñón *et. al.*, or do not distinguish different hydrophobicity levels as in Tuñón *et. al.* (1992). To get the correct values of the chemical potentials and geometric constants, commonly used method of training with data can be employed, in which one can also test the formulas' ability of discriminating native and nearby compact non-native structures. After that, a direct test is to predict the native structure from the amino acid sequence of a protein by minimizing the following:

$$G(\mathbf{X}_N) = \inf_{\mathbf{X} \in \mathfrak{X}} G(\mathbf{X}). \tag{57}$$

This is the first time that a theoretically derived formula of the Gibbs free energy is available. Before this, all *ab intitio* predictions are not really *ab initio.* A combined (theoretical and experimental) search for the values of chemical potentials will be the key for the success of the *ab initio* prediction of protein structure.

10.3. Gradient

With formula (2) as the Gibbs free energy, the minimization in equation (57) can be pursued by Newton's fastest descending method. To state the result, some definitions are necessary.

10.3.1. Molecular graphs

Given a molecule U, let V be the set of atoms in U and $N = |V|$ be the number of atoms and label the atoms as $\mathbf{a}_1, \mathbf{a}_2, \cdots, \mathbf{a}_N$. For $1 \leq i, j \leq N$, define $B_{ij} = n$ if atoms i and j are connected by a bond with valency n (one can imagine that n is not necessarily a whole number), if i and j do not form a bond, then $B_{ij} = 0$. The molecule formula of U in chemistry can be seen as a graph $G(U) = (V, E)$, where V acts as the vertex set of $G(U)$ and E is the edge set of $G(U)$. An edge in E is denoted by $\{i, j\}$, If two atoms $\mathbf{a}_i$ and $\mathbf{a}_j$ are connected by a covalent bond, i.e., $B_{ij} = n \geq 1$, then $\{i, j\} \in E$ is an edge. Call $G(U)$ the **molecular graph** of U. FIGURE 1 is a molecular graph if the side chain R consisting of only one atom, such as in the amino acid Glycine.

A graph G is connected if from any vertex v one can follow the edges in the graph to arrive any other vertex. If a graph is not connected, then it has several **connected components**, each is itself a connected graph. All molecular graphs are connected.

10.3.2. Rotatable bonds

Let $\mathbf{b} = \mathbf{a}_\alpha \mathbf{a}_\beta$ be a covalent bond in the molecule U connecting two atoms $\mathbf{a}_\alpha$ and $\mathbf{a}_\beta$. The bond $\mathbf{b}$ is rotatable if and only if: 1. the valency of $\mathbf{b}$ is not greater than 1; 2. in the molecular graph $G(U)$, if one deletes $\{\alpha, \beta\}$, the remaining graph $G(U) \backslash \{\alpha, \beta\} = (V, E \backslash \{\alpha, \beta\})$ has exactly two connected components and neither component has rotational symmetry around the bond $\mathbf{b}$.

10.3.3. Derivatives of $G(\mathbf{X})$

Let $\mathbf{x}_i = (x_i, y_i, z_i)$, write $\mathbf{F} = -\nabla_{\mathbf{x}_i} G(\mathbf{X}) = -(G_{x_i}, G_{y_i}, G_{z_i})(\mathbf{X})$. The calculation of $G_{x_i}(\mathbf{X})$, for example, is via Lie vector field induced by moving the atomic position $\mathbf{x}_i$. In fact, any infinitesimal change of structure $\mathbf{X}$ will induce a Lie vector field $\vec{L} : \mathbf{X} \to \mathbb{R}^3$. For example, moving $\mathbf{x}_i$ from $\mathbf{x}_i$ to $\mathbf{x}_i + (\Delta x_i, 0, 0)$ while keep other nuclear center fixed, will induce L_{x_i} : $\mathbf{X} \to \mathbb{R}^3$, such that $\vec{L}_{x_i}(\mathbf{x}_i) = (1, 0, 0)$ and $\vec{L}_{x_i}(\mathbf{x}_j) = (0, 0, 0)$ for $j \neq i$. Similarly $\vec{L}_{y_i}$ and $\vec{L}_{z_i}$ can be described as well. Then write $G_{x_i} = G_{\vec{L}_{x_i}}$, etc. and

$$\nabla_{\mathbf{x}_i} G(\mathbf{X}) = (G_{\vec{L}_{x_i}}, G_{\vec{L}_{y_i}}, G_{\vec{L}_{z_i}})(\mathbf{X}), \tag{58}$$

Rotating around a covalent bond b_{ij} also induce a Lie vector field $L_{b_{ij}} : \mathbf{X} \to \mathbb{R}^3$. In fact if $\mathbf{a}_i \mathbf{a}_j$ form the covalent bond b_{ij}, then the bond axis is

$$\mathbf{b}_{ij} = \frac{\mathbf{x}_j - \mathbf{x}_i}{|\mathbf{x}_j - \mathbf{x}_i|}. \tag{59}$$

If $\mathbf{b}_{ij}$ is rotatable, denoting all nuclear centers in one component by $R_{b_{ij}}$ and others in $F_{b_{ij}}$. One can rotate all centers in $R_{b_{ij}}$ around $\mathbf{b}_{ij}$ for certain angle while keep all centers in $F_{b_{ij}}$ fixed. The induced Lie vector field $\vec{L}_{b_{ij}}$ will be

$$\vec{L}_{b_{ij}}(\mathbf{x}_k) = (\mathbf{x}_k - \mathbf{x}_i) \wedge \mathbf{b}_{ij}, \text{ if } \mathbf{x}_k \in R_{b_{ij}}; \tag{60}$$

$$\vec{L}_{b_{ij}}(\mathbf{x}_k) = \vec{0}, \text{ if } \mathbf{x}_k \in F_{b_{ij}}. \tag{61}$$

Any such a Lie vector field $\vec{L}$ will generate a family of conformations $\mathbf{X}_t = (\mathbf{x}_{1t}, \cdots, x_{it}, \cdots, x_{Mt})$, where $\mathbf{x}_{kt} = \mathbf{x}_k + t\vec{L}(\mathbf{x}_k)$, $k = 1, \cdots, M$. Moreover, the Lie vector field $\vec{L}$ can be generated to the molecular surface $M_{\mathbf{X}}$, as shown in Appendix A.

The derivative $G_{\vec{L}}(\mathbf{X})$ is given by

$$G_{\vec{L}}(\mathbf{X}) = \nu_e \mu_e V_{\vec{L}}(\Omega_{\mathbf{X}}) + d_w \nu_e \mu_e A_{\vec{L}}(M_{\mathbf{X}}) + \sum_{i=1}^{H} \nu_i \mu_i A_{\vec{L}}(M_{\mathbf{X}i}), \tag{62}$$

with

$$V_{\vec{L}}(\Omega_{\mathbf{X}}) = -\int_{M_{\mathbf{X}}} \vec{L} \bullet \vec{N} d\mathcal{H}^2, \quad A_{\vec{L}}(M_{\mathbf{X}}) = -2\int_{M_{\mathbf{X}}} H(\vec{L} \bullet \vec{N}) d\mathcal{H}^2, \tag{63}$$

where $\vec{N}$ is the outer unit normal of $M_{\mathbf{X}}$, H the mean curvature of $M_{\mathbf{X}}$, and $\mathcal{H}^2$ the Hausdorff measure. Define $f_{t,i} : \mathbb{R}^3 \to \mathbb{R}$ as $f_{t\,i}(\mathbf{x}) = \text{dist}(\mathbf{x}, M_{\mathbf{X}_t\,i}) - \text{dist}(\mathbf{x}, M_{\mathbf{X}_t} \setminus M_{\mathbf{X}_t\,i})$, and define on $M_{\mathbf{X}}$

$$\nabla_{M_{\mathbf{X}}} f_{0,i} = \nabla f_{0,i} - (\nabla f_{0,i} \bullet \vec{N})\vec{N}, \quad f'_{0,i} = \left.\frac{\partial f_{t\,i}}{\partial t}\right|_{t=0}, \quad \frac{df_{0,i}}{dt} = \vec{L} \bullet \nabla f_{0,i} + f'_{0,i}, \tag{64}$$

then let $\vec{\eta}$ be the unit outward conormal vector of $\partial M_{\mathbf{X}i}$ (normal to $\partial M_{\mathbf{X}i}$ but tangent to $M_{\mathbf{X}}$),

$$A_{\vec{L}}(M_{\mathbf{X}i}) = -2\int_{M_{\mathbf{X}i}} H(\vec{L} \bullet \vec{N}) d\mathcal{H}^2 + \int_{\partial M_{\mathbf{X}i}} \left[\vec{L} \bullet \vec{\eta} - \frac{\frac{df_{0,i}}{dt}}{|\nabla_{M_{\mathbf{X}}} f_{0,i}|} \right] d\mathcal{H}^1. \tag{65}$$

The $\mathbf{X}_t$ is all the information needed in calculating the molecular surface $M_{\mathbf{X}_t}$, see Connolly (1983). To calculate, the above formulas have to be translated into formulas on the molecular surface $M_{\mathbf{X}}$. These translations are given in Appendix A, they are calculable (all integrals are integrable, i.e., can be expressed by analytic formulas with variables $\mathbf{X}$) and were calculated piecewise on $M_{\mathbf{X}}$.

10.3.4. The gradient

Let a protein $\mathfrak{U}$ have L rotatable bonds $(\mathbf{b}_1, \cdots, \mathbf{b}_i, \cdots, \mathbf{b}_L)$. Let θ_i denote the dihedral angle around the rotatable bond $\mathbf{b}_i$. A conformation $\mathbf{X}$ of $\mathfrak{U}$ can be expressed in terms of these rotatable dihedral angles $\Theta = (\theta_1, \cdots, \theta_i, \cdots, \theta_L)$, then

$$G(\mathbf{X}) = G(\Theta), \tag{66}$$

and the gradient of G can be written as

$$\nabla G(\Theta) = \left(\frac{\partial G}{\partial \theta_1}, \cdots, \frac{\partial G}{\partial \theta_i}, \cdots, \frac{\partial G}{\partial \theta_L}\right)(\Theta) = (G_{\vec{L}_{\mathbf{b}_1}}, \cdots, G_{\vec{L}_{\mathbf{b}_i}}, \cdots, G_{\vec{L}_{\mathbf{b}_L}})(\mathbf{X}). \quad (67)$$

If the rotation around $\mathbf{b}_i$ with rotating angle $-sG_{\vec{L}_{b_i}}(\mathbf{X})$ on R_{b_i} and fix atoms in F_{b_i} be denoted as M_i, new conformation $\mathbf{Y}_s = M_L \circ M_{L-1} \circ \ldots \circ M_1$ will be obtained, where $s > 0$ is a suitable step length. That is to say, the dihedral angles of $\mathbf{Y}_s$ are

$$[\theta_1 - sG_{\vec{L}_{b_1}}(\mathbf{X}), \cdots, \theta_i - sG_{\vec{L}_{b_i}}(\mathbf{X}), \cdots, \theta_L - sG_{\vec{L}_{b_L}}(\mathbf{X})].$$

The order of rotations in fact is irrelevant, i.e., by any order, the same conformation $\mathbf{Y}_s$ will always be obtained, as proved in Fang and Jing (2008) and Appendix A. This way one can fast change the structure by simultaneous rotate around all rotatable bonds.

This actually is the Newton's fastest descending method, it reduces the Gibbs free energy $G(\mathbf{X})$ most efficiently. Afore mentioned simulations of Fang and Jing (2010) used this method.

10.4. Kinetics

There are evidence that some protein's native structure is not the global minimum of the Gibbs free energy, but only a local minimum. If the native structure of a protein achieves the global minimal value of the Gibbs free energy, the folding process is **thermodynamic**; if it is only a local minimum, the folding process is **kinetic**, Lazaridis and Karplus (2003).

With the formula (2) and the gradient just obtained, one actually has the kinetic in hand. In fact, for any atomic position $\mathbf{x}_i$, the kinetic force is $\mathbf{F}_i(\mathbf{X}) = -\nabla_{\mathbf{x}_i} G(\mathbf{X})$, Dai (2007). With formula (2) these quantities are readily calculable as mentioned above. The resulting Newton's fastest descending method will help us find the native structure, either in the thermodynamic case or in the kinetic case, here the thermodynamic and kinetic cases are combined by the Gibbs free energy formula (2) and its derivatives.

The moving along $-\nabla G$ method was used in the simulation in Fang and Jing (2010).

11. Conclusion

A quantum statistical theory of protein folding for monomeric, single domain, self folding globular proteins is suggested. The assumptions of the theory fit all observed realities of protein folding. The resulting formulas (1) and (2) do not have any arbitrary parameters and all terms in them have clear physical meaning. Potential energies involving pairwise interactions between atoms do not appear in them.

Formulas (1) and (2) have explanation powers. They give unified explanation to folding and denaturation, to the hydrophobic effect in protein folding and its relation with the hydrogen bonding. The formulas also explain the relative successes of surface area protein folding models. Relation between kinetic and thermodynamic of protein folding is discussed, driving force formula comes from the Gibbs free energy formula (2) are also given. Energy surface theory will be much easier to handle. The concept of $\triangle G$ is clarified.

Appendix

A. Calculations on the molecular surface

A.1. Rotation order

Let $P_{\mathbf{X}} = \cup_{i=1}^{N} B(\mathbf{x}_i, r_i)$ and $\mathbf{x}_{\alpha 0}$ and $\mathbf{x}_{\alpha 1}$ be bonded by b_α, the rotation line of b_α is $\mathbf{x}_{\alpha 0} + t\frac{\mathbf{x}_{\alpha 1}-\mathbf{x}_{\alpha 0}}{|\mathbf{x}_{\alpha 1}-\mathbf{x}_{\alpha 0}|} = \mathbf{x}_{\alpha 0} + t\mathbf{b}_\alpha$. Each b_α divides $\{\mathbf{x}_1, \cdots, \mathbf{x}_M\}$ into two groups F_α and R_α, balls in R_α will be rotated while balls in F_α will be fixed. Note that these partitions are independent of $P_{\mathbf{X}}$, they only depend on the molecular graph of the protein molecule. Let M_α be this rotation-fixation, it will be shown that

$$M_\alpha \circ M_\beta(\mathbf{X}) = M_\beta \circ M_\alpha(\mathbf{X}), \mathbf{X} \in (\mathbf{x}_1, \mathbf{x}_2, \cdots, \mathbf{x}_M), 1 \leq \alpha, \beta \leq L. \tag{1}$$

The formula of rotating a point X around a line $L : \mathbf{y} = \mathbf{x} + t\mathbf{b}$ ($|\mathbf{b}| = 1$) by an angle ω is $R(X) = \mathbf{x} + \mathbf{A}(\omega)(X - \mathbf{x})$. Let I be the identity matrix, $B = \mathbf{b}\mathbf{b}^T$ and Z_b the matrix such that the outer product $\mathbf{b} \wedge X = Z_b X$, then the orthonormal matrix $\mathbf{A}(\omega) = (1 - \cos\omega)B + \cos\omega I + \sin\omega Z_b$.

The topology of a protein molecule guarantees that if two bonds b_α and b_β such that $R_\alpha \subset R_\beta$, then $\{\mathbf{x}_{\alpha 0}, \mathbf{x}_{\alpha 1}\} \subset R_\beta$. Let b_1 and b_2 be two bonds and $L_1 : \mathbf{x} = \mathbf{x}_1 + t\mathbf{b}_1$ and $L_2 : \mathbf{x} = \mathbf{x}_2 + t\mathbf{b}_2$ be the two rotating lines and $X \in (\mathbf{x}_1, \mathbf{x}_2, \cdots, \mathbf{x}_N)$. To prove equation (1), there are only two cases to consider: $R_1 \subset R_2$ and $R_1 \cap R_2 = \emptyset$. In any case, if $X \in F_1 \cap F_2$, then $M_1 \circ M_2(X) = M_2 \circ M_1(X) = X$. If $X \in R_1 \subset R_2$, then

$$M_2 \circ M_1(X) = \mathbf{x}_2 + \mathbf{A}_2(\omega_2)(\mathbf{x}_1 - \mathbf{x}_2) + \mathbf{A}_2(\omega_2)\mathbf{A}_1(\omega_1)(X - \mathbf{x}_1). \tag{2}$$

On the other hand b_1 and hence L_1 itself will be rotated by M_2, $L_3 = M_2(L_1) = \mathbf{x}_3 + t\mathbf{b}_3$, where $\mathbf{x}_3 = \mathbf{x}_2 + \mathbf{A}_2(\omega_2)(\mathbf{x}_1 - \mathbf{x}_2)$, $\mathbf{b}_3 = \mathbf{A}_2(\omega_2)\mathbf{b}_1$. Since $X \in R_1 \subset R_2$ and $M_2(X) \in R_1$ (in the new conformation $M_2(P)$ where rotation around b_1 is rotation around L_3), $M_1 \circ M_2(X)$ will be the rotation R_3 around L_3 of $M_2(X)$ by angle ω_1, thus

$$M_1 \circ M_2(X) = \mathbf{x}_2 + \mathbf{A}_2(\omega_2)(\mathbf{x}_1 - \mathbf{x}_2) + \mathbf{A}_3(\omega_1)\mathbf{A}_2(\omega_2)(X - \mathbf{x}_1). \tag{3}$$

Let $\mathbf{v} \in \mathbb{R}^3$ be an arbitrary vector, writing $\mathbf{A}_1(\omega_1) = \mathbf{A}_1$, $\mathbf{A}_2(\omega_2) = \mathbf{A}_2$, and $\mathbf{A}_3(\omega_1) = \mathbf{A}_3$, then

$$\mathbf{A}_2\mathbf{A}_1\mathbf{v} = (1 - \cos\omega_1)(\mathbf{b}_1 \bullet \mathbf{v})\mathbf{A}_2\mathbf{b}_1 + \cos\omega_1 \mathbf{A}_2\mathbf{v} + \sin\omega_1 \mathbf{A}_2(\mathbf{b}_1 \wedge \mathbf{v}). \tag{4}$$

For any orthonormal matrix O, $(O\mathbf{b}_1) \bullet (O\mathbf{v}) = \mathbf{b}_1 \bullet \mathbf{v}$, $O(\mathbf{b}_1 \wedge \mathbf{v}) = (O\mathbf{b}_1) \wedge (O\mathbf{v})$. Then by $\mathbf{b}_3 = \mathbf{A}_2(\omega_2)\mathbf{b}_1$,

$$\begin{aligned} \mathbf{A}_3\mathbf{A}_2\mathbf{v} &= (1 - \cos\omega_1)[\mathbf{b}_3 \bullet (\mathbf{A}_2\mathbf{v})]\mathbf{b}_3 + \cos\omega_1(\mathbf{A}_2\mathbf{v}) + \sin\omega_1 \mathbf{b}_3 \wedge (\mathbf{A}_2\mathbf{v}) \\ &= (1 - \cos\omega_1)(\mathbf{b}_1 \bullet \mathbf{v})\mathbf{A}_2\mathbf{b}_1 + \cos\omega_1 \mathbf{A}_2\mathbf{v} + \sin\omega_1 \mathbf{A}_2(\mathbf{b}_1 \wedge \mathbf{v}). \end{aligned} \tag{5}$$

Since $\mathbf{v}$ was arbitrary, equations (2) to (5) show equation (1) is true.

If $R_1 \cap R_2 = \emptyset$ and $X \in R_2$, then X and $M_2(X) \in F_1$ hence $M_1 \circ M_2(X) = M_2(X) = M_2 \circ M_1(X)$.

The molecular surface is consisted of faces. Thus all integrals can be integrated piecewise on faces. There are three kinds of faces, convex, concave, and saddle, Connolly (1983). The formulas on each kind of face are given below. The notation $\dot{\mathbf{x}}$ means $L(\mathbf{x})$ with L the corresponding Lie vector field. All van der Waals radii r_i, as well as the probe radius r_p, are constants.

A.2. Convex face

A convex face is a piece of spherical surface lying on some $S_i = \partial B(\mathbf{x}_i, r_i)$ and bounded by circular arcs $\gamma_\nu, \nu = 1, \cdots, n_F$, let $\mathbf{v}_\nu^0$ and $\mathbf{v}_\nu^1$ be γ_ν's vertices and $\mathbf{c}_\nu$ and r_ν the center and radius of γ_ν's circle, $r_\nu\phi_\nu$ the arc length of γ_ν, $\mathbf{e}_3^\nu = (z_{\nu 1}, z_{\nu 2}, z_{\nu 3})$ be the unit vector in the direction of $(\mathbf{v}_\nu^0 - \mathbf{c}_\nu) \wedge (\mathbf{v}_\nu^1 - \mathbf{c}_\nu)$, $d_\nu = \mathbf{e}_3^\nu \bullet (\mathbf{c}_\nu - \mathbf{x}_i)$, $\mathbf{e}_1^\nu = \frac{\mathbf{v}_\nu^0 - \mathbf{c}_{ij\nu}}{r_\nu} = (x_{\nu 1}, x_{\nu 2}, x_{\nu 3})$, $\mathbf{e}_2^\nu = \mathbf{e}_3^\nu \wedge \mathbf{e}_1^\nu = (y_{\nu 1}, y_{\nu 2}, y_{\nu 3})$, $1 \le \nu \le n_F$. A point $\mathbf{x}$ on F has the form $\mathbf{x} = \mathbf{x}_i - r_i N$ and $X_\alpha(\mathbf{x}) = \dot{\mathbf{x}} - r_i \dot{N}$, by $N \bullet \dot{N} \equiv 0$ and the general divergence formula on sphere,

$$r_i \int_F (X_\alpha \bullet N) H \, d\mathcal{H}^2 = \int_F X_\alpha \bullet N \, d\mathcal{H}^2 = \frac{-1}{r_i} \dot{\mathbf{x}}_i \bullet \sum_{\nu=1}^{M} (X_\nu, Y_\nu, Z_\nu), \tag{6}$$

where

$$\begin{aligned} X_\nu = & \frac{r_\nu^2}{2} \{\phi_\nu z_{\nu 1} + \sin\phi_\nu [\cos\phi_\nu (x_{\nu 2} y_{\nu 3} + x_{\nu 3} y_{\nu 2}) + \sin\phi_\nu (y_{\nu 2} y_{\nu 3} - x_{\nu 2} x_{\nu 3})\} \\ & + r_\nu d_\nu z_{\nu 2} [y_{\nu 3} \sin\phi_\nu - x_{\nu 3} (1 - \cos\phi_\nu)], \end{aligned} \tag{7}$$

$$\begin{aligned} Y_\nu = & \frac{r_\nu^2}{2} \{\phi_\nu z_{\nu 2} + \sin\phi_\nu [\cos\phi_\nu (x_{\nu 3} y_{\nu 1} + x_{\nu 1} y_{\nu 3}) + \sin\phi_\nu (y_{\nu 1} y_{\nu 3} - x_{\nu 1} x_{\nu 3})]\} \\ & + r_\nu d_\nu z_{\nu 3} [y_{\nu 1} \sin\phi_\nu - x_{\nu 1} (1 - \cos\phi_\nu)], \end{aligned} \tag{8}$$

$$\begin{aligned} Z_\nu = & \frac{r_\nu^2}{2} \{\phi_\nu z_{\nu 3} + \sin\phi_\nu [\cos\phi_\nu (x_{\nu 1} y_{\nu 2} + x_{\nu 2} y_{\nu 1}) + \sin\phi_\nu (y_{\nu 1} y_{\nu 2} - x_{\nu 1} x_{\nu 2})]\} \\ & + r_\nu d_\nu z_{\nu 1} [y_{\nu 2} \sin\phi_\nu - x_{\nu 2} (1 - \cos\phi_\nu)]. \end{aligned} \tag{9}$$

A.3. Concave face

A concave face F is a spherical polygon on the probe sphere S when S is simultaneously tangent to 3 balls $B(\mathbf{x}_i, r_i), 1 \le i \le 3$. F is expressed by parameters $t_i \ge 0, i = 1, 2, 3$,

$$\mathbf{x} = \mathbf{p} + rN = \mathbf{p} + r \frac{t_1 \mathbf{x}_1 + t_2 \mathbf{x}_2 + t_3 \mathbf{x}_3 - \mathbf{p}}{|t_1 \mathbf{x}_1 + t_2 \mathbf{x}_2 + t_3 \mathbf{x}_3 - \mathbf{p}|}, \quad t_1 + t_2 + t_3 = 1. \tag{10}$$

$$\phi_t(\mathbf{x}) = \mathbf{p}(t) + rN(t) = \mathbf{p}(t) + r \frac{t_1 \mathbf{x}_1(t) + t_2 \mathbf{x}_2(t) + t_3 \mathbf{x}_3(t) - \mathbf{p}(t)}{|t_1 \mathbf{x}_1(t) + t_2 \mathbf{x}_2(t) + t_3 \mathbf{x}_3(t) - \mathbf{p}(t)|}, \tag{11}$$

$X_\alpha(\mathbf{x}) = \left.\frac{d\phi_t(\mathbf{x})}{dt}\right|_{t=0} = \dot{\mathbf{p}} + r\dot{N}$. Using $|\mathbf{p}(t) - \mathbf{x}_i(t)| = r_i + r = \text{constant}$, let $b_i = (\mathbf{x}_i - \mathbf{p})\bullet\dot{\mathbf{x}}_i$, $\mathbf{b} = (b_1, b_2, b_3)^T$, $A = \begin{pmatrix} \mathbf{x}_1 - \mathbf{p} \\ \mathbf{x}_2 - \mathbf{p} \\ \mathbf{x}_3 - \mathbf{p} \end{pmatrix}$, then $\det A \neq 0$, $\dot{\mathbf{p}} = A^{-1}\mathbf{b}$. By $X_\alpha \bullet N = \dot{\mathbf{p}} \bullet N$,

$$r\int_F (X_\alpha \bullet N) H \, d\mathcal{H}^2 = -\int_F X_\alpha \bullet N \, d\mathcal{H}^2 = \frac{1}{r}\dot{\mathbf{p}} \bullet \sum_{i=1}^{3} (X_i, Y_i, Z_i). \tag{12}$$

Here the X_i, Y_i, and Z_i are the same as in equations (7) to (9).

Assume that $\mathbf{x}_1$ has different water association with $\mathbf{x}_2$ and $\mathbf{x}_3$, let $R_i = r_i + r$, $d_{ij} = |\mathbf{x}_i - \mathbf{x}_j|$, $y_{ij} = (R_i^2 - R_j^2)/2d_{ij}$. Then $f_P(\mathbf{x}) = (\mathbf{x} - \mathbf{p}) \bullet \mathbf{n}_k$, where $\mathbf{n}_k = (\mathbf{x}_k - \mathbf{x}_1)/d_{1k}$ is the directed unit normal of the dividing plane P_k (passing through $\mathbf{p}$ and $\mathbf{t}_{1k} = \frac{1}{2}(\mathbf{x}_1 + \mathbf{x}_k) + y_{1k}\mathbf{n}_k$ and perpendicular to it), $k = 2, 3$. The projection of $\partial W \cap F$ on the $\mathbf{x}_1\mathbf{x}_2\mathbf{x}_3$ plane is in the form of one or two curves γ_k, $(\{j,k\} = \{2,3\})$

$$t_k = A_k t_j + B_k, \; 0 \leq t_j \leq z_j, \; A_k = \frac{d_{1j}\cos\omega}{-d_{1k}}, \; B_k = \frac{d_{1k} + 2y_{1k}}{2d_{1k}}, \tag{13}$$

where $\cos\omega = \frac{(\mathbf{x}_2 - \mathbf{x}_1)\bullet(\mathbf{x}_3 - \mathbf{x}_1)}{d_{12}d_{13}}$. $F \cap W$ is a spherical polygon with arcs γ_v, $1 \leq v \leq n$, including some γ_k as above, so $\int_{W\cap F}(X_\alpha \bullet N) H d\mathcal{H}^2$ has the similar form as that in equation (12).

Let $\mathbf{A}_k = \mathbf{x}_j - \mathbf{x}_1 + A_k(\mathbf{x}_k - \mathbf{x}_1)$, $\mathbf{B}_k = B_k(\mathbf{x}_k - \mathbf{x}_1) + (\mathbf{x}_1 - \mathbf{p})$, $\mathbf{C}_k = \mathbf{B}_k \wedge \mathbf{A}_k$. Treating A_k and B_k as constants and setting $H_k = \dot{\mathbf{p}} \bullet \mathbf{C}_k$, $J_k = \dot{\mathbf{A}}_k \bullet \mathbf{C}_k$, and $K_k = \dot{\mathbf{B}}_k \bullet \mathbf{C}_k$. Let $a_k t_j^2 + b_k t_j + c_k = |\mathbf{A}_k t_j + \mathbf{B}_k|^2 > 0$, then $\Delta_k = 4a_k c_k - b_k^2 > 0$. By $\eta = N'_{t_j} \wedge N / |N'_{t_j}|$ and $d\mathcal{H}^1 = r|N'_{t_j}| dt_j$,

$$\begin{aligned}\int_{(\partial W\cap F)\cap\gamma_k} X_\alpha \bullet \eta \, d\mathcal{H}^1 = & \frac{2rH_k}{\sqrt{\Delta_k}}\left(\arctan\frac{2a_k z_j + b_k}{\sqrt{\Delta_k}} - \arctan\frac{b_k}{\sqrt{\Delta_k}}\right) \\ & + \frac{2r^2 J_k}{\Delta_k}\left(2\sqrt{c_k} - \frac{b_k z_j + 2c_k}{\sqrt{a_k z_j^2 + b_k z_j + c_k}}\right) \\ & + \frac{2r^2 K_k}{\Delta_k}\left(\frac{2a_k z_j + b_k}{\sqrt{a_k z_j^2 + b_k z_j + c_k}} - \frac{b_k}{\sqrt{c_k}}\right).\end{aligned} \tag{14}$$

Let $U_k = (\mathbf{A}_k \dot{\bullet} \mathbf{n}_k)$, $V_k = (\mathbf{B}_k \dot{\bullet} \mathbf{n}_k)$, $W_k = |\mathbf{C}_k \bullet \mathbf{n}_k| > 0$, then

$$\begin{aligned}\int_{\gamma_k} \frac{\frac{df_P}{dt}}{|\nabla_{M_P} f_P|} d\mathcal{H}^1 = & \frac{\pm 2r^2}{W_k}\left(\frac{(2a_k z_j + b_k)V_k}{\sqrt{a_k z_j^2 + b_k z_j + c_k}} - \frac{b_k V_k}{\sqrt{c_k}}\right. \\ & \left. + \; 2\sqrt{c_k}U_k - \frac{(b_k z_j + 2c_k)U_k}{\sqrt{a_k z_j^2 + b_k z_j + c_k}}\right),\end{aligned} \tag{15}$$

where the sign is determined by orientation.

A.4. Saddle face

A saddle face F is generated when the probe S simultaneously tangents to two balls $B(\mathbf{x}_1, r_1)$ and $B(\mathbf{x}_2, r_2)$, and rolls around the axis $\mathbf{e}_2 = \frac{\mathbf{x}_2 - \mathbf{x}_1}{d_{12}}$. The starting and stopping positions of the probe center is $\mathbf{p}$ and $\mathbf{q}$. Let $y = [(r_1 + r)^2 - (r_2 + r)^2]/2d_{12}$ and $\mathbf{t} = \frac{1}{2}(\mathbf{x}_1 + \mathbf{x}_2) + y\mathbf{e}_2$, $R = |\mathbf{p} - \mathbf{t}|$, $\mathbf{e}_1 = (\mathbf{p} - \mathbf{t})/R$, $\mathbf{e}_3 = \mathbf{e}_1 \wedge \mathbf{e}_2$, then F is parametrized by $0 \le \psi \le \psi_s$, $\theta_1 \le \theta \le \theta_2$,

$$\mathbf{x}(\psi, \theta) = \mathbf{t} + (R - r\cos\theta)(\cos\psi\mathbf{e}_1 + \sin\psi\mathbf{e}_3) + r\sin\theta\mathbf{e}_2, \tag{16}$$

where let $\omega_s = \arccos[(\mathbf{p} - \mathbf{t}) \bullet (\mathbf{q} - \mathbf{t})/R^2]$, then $\psi_s = \omega_s$ or $2\pi - \omega_s$. $\theta_1 = \arctan[-(d_{12} + 2y)/2R]$, $\theta_2 = \arctan[(d_{12} - 2y)/2R]$. These data are uniquely determined by the conformation P, see Connolly (1983). Let $\theta_k(t)$ and $\phi_s(t)$ be similarly defined for the conformation P_t, one can define $\phi_t(\psi) = \frac{\psi\psi_s(t)}{\psi_s}$, $\phi_t(\theta) = \frac{\theta_1(t)(\theta_2 - \theta) + \theta_2(t)(\theta - \theta_2)}{\theta_2 - \theta_1}$, and $U(\psi) = \cos\psi\mathbf{e}_1 + \sin\psi\mathbf{e}_3$, then for the same $0 \le \phi \le \phi_s$ and $\theta_1 \le \theta \le \theta_2$,

$$\phi_t(\mathbf{x}) = \mathbf{t}(t) + [R - r\cos\phi_t(\theta)]U(\phi_t(\psi)) + r\sin\phi_t(\theta)\mathbf{e}_2(t). \tag{17}$$

Let $\dot{U} = \cos\psi\dot{\mathbf{e}}_1 + \sin\psi\dot{\mathbf{e}}_3$, $U' = -\sin\psi\mathbf{e}_1 + \cos\psi\mathbf{e}_3$, then

$$\begin{aligned} X_\alpha(\mathbf{x}) &= \dot{\mathbf{t}} + (\dot{R} + r\dot{\theta}\sin\theta)U + (R - r\cos\theta)(\dot{U} + \dot{\psi}U') \\ &\quad + r\dot{\theta}\cos\theta\mathbf{e}_2 + r\sin\theta\dot{\mathbf{e}}_2. \end{aligned} \tag{18}$$

On F, $N = -\cos\theta U(\psi) + \sin\theta\mathbf{e}_2$, $\mathrm{d}\mathcal{H}^2 = r(R - r\cos\theta)\mathrm{d}\theta\mathrm{d}\psi$, $2H = \frac{R - 2r\cos\theta}{r(R - r\cos\theta)}$. Let $J = J(\psi_s) = \int_0^{\psi_s} U(\psi)\mathrm{d}\psi$, then

$$\begin{aligned} 4\int_F X_\alpha \bullet N\mathrm{d}\mathcal{H}^2 &= 4rR(\phi_s\dot{\mathbf{t}} \bullet \mathbf{e}_2 - RJ \bullet \dot{\mathbf{e}}_2)(\cos\theta_1 - \cos\theta_2) \\ &+ 4rR(\phi_s\dot{R} + \dot{\mathbf{t}} \bullet J)(\sin\theta_1 - \sin\theta_2) \\ &- r^2(\phi_s\dot{\mathbf{t}} \bullet \mathbf{e}_2 - RJ \bullet \dot{\mathbf{e}}_2)(\cos 2\theta_1 - \cos 2\theta_2) \\ &+ r^2(\phi_s\dot{R} + \dot{\mathbf{t}} \bullet J)[2(\theta_1 - \theta_2) + \sin 2\theta_1 - \sin 2\theta_2], \end{aligned} \tag{19}$$

$$\begin{aligned} 2\int_F (X_\alpha \bullet N)H\mathrm{d}\mathcal{H}^2 &= 2R(\phi_s\dot{\mathbf{t}} \bullet \mathbf{e}_2 - R \bullet \dot{\mathbf{e}}_2)(\cos\theta_1 - \cos\theta_2) \\ &+ 2R(\phi_s\dot{R} + \dot{\mathbf{t}} \bullet J)(\sin\theta_1 - \sin\theta_2) \\ &- r(\phi_s\dot{\mathbf{t}} \bullet \mathbf{e}_2 - RJ \bullet \dot{\mathbf{e}}_2)(\cos 2\theta_1 - \cos 2\theta_2) \\ &+ r(\phi_s\dot{R} + \dot{\mathbf{t}} \bullet J)[2(\theta_1 - \theta_2) + \sin 2\theta_1 - \sin 2\theta_2]. \end{aligned} \tag{20}$$

Assume that $\mathbf{x}_1$ is hydrophobic and $\mathbf{x}_2$ is not, then the dividing plane P passing through $\mathbf{p}$ and $\mathbf{t}$ and is perpendicular to $\mathbf{e}_2$. The curve $\partial W \cap F$ is given by $\mathbf{x}(\psi) = \mathbf{t} + (R - r)U(\psi)$, $0 \le \phi \le \phi_s$, on which $\mathrm{d}\mathcal{H}^1 = (R - r)\mathrm{d}\phi$. The hydrophobic surface integral on F then is the same as in equation (20), except $\theta_1 = 0$. Since on $\partial W \cap F$, $\eta = N' \wedge N = \mathbf{e}_2$, $\frac{\mathrm{d}\theta(t)}{\mathrm{d}t}|_{t=0} = \dot{\theta}_0 = \frac{\dot{\theta}_1\theta_2 - \dot{\theta}_2\theta_1}{\theta_2 - \theta_1}$, by equation (18),

$$\int_{\partial W \cap F} X_\alpha \bullet \eta\,\mathrm{d}\mathcal{H}^1 = (R - r)\phi_s(r\dot{\theta}_0 + \dot{\mathbf{t}} \bullet \mathbf{e}_2) - (R - r)^2\dot{\mathbf{e}}_2 \bullet J, \tag{21}$$

Let $\mathbf{n}_j = \mathbf{e}_2$, then $f_{P_t}(\phi_t(\mathbf{x})) = [\phi_t(\mathbf{x}) - \mathbf{t}(t)] \bullet \mathbf{n}_j(t)$, $|\nabla_{M_P} f_P| = \mathbf{n}_j \bullet \eta = 1$, and $\frac{\mathrm{d} f_{P_t}(\phi_t(X))}{\mathrm{d}t}|_{t=0} = [\dot{\phi_t}(X) - \dot{\mathbf{t}}(t)] \bullet \mathbf{e}_2 + [(R-r)U] \bullet \dot{\mathbf{e}}_2 = r\dot{\theta}_0$.

$$\int_{\partial W_P \cap F} \frac{\frac{\mathrm{d} f_P}{\mathrm{d}t}}{|\nabla_{M_P} f_P|} \mathrm{d}\mathcal{H}^1 = (R-r)\phi_s r\dot{\theta}_0. \tag{22}$$

Author details

Yi Fang

Department of Mathematics, Nanchang University, 999 Xuefu Road, Honggutan New District, Nanchang, 330031, China

12. References

[1] Anfinsen, C. B. (1973) Principles that govern the folding of protein chains. *Science* 181, 223-230.

[2] Bader, R. F. W. (1990) *Atoms in Molecules: A Quantum Theory.* (Clarendon Press · Oxford).

[3] Bailyn, M. (1994) *A Survey of Thermodynamics.* American Institute of Physics New York.

[4] Ben-Naim, A. (2012) Levinthal's question revisited, and answered. *Journal of Biomolecular Structure and Dynamics* 30(1), 113-124 (2012).

[5] Branden, C. and J. Tooze, J. (1999) *Introduction to Protein Structure.* (Second Edition, Garland).

[6] Connolly, M. L. (1983) Analytical molecular surface calculation. *J. Appl. Cryst.*, 16:548-558.

[7] Dai, X. (2007) *Advanced Statistical Physics.* (Fudan University Press, Shanghai).

[8] Dill, K. A. (1990) Dominant forces in protein folding. *Biochemistry,* 29 7133-7155.

[9] Dyson, F. (2004) A meeting with Enrico Fermi: How an intuitive physicist rescued a team from fruitless research. *Nature* 427, 297.

[10] Eisenberg, D and McLachlan, A. D. Solvation energy in protein folding and binding. *Nature* 319, 199-203.

[11] ExPASy Proteomics Server. http://au.expasy.org/sprot/relnotes/relstat.html

[12] Fang, Y. (2005) Mathematical protein folding problem. In: D. Hoffman, Ed, *Global Theory of Minimal Surfaces.* (*Proceedings of the Clay Mathematical Proceedings,* 2 2005) pp. 611-622.

[13] Fang, Y. and Jing, J. (2008) Implementation of a mathematical protein folding model. *International Journal of Pure and Applied Mathematics,* 42(4), 481-488.

[14] Fang, Y. and Jing, J. (2010) Geometry, thermodynamics, and protein. *Journal of Theoretical Biology,* 262, 382-390.

[15] Finkelstein, A. V. and Ptitsyn, O. B. (2002) *Protein Physics: A Course of Lectures.* Academic Press, An imprint of Elsevier Science, Amsterdam .... .

[16] Greiner, W., Neise, L., and Stöker, H. (1994) *Thermodynamics and Statistical Mechanics.* (Spriger-Verlag, New York, Berlin, ...).

[17] Jackson, R. M., and Sternberg, M. J. E. (1993) Protein surface area defined. *Nature* 366, 638.

[18] Kang, H. J. and Baker, E. N. (2011) Intramolecular isopeptide bonds: protein crosslinks built for stress? *TIBS,* 36(4), 229-237.

[19] Kauzmann, W. (1959) Some factors in the interpretation of protein denaturation. *Adv. Protein Chem.* 14, 1-63 (1959).
[20] Lazaridis, T., and Karplus, M. (2003) Thermodynamics of protein folding: a microscopic view. *Biophysical Chemistry,* 100, 367-395.
[21] Lee, B., and Richards, F. M. (1971) The interpretation of protein structures: estimation of static accessibility. *J. Mol. Biol.* 55:379-400.
[22] Novotny, J., Bruccoleri, R., and Karplus, M. (1984) An analysis of incorrectly folded protein models. Implications for structure predictions. *J. Mol. Biol.* 177, 787-818.
[23] Popelier, P. (2000) *Atoms in Molecules: An Introduction.* (Prentice Hall).
[24] Reynolds, J. A., Gilbert, D. B., and Tanford, C. (1974) Empirical correlation between hydrophobic free energy and aqueous cavity surface area. *Proc. Natl. Acad. Sci. USA* 71(8), 2925-2927.
[25] Richards, F. M. (1977) Areas, volumes, packing, and protein structure. *Ann. Rev. Biophys. Bioeng.* 6, 151-176.
[26] Rose, G. D., Fleming, P. J., Banavar, J. R., and Maritan, A. (2006) A backbone based theory of protein folding.
PNAS 103(45), 16623-16633.
[27] Tuñón, I., Silla, E., and Pascual-Ahuir, J. L. (1992) Molecular surface area and hydrophobic effect. Protien Engineering 5(8), 715-716.
[28] Wang, Y., Zhang, H., Li, W., and Scott, R. A. (1995) Discriminating compact nonnative structures from the native structure of globular proteins. *PNAS,* 92, 709-713.
[29] Wu, H. (1931) Studies on denaturation of proteins XIII. A theory of denaturation. *Chinese Journal nof Physiology* 4, 321-344 (1931). A preliminary report was read before the XIIIth International Congress of Physiology at Boston, Aug. 19-24, 1929 and published in the *Am. J Physiol.* for Oct. 1929. Reprinted in *Advances in Protein Chemicsty* 46, 6-26 (1995).

Thermodynamics' Microscopic Connotations

A. Plastino, Evaldo M. F. Curado and M. Casas

Additional information is available at the end of the chapter

1. Introduction

Thermodynamics is the science of energy conversion. It involves heat and other forms of energy, mechanical one being the foremost one. Potential energy is the capacity of doing work because of the position of something. Kinetic energy is due to movement, depending upon mass and speed. Since all objects have structure, they possess some internal energy that holds such structure together, a kind of strain energy. As for work, there are to kinds of it: internal and external. The later is work done on "something". The former is work effected within something, being a capacity. Heat is another king of energy, the leit-motif of thermodynamics. Thermodynamics studies and interrelates the macroscopic variables, such as temperature, volume, and pressure that are employed to describe thermal systems and concerns itself with phenomena that can be experimentally reproducible.

In thermodynamics one is usually interested in special system's states called equilibrium ones. Such states are steady ones reached after a system has stabilized itself to such an extent that it no longer keeps changing with the passage of time, as far as its macroscopic variables are concerned. From a thermodynamics point of view a system is defined by its being prepared in a certain, specific way. The system will always reach, eventually, a unique state of thermodynamic equilibrium, univocally determined by the preparation-manner. Empiric reproducibility is a fundamental requirement for physics in general an thermodynamics in particular. The main source of the strength, or robustness, of thermodynamics, lies on the fact does it deals only with phenomena that are experimentally reproducible.

Historically, thermodynamics developed out the need for increasing the efficiency of early steam engines, particularly through the work of the French physicist Nicolas Sadi-Carnot (1824) who believed that a heat engine's efficiency was to play an important role in helping France win the Napoleonic Wars. Scottish physicist Lord Kelvin was the first to formulate a succint definition of thermodynamics in 1854: "Thermodynamics is the subject of the relation of heat to forces acting between contiguous parts of bodies, and the relation of heat to electrical agency". Chemical thermodynamics studies the role of entropy in the process of chemical reactions and provides the main body of knowledge of the field. Since Boltzmann in the 1870's, statistical thermodynamics, or statistical mechanics, that are microscopic theories, began to explain macroscopic thermodynamics via statistical predictions on the collective motion of atoms.

1.1. Thermodynamics' laws

The laws of physics are established scientific regularities regarded as universal and invariable facts of the universe. A "law" differs from hypotheses, theories, postulates, principles, etc., in that it constitutes an analytic statement. A theory starts from a set of axioms from which all laws and phenomena should arise via adequate mathematical treatment. The principles of thermodynamics, often called "its laws", count themselves amongst the most fundamental regularities of Nature [1]. These laws define fundamental physical quantities, such as temperature, energy, and entropy, to describe thermodynamic systems and they account for the transfer of energy as heat and work in thermodynamic processes. An empirically reproducible distinction between heat and work constitutes the "hard-core" of thermodynamics. For processes in which this distinction cannot be made, thermodynamics remains silent. One speaks of four thermodynamics' laws:

- The zeroth law of thermodynamics allows for the assignment of a unique temperature to systems that are in thermal equilibrium with each other.
- The first law postulates the existence of a quantity called the internal energy of a system and shows how it is related to the distinction between energy transfer as work and energy transfer as heat. The internal energy is conserved but work and heat are not defined as separately conserved quantities. Alternatively, one can reformulate the first law as stating that perpetual motion machines of the first kind can not exist.
- The second law of thermodynamics expresses the existence of a quantity called the entropy S and states that for an isolated macroscopic system S never decreases, or, alternatively, that perpetual motion machines of the second kind are impossible.
- The third law of thermodynamics refers to the entropy of a system at absolute zero temperature ($T = 0$) and states that it is impossible to lower T in such a manner that reaches the limit $T = 0$.

Classical thermodynamics accounts for the exchange of work and heat between systems with emphasis in states of thermodynamic equilibrium. Thermal equilibrium is a condition *sine qua non* for *macroscopically specified systems* only. It shoul be noted that, at the microscopic (atomic) level all physical systems undergo random fluctuations. Every finite system will exhibit statistical fluctuations in its thermodynamic variables of state (entropy, temperature, pressure, etc.), but these are negligible for macroscopically specified systems. Fluctuations become important for microscopically specified systems. Exceptionally, for macroscopically specified systems found at critical states, fluctuations are of the essence.

1.2. The Legendre transform

The Legendre transform is an operation that transforms one real-valued function of f a real variable x into another f_T, of a different variable y, maintaining constant its information content. The derivative of the function f becomes the argument to the function f_T.

$$f_T(y) = xy - f(x); \; y = f'(x) \Rightarrow \text{ reciprocity.} \tag{1}$$

The Legendre transform its own inverse. It is used to get from Lagrangians the Hamiltonian formulation of classical mechanics.

Legendre' reciprocity relations constitute thermodynamics' essential formal ingredient [2]. In general, for two functions I and α one has

$$I(A_1, \ldots, A_M) = \alpha + \sum_{k=1}^{M} \lambda_k A_k, \tag{2}$$

with the A_i extensive variables and the λ_i intensive ones. Obviously, the Legendre transform main goal is that of changing the identity of our relevant independent variables. For α we have

$$\alpha(\lambda_1, \ldots, \lambda_M) = I - \sum_{k=1}^{M} \lambda_k \langle A_k \rangle \, . \tag{3}$$

The three operative reciprocity relations become [2]

$$\frac{\partial \alpha}{\partial \lambda_k} = -\langle A_k \rangle \, ; \qquad \frac{\partial I}{\partial \langle A_k \rangle} = \lambda_k \, ; \qquad \frac{\partial I}{\partial \lambda_i} = \sum_k^M \lambda_k \frac{\partial \langle A_k \rangle}{\partial \lambda_i}, \tag{4}$$

the last one being the so-called Euler theorem.

1.3. The axioms of thermodynamics

Thermodynamics can be regarded as a formal logical structure whose *axioms* are empirical facts [2], which gives it a unique status among the scientific disciplines [1]. The four axioms given below are equivalent to the celebrated laws of thermodynamics of the prevous Subsection [2].

- For every system there exists a quantity E, the internal energy, such that a unique E_s—value is associated to each and every state s. The difference $E_{s1} - E_{s2}$ for two different states s_1 and s_2 in a closed system is equal to the work required to bring the system, while adiabatically enclosed, from one state to the other.
- There exist particular states of a system, the equilibrium ones, that are uniquely determined by E and by a set of extensive macroscopic parameters A_ξ, $\xi = 1, \ldots, M$. The number and characteristics of the A_ξ depends on the nature of the system.
- For every system there exists a state function $S(E, \forall A_\xi)$ that (i) always grows if internal constraints are removed and (ii) is a monotonously (growing) function of E. S remains constant in quasi-static adiabatic changes.

- S and the temperature $T = [\frac{\partial E}{\partial S}]_{A_1,\ldots,A_M}$ vanish for the state of minimum energy and are non-negative for all other states.

From the second and 3rd. Postulates one extracts the following two essential assertions

1. **Statement 3a)** for every system there exists a state function S, a function of E and the A_ξ

$$S = S(E, A_1, \ldots, A_M). \tag{5}$$

2. **Statement 3b)** S is a monotonous (growing) function of E, so that one can interchange the roles of E and S in (5) and write

$$E = E(S, A_1, \ldots, A_M), \tag{6}$$

Eq. (6) clearly indicates that

$$dE = \frac{\partial E}{\partial S} dS + \sum_\xi \frac{\partial E}{\partial A_\xi} dA_\xi \Rightarrow dE = TdS + \sum_\xi P_\xi dA_\xi, \tag{7}$$

with P_ξ generalized pressures and the temperature T defined as [2]

$$T = \left(\frac{\partial E}{\partial S}\right)_{[\forall A_\xi]}. \tag{8}$$

Eq. (7) will play a key-role in our future considerations. If we know $S(E, A_1, \ldots, A_n)$ or, equivalently because of monotonicity, $E(S, A_1, \ldots, A_n))$ we have a *complete* thermodynamic description of a system [2]. For experimentalists, it is often more convenient to work with *intensive* variables defined as follows [2].

Let $S \equiv A_0$. The intensive variable associated to the extensive A_i, to be called P_i are the derivatives

$$P_0 \equiv T = [\frac{\partial E}{\partial S}]_{A_1,\ldots,A_n}, \;\; 1/T = \beta. \tag{9}$$

$$P_j \equiv \lambda_j/T = [\frac{\partial E}{\partial A_j}]_{S,A_1,\ldots,A_{j-1},A_{j+1},\ldots,A_n}. \tag{10}$$

Any one of the Legendre transforms that replaces any s extensive variables by their associated intensive ones (β, λ's will be Lagrange multipliers in SM)

$$L_{r_1,\ldots,r_s} = E - \sum_j P_j A_j, \;\; (j = r_1, \ldots, r_s)$$

contains the same information as either S or E. The transform $L_{r_1,\ldots,r_s}$ is a function of $n - s$ extensive and s intensive variables. This is called the *Legendre invariant structure of thermodynamics*. As we saw above, this implies certain relationships amongst the relevant system's variables, called the *reciprocity relations* (RR), that are crucial for the microscopic discussion of Thermodynamics.

2. Classical statistical mechanics

In 1903 Gibbs formulated the first axiomatic theory for statistical mechanics [1, 3], revolving around the concept of phase space. The phase space (PS) precise location is given by generalized coordinates and momenta. Gibbs' postulates properties of an imaginary (Platonic) ad-hoc notion: the "ensemble" (a mental picture). The ensemble consists of extremely many (N) independent systems, all identical in nature with the one of actual physical interest, but differing in PS-location. That is, the original system is to be mentally repeated many times, each with a different arrangement of generalized coordinates and momenta. Here Liouville's theorem of volume conservation in phase space for Hamiltonian motion plays a crucial role. The ensemble amounts to a distribution of N PS-points, representative of the actual system. N is large enough that one can properly speak of a density D at any PS-point $\phi = q_1, \ldots, q_N; p_1, \ldots, p_N$, with $D = D(q_1, \ldots, q_N; p_1, \ldots, p_N, t) \equiv D(\phi)$, with t the time, and, if we call $d\phi$ the volume element,

$$N = \int d\phi\, D; \quad \forall t. \tag{11}$$

Randomly extracting a system from the ensemble, the probability of selecting it being located in a neighborhood of ϕ would yield

$$P(\phi) = D(\phi)/N. \tag{12}$$

Consequently,

$$\int P\, d\phi = 1. \tag{13}$$

Liouville's theorem follows from the fact that, since phase-space points can not be "destroyed", if

$$N_{12} = \int_{\phi_1}^{\phi_2} D\, d\phi, \tag{14}$$

then

$$\frac{dN_{12}}{dt} = 0. \tag{15}$$

An appropriate analytical manipulation involving Hamilton's canonical equations of motion then yields the theorem in the form [1]

$$\dot{D} + \sum_{i}^{N} \frac{\partial D}{\partial p_i} \dot{p}_i + \sum_{i}^{N} \frac{\partial D}{\partial q_i} \dot{q}_i = 0, \tag{16}$$

entailing the PS-conservation of density.

Equilibrium means simply $\dot{D} = 0$, i. e.,

$$\sum_{i}^{N} \frac{\partial D}{\partial p_i} \dot{p}_i + \sum_{i}^{N} \frac{\partial D}{\partial q_i} \dot{q}_i = 0. \tag{17}$$

2.1. The classical axioms

Gibbs refers to PS-location as the "phase" of the system [1, 3]. The following statements completely explain in microscopic fashion the corpus of classical equilibrium thermodynamics [1].

- The probability that at time t the system will be found in the dynamical state characterized by ϕ equals the probability $P(\phi)$ that a system randomly selected from the ensemble shall possess the phase ϕ will be given by Eq. (12) above.
- All phase-space neighborhoods (cells) have the same a priori probability.
- D depends only upon the system's Hamiltonian.
- The time-average of a dynamical quantity F equals its average over the ensemble, evaluated using D.

3. Information

Information theory (IT) treats information as data communication, with the primary goal of concocting efficient manners of encoding and transferring data. IT is a branch of applied mathematics and electrical engineering, involving the quantification of information, developed by Claude E. Shannon [4] in order to i) find fundamental limits on signal processing operations such as compressing data and ii) finding ways of reliably storing and communicating data. Since its 1948-inception it has considerably enlarged its scope and found applications in many areas that include statistical inference, natural language processing, cryptography, and networks other than communication networks. A key information-measure (IM) was originally called (by Shannon) entropy, in principle unrelated to thermodynamic entropy. It is usually expressed by the average number of bits needed to store or communicate one symbol in a message and quantifies the uncertainty involved in predicting the value of a random variable.Thus, a degree of knowledge (or ignorance) is associated to any normalized probability distribution $p(i)$, $(i = 1, \ldots, N)$, determined by a functional $I[\{p_i\}]$ of the $\{p_i\}$ [4–7] which is precisely Shannon's entropy. IT was la axiomatized in 1950 by Kinchin [8], on the basis of four axioms, namely,

- I is a function ONLY of the $p(i)$,
- I is an absolute maximum for the uniform probability distribution,
- I is not modified if an $N+1$ event of probability zero is added,
- Composition law.

As for the last axiom, consider two sub-systems $[\Sigma^1, \{p^1(i)\}]$ and $[\Sigma^2, \{p^2(j)\}]$ of a composite system $[\Sigma, \{p(i,j)\}]$ with $p(i,j) = p^1(i)\, p^2(j)$. Assume further that the conditional probability distribution (PD) $Q(j|i)$ of realizing the event j in system 2 for a fixed $i-$event in system 1. To this PD one associates the information measure $I[Q]$. Clearly,

$$p(i,j) = p^1(i)\, Q(j|i). \tag{18}$$

Then Kinchin's fourth axiom states that

$$I(p) = I(p^1) + \sum_i p^1(i)\, I(Q(j|i)). \tag{19}$$

An important consequence is that, out of the four Kinchin axioms one finds that Shannons's measure

$$S = -\sum_{i=1}^{N} p(i) \ln [p(i)], \tag{20}$$

gives us the only way of complying with Kinchin's axioms.

4. Statistical mechanics and information theory

It has been argued [9] that the statistical mechanics (SM) of Gibbs is a juxtaposition of subjective, probabilistic ideas on the one hand and objective, mechanical ideas on the other. From the mechanical viewpoint, the vocables "statistical mechanics" suggest that for solving physical problems we ought to acknowledge a degree of uncertainty as to the experimental conditions. Turning this problem around, it also appears that the purely statistical arguments are incapable of yielding any physical insight unless some mechanical information is a priori assumed [9]. This is the conceptual origin of the link SM-IT pioneered by Jaynes in 1957 via his Maximum Entropy Principle (MaxEnt) [5, 6, 10] which allowed for reformulating SM in information terms. Since IT's central concept is that of information measure (IM)

Descartes' scientific methodology considers that truth is established via the agreement between two *independent* instances that can neither suborn nor bribe each other: analysis (purely mental) and experiment [11]. The analytic part invokes mathematical tools and concepts: Mathematics' world ⇔ Laboratory. The mathematical realm is called Plato's Topos Uranus (TP). Science in general, and physics in particular, may thus be seen as a [TP ⇔ "Experiment"] two-way bridge. TP concepts are related to each other in the form of "laws" that adequately describe the relationships obtaining among suitable chosen variables that describe the phenomenon at hand. In many cases these laws are integrated into a comprehensive theory (e.g., classical electromagnetism, based upon Maxwell's equations) [1, 12–15].

Jaynes' MaxEnt ideas describe thermodynamics via the link [IT as a part of TP]⇔ [Thermal experiment], or in a more general scenario: [IT] ⇔ [Phenomenon at hand]. It is clear that the relation between an information measure and entropy is [IM] ⇔ [Entropy S]. One can then assert that an IM is not necessarily an entropy, since the first belongs to the Topos Uranus and the later to the laboratory. Of course, in some special cases an association IM ⇔ entropy S can be established. Such association is both useful and proper in very many situations [5].

If, in a given scenario, N distinct outcomes ($i = 1, \ldots, N$) are possible, three alternatives are to be considered [6]:

1. Zero ignorance: predict with certainty the actual outcome.
2. Maximum ignorance: Nothing can be said in advance. The N outcomes are equally likely.

3. Partial ignorance: we are given the probability distribution $\{P_i\}$; $i = 1, \ldots, N$.

If our state of knowledge is appropriately represented by a set of, say, M expectation values, then the "best", least unbiased probability distribution is the one that [6]

- reflects just what we know, without "inventing" unavailable pieces of knowledge [5, 6] and, additionally,
- maximizes ignorance: the truth, all the truth, *nothing but* the truth [6].

Such is the MaxEnt rationale. In using MaxEnt, one is not maximizing a physical entropy, but only maximizing ignorance in order to obtain the least biased distribution compatible with the a priori knowledge.

Statistical mechanics and thereby thermodynamics can be formulated on an information theory basis if the density operator $\hat{\rho}$ is obtained by appealing to Jaynes' maximum entropy principle (MaxEnt), that can be stated as follows:

Assume that your prior knowledge about the system is given by the values of M expectation values $< A_1 >, \ldots, < A_M >$. In such circumstances $\hat{\rho}$ is uniquely determined by extremizing $I(\hat{\rho})$ subject to M constraints given, namely, the M conditions $< A_j >= Tr[\hat{\rho}\, \hat{A}_j]$, a procedure that entails introducing M Lagrange multipliers λ_i. Additionally, since normalization of $\hat{\rho}$ is necessary, a normalization Lagrange multiplier ξ should be invoked. The procedure immediately leads one [6] to realizing that $I \equiv S$, the equilibrium Boltzmann's entropy, if the a priori knowledge $< A_1 >, \ldots, < A_M >$ refers only to extensive quantities. Of course, I, once determined, *affords for complete thermodynamical information for the system of interest* [6].

5. A new micro-macroscopic way of accounting for thermodynamics

Gibbs' and MaxEnt approaches satisfactorily describe equilibrium thermodynamics. We will here search for a new, different alternative able to account for thermodynamics from first principles. Our idea is to give axiom-status to Eq. (7), *which is an empirical statement*. Why? Because neither in Gibbs' nor in MaxEnt's axioms we encounter a direct connection with actual thermal data. By appealing to Eq. (7) we would instead be actually employing empirical information. This is our rationale.

Consequently, we will concoct a new SM-axiomatics by giving postulate status to the following macroscopic statement:

Axiom (1)

$$dE = TdS + \sum_{\nu} P_\nu dA_\nu. \tag{21}$$

This is a macroscopic postulate to be inserted into a microscopic axiomatics' corpus.

We still need *some* amount of microscopic information, since we are building up a microscopic theory. We wish to add as little as possible, of course (Ockham's razor). At this point it is useful to remind the reader of Kinchin's postulates, recounted above. We will content ourselves with borrowing for our theoretical concerns just his his first axiom. Thus, we conjecture at this point, and will prove below, that the following assertion suffices for our theoretical purposes:

Axiom (2) If there are W microscopic accessible states labelled by i, whose microscopic probability we call p_i, then

$$S = S(p_1, p_2, \ldots, p_W). \tag{22}$$

Thus, we are actually taking as a postulate something that is actually known from both quantum and classical mechanics.

Axiom (3) The internal energy E and the external parameters A_ν are to be considered as the expectation values of suitable operators, that is, the hamiltonian H and the hermitian operators $\mathcal{R}_\nu$ (i.e., $A_\nu \equiv <\mathcal{R}_\nu>$). Thus, the A_ν (and also E) will depend on the eigenvalues of these operators *and* on the probability set. (Note that energy eigenvalues depend of course upon the $\mathcal{R}_\nu$.

The reader will immediately realize that Axiom (2) is just a way of re-expressing Boltzmann's "atomic" conjecture. Thus, macroscopic quantities become statistical averages evaluated using a microscopic probability distribution [16]. Our present three new axioms are statements of fact. What do we mean? That they are borrowed from either experiment or pre-existent theories. Somewhat surprisingly, our three axioms do not actually incorporate any knew knowledge at all. The merely re-express known previous notions. Ockham's razor at its best! Our theory could no be more economical.

We need now to prove that the above three postulates allow one to reconstruct the imposing edifice of statistical mechanics. We will tackle this issue by showing below that they our axioms are equivalent to those of Jaynes' [17]. At this point we need to recall the main goal of statistical mechanics, namely, finding the probability distribution (or the density operator) that best describes our physical system. In order to do so Jaynes appealed to his MaxEnt postulate, that we restate below for the sake of fixing notation.

MaxEnt axiom: assume your prior knowledge about the system is given by the values of M expectation values

$$A_1 \equiv <\mathcal{R}_1>, \ldots, A_R \equiv <\mathcal{R}_M>. \tag{23}$$

Then, ρ is uniquely fixed by extremizing the information measure $I(\rho)$ subject to $\rho-$normalization plus the constraints given by the M conditions constituting our assumed foreknowledge

$$A_\nu = <\mathcal{R}_\nu> = Tr[\rho\,\mathcal{R}_\nu]. \tag{24}$$

This leads, after a Lagrange-constrained extremizing process, to the introduction of M Lagrange multipliers λ_ν, that one assimilates to the generalized pressures P_ν. The truth, the whole truth, nothing but the truth [6]. Jaynes rationale asserts that if the entropic measure that reflects our ignorance were not of maximal character, we would actually be *inventing* information not at hand.

While working through his variational process, Jaynes discovers that, after multiplying by Boltzmann's constant k_B the right-hand-side of his expression for the information measure, it converts itself into an entropy, $I \equiv S$, the equilibrium thermodynamic one, with the caveat that $A_1 = <\mathcal{R}_1>, \ldots, A_M = <\mathcal{R}_M>$ refer to extensive quantities. Having ρ, his universal form $I(\rho)$ *yields complete microscopic information with respect to the system of interest.* To achieve our ends one needs now just to prove that the new axiomatics, with (21) and (22), is equivalent to MaxEnt.

6. New connection between macroscopic and microscopic approaches

In establishing our new connections between the micro- and macro-scenarios we shall work with the classical instance only, since the corresponding quantum treatment constitute in this sense just a straightforward extension.

Our main idea is to pay attention to the generic change $p_i \rightarrow p_i + dp_i$ as constrained by Eq. (21). In other word, we insist on studying the change dp_i that takes place in such a manner that (21) holds. Our main macroscopic quantities S, A_j, and E will vary with dp_i. These changes are not arbitrary but are constrained by (21). Note here an important advantage to be of our approach. We need *not* specify beforehand the information measure employed.

Since several possibilities exist (see for instance Gell-Mann and Tsallis [18]), this entails that the choice of information nature is not predetermined by macroscopic thermodynamics. For a detailed discussion of this issue see Ferri, Martinez, and Plastino [19].

The pertinent ingredients at hand are

- an arbitrary, smooth function $f(p)$ permitting one expressing the information measure via

$$I \equiv S(\{p_i\}) = \sum_i p_i f(p_i), \tag{25}$$

  such that $S(\{p_i\})$ is a concave function,

- M quantities A_ν representing values of extensive quantities $\langle \mathcal{R}_\nu \rangle$, that adopt, for a micro-state i, the value a_i^ν with probability p_i,
- still another arbitrary smooth, monotonic function $g(p_i)$ ($g(0) = 0$; $g(1) = 1$). With the express purpose of employing generalized, non-Shannonian entropies, we slightly generalize here the expectation-value definitions by recourse to g via (26):

$$A_\nu \equiv \langle \mathcal{R}_\nu \rangle = \sum_i^W a_i^\nu \, g(p_i); \;\; \nu = 2, \ldots, M, \tag{26}$$

$$E = \sum_i^W \epsilon_i \, g(p_i), \tag{27}$$

  where ϵ_i is the energy associated to the microstate i.

We take $A_1 \equiv E$ and pass to a consideration of the probability variations dp_i that should generate accompanying changes dS, dA_ν, and dE in, respectively, S, the A_ν, and E.

The essential issue at hand is that of enforcing compliance with

$$dE - TdS + \sum_{\nu=1}^W dA_\nu \lambda_\nu = 0, \tag{28}$$

with T the temperature and λ_ν generalized pressures. By recourse to (25), (26), and (27) we i) recast now (28) for

$$p_i \to p_i + dp_i, \tag{29}$$

and ii) expand the resulting equation up to first order in the dp_i.

Remembering that the Lagrange multipliers λ_ν are identical to the generalized pressures P_ν of Eq. (7), one thus encounter, after a little algebra [20–26],

$$C_i^{(1)} = [\Sigma_{\nu=1}^{M} \lambda_\nu \, a_i^\nu + \epsilon_i]$$

$$C_i^{(2)} = -T \frac{\partial S}{\partial p_i}$$

$$\Sigma_i [C_i^{(1)} + C_i^{(2)}] dp_i \equiv \Sigma_i K_i dp_i = 0, \tag{30}$$

so that, appropriately rearranging things

$$T_i^{(1)} = f(p_i) + p_i f'(p_i)$$
$$T_i^{(2)} = -\beta[(\Sigma_{\nu=1}^{M} \lambda_\nu \, a_i^\nu + \epsilon_i) \, g'(p_i) - K],$$
$$(\beta \equiv 1/kT), \tag{31}$$

and we are in a position to recast (30) in the fashion

$$T_i^{(1)} + T_i^{(2)} = 0; \;\; (for \;\; any \;\; i), \tag{32}$$

an expression whose importance will become manifest later on.

Eqs. (30) or (32) yield one and just one p_i–expression, as demonstrated in Refs. [20–26]. However, it will be realized below that, at this stage, an explicit expression for this probability distribution is not required.

We pass now to traversing the opposite road that leads from Jaynes' MaxEnt procedure and ends up with *our* present equations. This entails extremization of S subject to constraints in E, A_ν, and normalization. For details see [20–26].

Setting $\lambda_1 \equiv \beta = 1/T$ one has

$$\delta_{p_i}[S - \beta\langle H\rangle - \sum_{\nu=2}^{M} \lambda_\nu \langle \mathcal{R}_\nu \rangle - \xi \sum_i p_i] = 0, \tag{33}$$

(normalization Lagrange multiplier ξ) is easily seen in the above cited references to yield as a solution the very set of Eqs. (30). The detailed proof is given in the forthcoming Section. Eqs. (30) arise then from two different approaches:

- our methodology, based on Eqs. (21) and (22), and
- following the well known MaxEnt route.

Accordingly, we see that both MaxEnt and our axiomatics co-imply one another. They are indeed equivalent ways of constructing equilibrium statistical mechanics. As a really relevant fact

One does not need to know the analytic form of $S[p_i]$ neither in Eqs. (30) nor in (33).

7. Proof

Here we prove that Eqs. (30) can be derived from the MaxEnt approach (33). One wishes to extremize S subject to the constraints of fixed valued for i) U, ii) the M values A_ν (entailing Lagrange multipliers (1) β and (2) M γ_ν), and iii) normalization (Lagrange multiplier ξ). One has also

$$A_\nu = \langle \mathcal{R}_\nu \rangle = \sum_i p_i \, a_i^\nu, \tag{34}$$

with $a_i^\nu = \langle i | \mathcal{R}_\nu | i \rangle$ the matrix elements in the basis $\langle i \rangle$ of $\mathcal{R}_\nu$. The ensuing variational problem one faces, with $U = \sum_i p_i \epsilon_i$, is

$$\delta_{\{p_i\}} \left[S - \beta U - \sum_{\nu=1}^{M} \gamma_\nu A_\nu - \xi \sum_i p_i \right] = 0, \tag{35}$$

that immediately leads, for $\gamma_\nu = \beta \lambda_\nu$, to

$$\delta_{p_m} \sum_i \left(p_i f(p_i) - [\beta p_i (\sum_{\nu=1}^{M} \lambda_\nu \, a_i^\nu + \epsilon_i) + \xi p_i] \right) = 0, \tag{36}$$

so that the the following two quantities vanish

$$\begin{aligned} & f(p_i) + p_i f'(p_i) - [\beta(\textstyle\sum_{\nu=1}^{M} \lambda_\nu \, a_i^\nu + \epsilon_i) + \xi] \\ & \quad \Rightarrow \text{ if } \xi \equiv \beta K, \\ & f(p_i) + p_i f'(p_i) - \beta(\textstyle\sum_{\nu=1}^{M} \lambda_\nu a_i^\nu + \epsilon_i) + K] \\ & \quad \Rightarrow 0 = T_i^{(1)} + T_i^{(2)}. \end{aligned} \tag{37}$$

We realize now that (32) and the last equality of (37) are one and the same equation. MaxEnt does lead to (32).

8. Conclusions

We have formally proved above that our axiomatics allows one to derive MaxEnt equations and viceversa. Thus, our treatment provides an alternative foundation for equilibrium statistical mechanics. We emphasized that, opposite to what happens with both Gibbs' and Jaynes' axioms, our postulates have zero new informational content. Why? Because they are borrowed either from experiment or from pre-existing theories, namely, information theory and quantum mechanics.

The first and second laws of thermodynamics are two of physics' most important empirical facts, constituting pillars to our present view of Nature. Statistical mechanics (SM) adds an underlying microscopic substratum able to explain not only these two laws but the whole of thermodynamics itself [2, 6, 27–30]. Basic SM-ingredient is a microscopic probability distribution (PD) that controls microstates-population [27]. Our present ideas yield a detailed picture, from a new perspective [20–26], of how changes in the independent external thermodynamic parameters affect the micro-state population and, consequently, the entropy and the internal energy.

Acknowledgement

This work was partially supported by the MEC Grant FIS2005-02796 (Spain).

Author details

A. Plastino
Universidad Nacional de La Plata, Instituto de Física (IFLP-CCT-CONICET), C.C. 727, 1900 La Plata, Argentina
Physics Departament and IFISC-CSIC, University of Balearic Islands, 07122 Palma de Mallorca, Spain

Evaldo M. F. Curado
Centro Brasileiro de Pesquisas Fisicas, Rio de Janeiro, Brazil

M. Casas
Physics Departament and IFISC-CSIC, University of Balearic Islands, 07122 Palma de Mallorca, Spain

9. References

[1] R. B. Lindsay and H. Margenau, *Foundations of physics*, NY, Dover, 1957.
[2] E. A. Desloge, *Thermal physics* NY, Holt, Rhinehart and Winston, 1968.
[3] J. Willard Gibbs, *Elementary Principles in Statistical Mechanics*, New Haven, Yale University Press, 1902.
[4] C. E. Shannon, Bell System Technol. J. 27 (1948) 379-390.
[5] E. T. Jaynes *Papers on probability, statistics and statistical physics*, edited by R. D. Rosenkrantz, Dordrecht, Reidel, 1987.
[6] A. Katz, *Principles of Statistical Mechanics, The information Theory Approach*, San Francisco, Freeman and Co., 1967.
[7] T. M. Cover and J. A. Thomas, *Elements of information theory*, NY, J. Wiley, 1991.
[8] A. Plastino and A. R. Plastino in *Condensed Matter Theories*, Volume 11, E. Ludeña (Ed.), Nova Science Publishers, p. 341 (1996).
[9] D. M. Rogers, T. L. Beck, S. B. Rempe, *Information Theory and Statistical Mechanics Revisited*, ArXiv 1105.5662v1.
[10] D. J. Scalapino in *Physics and probability. Essays in honor of Edwin T. Jaynes* edited by W. T. Grandy, Jr. and P. W. Milonni (Cambridge University Press, NY, 1993), and references therein.
[11] B. Russell, *A history of western philosophy* (Simon & Schuster, NY, 1945).
[12] P. W. Bridgman *The nature of physical theory* (Dover, NY, 1936).
[13] P. Duhem *The aim and structure of physical theory* (Princeton University Press, Princeton, New Jersey, 1954).
[14] R. B. Lindsay *Concepts and methods of theoretical physics* (Van Nostrand, NY, 1951).

[15] H. Weyl *Philosophy of mathematics and natural science* (Princeton University Press, Princeton, New Jersey, 1949).
[16] D. Lindley, *Boltzmann's atom*, NY, The free press, 2001.
[17] W.T. Grandy Jr. and P. W. Milonni (Editors), *Physics and Probability. Essays in Honor of Edwin T. Jaynes*, NY, Cambridge University Press, 1993.
[18] M. Gell-Mann and C. Tsallis, Eds. *Nonextensive Entropy: Interdisciplinary applications*, Oxford, Oxford University Press, 2004.
[19] G. L. Ferri, S. Martinez, A. Plastino, Journal of Statistical Mechanics, P04009 (2005).
[20] E. Curado, A. Plastino, Phys. Rev. E 72 (2005) 047103.
[21] A. Plastino, E. Curado, Physica A 365 (2006) 24
[22] A. Plastino, E. Curado, International Journal of Modern Physics B 21 (2007) 2557
[23] A. Plastino, E. Curado, Physica A 386 (2007) 155
[24] A. Plastino, E. Curado, M. Casas, Entropy A 10 (2008) 124
[25] International Journal of Modern Physics B 22, (2008) 4589
[26] E. Curado, F. Nobre, A. Plastino, Physica A 389 (2010) 970.
[27] R.K. Pathria, *Statistical Mechanics* (Pergamon Press, Exeter, 1993).
[28] F. Reif, *Statistical and thermal physics* (McGraw-Hill, NY, 1965).
[29] J. J.Sakurai, *Modern quantum mechanics* (Benjamin, Menlo Park, Ca., 1985).
[30] B. H. Lavenda, *Statistical Physics* (J. Wiley, New York, 1991); B. H. Lavenda, *Thermodynamics of Extremes* (Albion, West Sussex, 1995).

Chapter 5

Information Capacity of Quantum Transfer Channels and Thermodynamic Analogies

Bohdan Hejna

Additional information is available at the end of the chapter

1. Introduction

We will begin with a simple type of *stationary stochastic*[1] systems of quantum physics using them within a frame of the **Shannon** Information Theory and Thermodynamics but starting with their *algebraic representation*. Based on this algebraic description a model of information transmission in those systems by defining the Shannon information will be stated in terms of variable about the system state. Measuring on these system is then defined as a spectral decomposition of measured quantities - *operators*. The information capacity formulas, now of the *narrow-band* nature, are derived consequently, for the simple system governed by the **Bose–Einstein** (B–E) Law [bosonic (photonic) channel] and that one governed by the **Fermi-Dirac** (F–D) Law [fermionic (electron) channel]. *The not-zero value for the average input energy needed for information transmission existence in F–D systems* is stated [11, 12].

Further the *wide–band* information capacity formulas for B–E and F–D case are stated. Also the original *thermodynamic* capacity derivation for the wide–band photonic channel as it was stated by **Lebedev–Levitin** in 1966 is revised. This revision is motivated by apparent relationship between the B–E (photonic) wide–band information capacity and the *heat efficiency* for a certain *heat cycle*, being further considered as the demonstrating model for processes of information transfer in the original wide–band photonic channel. The information characteristics of a *model reverse* heat cycle and, by this model are analyzed, the information arrangement of which is set up to be most analogous to the structure of the photonic channel considered, we see the necessity of returning the transfer medium (the channel itself) to its initial state as a condition for a *sustain, repeatable* transfer. It is not regarded in [12, 30] where a single information transfer act only is considered. Or the return is

[1] We deal with such a system which is taking on at time $t = 0, 1, \ldots$ states $\boldsymbol{\theta}_t$ from a state space $\boldsymbol{\Theta}$. If for any t_0 the relative frequencies I_B of events $B \subset \boldsymbol{\Theta}$ is valid that $\frac{1}{T} \sum_{t=t_0+1}^{t_0+T} I_B(\boldsymbol{\theta}_t)$ tends for $T \to \infty$ to probabilities $p_{t_0}(B)$ we speak about a *stochastic system*. If these probabilities do not depend on the beginning t_0, a *stationary stochastic system* is spoken about.

regarded, but by opening the whole transfer chain for covering these *return energy needs from its environment* not counting them in - *not within the transfer chain only* as we do now. *The result is the corrected capacity formula for wide–band photonic (B–E) channels being used for an information transfer organized cyclicaly.*

2. Information transfer channel

An *information, transfer* channel $\mathcal{K}$ is defined as an arranged *tri–partite* structure [5]

$$\mathcal{K} \stackrel{\text{Def}}{=} [X,\ \varepsilon,\ Y] \ \text{ where } \ X \stackrel{\text{Def}}{=} [A,\ p_X(\cdot)],\ \ Y \stackrel{\text{Def}}{=} [B,\ p_Y(\cdot)] \ \text{ and} \tag{1}$$

- X is an *input* stochastic quantity, a *source* of *input* messages $a \in A^+ \stackrel{\triangle}{=} \mathcal{X}$, a *transceiver,*
- Y is an *output* stochastic quantity, a *source* of *output* messages $b \in B^+ \stackrel{\triangle}{=} \mathcal{Y}$, a *receiver,*
- output messages $b \in \mathcal{Y}$ are *stochastic* dependent on input messages $a \in \mathcal{X}$ and they are received by the receiver of messages, Y,
- ε is the *maximal* probability of an error in the transfer of any symbol $x \in A$ in an input message $a \in \mathcal{X}$
[the maximal probability of erroneously receiving $y \in B$ (inappropriate for x) in an output message $b \in \mathcal{Y}$],
- A denotes a finite *alphabet* of elements x of the source of input messages,
- B denotes a finite alphabet of elements y of the source of output messages,
- $p_X(\cdot)$ is the *probability distribution* of evidence of any symbol $x \in A$ in an input message,
- $p_Y(\cdot)$ is the probability distribution of evidence of any symbol $y \in B$ in an output message.

The structure $(X,\ \mathcal{K},\ Y)$ or $(\mathcal{X},\ \mathcal{K},\ \mathcal{Y})$ is termed a *transfer (Shannon) chain*. The symbols $H(X)$ and $H(Y)$ respectively denote the *input information (Shannon) entropy* and the *output information (Shannon) entropy* of channel $\mathcal{K}$, *discrete* for this while,

$$H(X) \stackrel{\text{Def}}{=} -\sum_{x \in A} p_X(x)\ \ln p_X(x),\quad H(Y) \stackrel{\text{Def}}{=} -\sum_{y \in B} p_Y(y)\ \ln p_Y(y) \tag{2}$$

The symbol $H(X|Y)$ denotes the *loss entropy* and the symbol $H(Y|X)$ denotes the *noise entropy* of channel $\mathcal{K}$. These entropies are defined as follows,

$$H(X|Y) \stackrel{\text{Def}}{=} -\sum_A \sum_B p_{X,Y}(x,y)\ \ln p_{X|Y}(x|y),\quad H(Y|X) \stackrel{\text{Def}}{=} -\sum_A \sum_B p_{X,Y}(x,y) \ln p_{Y|X}(y|x) \tag{3}$$

where the symbol $p_{\cdot|\cdot}(\cdot|\cdot)$ denotes the *condition* and the symbol $p_{\cdot,\cdot}(\cdot,\cdot)$ denotes the *simultaneous* probabilities. For *mutual (transferred) usable* information, *transinformation* $T(X;Y)$ or $T(Y;X)$ is valid that

$$T(X;Y) = H(X) - H(X|Y) \ \text{ and } \ T(Y;X) = H(Y) - H(Y|X) \tag{4}$$

From (2) and (3), together with the definitions of $p_{\cdot|\cdot}(\cdot|\cdot)$ and $p_{\cdot,\cdot}(\cdot,\cdot)$, is provable prove that the transinformation is symmetric. Then the *equation of entropy (information) conservation* is valid

$$H(X) - H(X|Y) = H(Y) - H(Y|X) \tag{5}$$

The *information capacity* of the channel $\mathcal{K}$ (both discrete an continuous) is defined by the equation

$$C \stackrel{\text{Def}}{=} \sup I(X;Y) \tag{6}$$

over all possible probability distributions $q(\cdot)$, $p(\cdot)$.It is the maximum (supremum) of the medium value of the usable amount of information about the input message x within the *output* message y.

Remark: For continuous distributions (densities) $p_{[\cdot]}(\cdot)$, $p_{[\cdot|\cdot]}(\cdot|\cdot)$ on intervals $\mathcal{X}$, $\mathcal{Y} \in \mathbb{R}$, $x \in \mathcal{X}, y \in \mathcal{Y}$ is

$$H(X) = -\int_{\mathcal{X}} p_X(x) \ln p_X \,\mathrm{d}x, \tag{7}$$

$$H(Y) = -\int_{\mathcal{Y}} p_Y(y) \ln p_Y(y) \,\mathrm{d}y \tag{8}$$

$$H(X|Y) = -\int_{\mathcal{X}}\int_{\mathcal{Y}} p_{X,Y}(x,y) \ln p_{X|Y}(x|y) \,\mathrm{d}x\mathrm{d}y,$$

$$H(Y|X) = -\int_{\mathcal{X}}\int_{\mathcal{Y}} p_{X,Y}(x,y) \ln p(y|x) \,\mathrm{d}x\mathrm{d}y$$

Equations (4), (5 are valid for both the quantities $H(\cdot)$ and $H(\cdot|\cdot)$, as well as for their respective changes $\Delta H(\cdot)[= H(\cdot)]$ and $\Delta H(\cdot|\cdot)[= H(\cdot|\cdot)]$.

3. Representation of physical transfer channels

The most simple way of description of *stationary* physical systems is an *eucleidian* space Ψ of their states expressed as *linear operators*.[2] This way enables the *mathematical* formulation of the term *(physical) state* and, generally, the term *(physical) quantity*.

Physical quantities $\boldsymbol{\alpha}$, associated with a physical system $\boldsymbol{\Psi}$ represented by the Eucleidian space Ψ are expressed by *symmetric* operators from the linear space $L(\Psi)$ of operators on Ψ, $\boldsymbol{\alpha} \in \mathbf{A} \subset L(\Psi)$ [7]. The supposition is that any physical quantity can achieve only those *real* values α which are the *eigenvalues* of the associated *symmetric operator* $\boldsymbol{\alpha}$ (symmetric *matrix* $[\alpha_{i,j}]_{n,n}$, $n = \dim \Psi$). They are elements of the *spectrum* $\mathbf{S}(\boldsymbol{\alpha})$ of the *operator* $\boldsymbol{\alpha}$. The eigenvalues $\alpha \in \mathbf{S}(\boldsymbol{\alpha}) \subset \mathbb{R}$ of the quantity $\boldsymbol{\alpha}$ being measured on the system $\boldsymbol{\Psi} \cong \Psi$ depend on the (inner) states $\boldsymbol{\theta}$ of this system $\boldsymbol{\Psi}$.

- The *pure states* of the system $\boldsymbol{\Psi} \cong \Psi$ are represented by *eigenvectors* $\psi \in \Psi$. It is valid that the *scalar project* $(\psi, \psi) = 1$; in *quantum physics* they are called *normalized wave functions.*
- The *mixed states* are nonnegative quantities $\boldsymbol{\theta} \in \mathbf{A}$; their *trace* [of an *square* matrix (operator) $\boldsymbol{\alpha}$] is defined

$$\mathrm{Tr}(\boldsymbol{\alpha}) \stackrel{\text{Def}}{=} \sum_{i=1}^{n} \alpha_{i,i} \quad \text{and, for } \boldsymbol{\alpha} \equiv \boldsymbol{\theta} \text{ is valid that } \mathrm{Tr}(\boldsymbol{\theta}) = 1. \tag{9}$$

The *symmetric projector* $\boldsymbol{\pi}\{\psi\} = \boldsymbol{\pi}[\Psi(\{\psi\})]$ (orthogonal) on the one–dimensional *subspace* $\Psi(\{\psi\})$ of the space Ψ is nonnegative quantity for which $\mathrm{Tr}(\boldsymbol{\pi}\{\psi\}) = 1$ is valid. The projector $\boldsymbol{\pi}\{\psi\}$ *represents* [on the set of quantities $\mathbf{A} \subset L(\Psi)$] the pure state ψ of the system $\boldsymbol{\Psi}$. Thus, an arbitrary state of the system $\boldsymbol{\Psi}$ can be defined as a nonnegative quantity $\boldsymbol{\theta} \in \boldsymbol{\Theta} \subset \mathbf{A}$ for which $\mathrm{Tr}(\boldsymbol{\theta}) = 1$ is valid. For the *pure state* $\boldsymbol{\theta}$ is then valid that $\boldsymbol{\theta}^2 = \boldsymbol{\theta}$ and the *state space* $\boldsymbol{\Theta}$ of the system $\boldsymbol{\Psi}$ is defined as the set of all states $\boldsymbol{\theta}$ of the system $\boldsymbol{\Psi}$.

[2] The motivation is the *axiomatic theory of algebraic representation of physical systems* [7].

3.1. Probabilities and information on physical systems

3.1.0.1. Theorem:

For any state $\boldsymbol{\theta} \in \boldsymbol{\Theta}$ the pure states $\boldsymbol{\theta}_i = \pi\{\psi_i\}$ and numbers $q(i|\boldsymbol{\theta}) \geq 0$ exists, that

$$q(i|\boldsymbol{\theta}) \geq 0,\ \boldsymbol{\theta} = \sum_{i=1}^{n} q(i|\boldsymbol{\theta})\,\boldsymbol{\theta}_i \quad \text{where} \quad \sum_{i=1}^{n} q(i|\boldsymbol{\theta}) = 1 \tag{10}$$

3.1.0.2. Proof:

Let $D(\boldsymbol{\theta}) = \{D_\theta : \theta \in \mathbf{S}(\boldsymbol{\theta})\}$ is a *disjoint decomposition* of the set $\{1, 2, ..., n\}$ of indexes of the base $\{\psi_1, \psi_2, ..., \psi_n\}$ of Ψ. The set $\{\psi_i : i \in D_\theta\}$ is an orthogonal basis of the *eigenspace* $\Psi(\theta|\boldsymbol{\theta}) \subset \Psi$ of the operator $\boldsymbol{\theta}$ [for its eigenvalue $\theta \in \mathbf{S}(\boldsymbol{\theta})$]. Then

$$\text{card}\, D_\theta = \dim \Psi(\theta|\boldsymbol{\theta}) \geq 1 \ \text{ and } \ \pi[\Psi(\theta|\boldsymbol{\theta})] = \sum_{i\in D_\theta} \pi\{\psi_i\},\ \forall\theta \in \mathbf{S}(\boldsymbol{\theta}) \tag{11}$$

Let $q(i|\boldsymbol{\theta}) = \theta$ is taken for all $i \in D_\theta$, $\theta \in \mathbf{S}(\boldsymbol{\theta})$. By the *spectral decomposition theorem* [9],

$$\boldsymbol{\theta} = \sum_{\theta\in\mathbf{S}(\boldsymbol{\theta})} \theta\,\pi_\theta = \sum_{\theta\in\mathbf{S}(\boldsymbol{\theta})} \theta\pi[\Psi(\theta|\boldsymbol{\theta})] = \sum_{\theta\in\mathbf{S}(\boldsymbol{\theta})} \theta \sum_{i\in D_\theta} \pi\{\psi_i\} = \sum_{i=1}^{n} q(i|\boldsymbol{\theta})\,\pi\{\psi_i\} \tag{12}$$

$$\text{Tr}(\boldsymbol{\theta}) = \sum_{i=1}^{n} q(i|\boldsymbol{\theta}) \cdot \text{Tr}(\pi\{\psi_i\}) = \sum_{i=1}^{n} q(i|\boldsymbol{\theta}) = 1$$

The symbol $q(\cdot|\boldsymbol{\theta})$ denotes the *probability distribution* into *pure, canonic components* $\boldsymbol{\theta}_i$ of $\boldsymbol{\theta}$ is called:

- *canonic distribution (q-distribution)* of the state $\boldsymbol{\theta} \in \boldsymbol{\Theta}$.[3]

Further the two distribution defined on spectras of $\boldsymbol{\alpha} \in \mathbf{A}$ and $\boldsymbol{\theta} \in \boldsymbol{\Theta} \subset \mathbf{A}$ will be dealt:

- *dimensional distribution (d-distribution)* of the state $\boldsymbol{\theta} \in \boldsymbol{\Theta}$[4]

$$d(\theta|\boldsymbol{\theta}) \stackrel{\text{Def}}{=} \frac{\dim \Psi(\theta|\boldsymbol{\theta})}{\dim \Psi} = \frac{\dim \Psi(\theta|\boldsymbol{\theta})}{n}, \quad \theta \in \mathbf{S}(\boldsymbol{\theta}) \tag{13}$$

- *distribution of measuring (p-distribution)* of the quantity $\boldsymbol{\alpha} \in \mathbf{A}$ in the state $\boldsymbol{\theta} \in \boldsymbol{\Theta}$,

$$p(\alpha|\boldsymbol{\alpha}|\boldsymbol{\theta}) \stackrel{\text{Def}}{=} \text{Tr}(\boldsymbol{\theta}\pi_\alpha), \quad \alpha \in \mathbf{S}(\boldsymbol{\alpha}) \tag{14}$$

where $\{\pi_\alpha : \alpha \in \mathbf{S}(\boldsymbol{\alpha})\}$ is the *spectral decomposition* of the *unit operator* $\mathbf{1}$. Due the nonnegativity of $\boldsymbol{\theta}\pi_\alpha$ is $p(\alpha|\boldsymbol{\alpha}|\boldsymbol{\theta}) = \text{Tr}(\boldsymbol{\alpha}\pi_\alpha) \geq 0$. By spectral decomposition of $\mathbf{1}$ and by definition of the *trace* $\text{Tr}(\cdot)$ is

$$\sum_{\alpha\in\mathbf{S}(\boldsymbol{\alpha})} p(\alpha|\boldsymbol{\alpha}|\boldsymbol{\theta}) = \text{Tr}\left(\boldsymbol{\theta} \sum_{\alpha\in\mathbf{S}(\boldsymbol{\alpha})} \pi_\alpha\right) = \text{Tr}(\boldsymbol{\theta}\mathbf{1}) = 1 \tag{15}$$

Thus the relation (14) defines the probability distribution on the spectrum of the operator $\mathbf{S}(\boldsymbol{\alpha})$.

[3] Or, the *system spectral distribution* SSD [11].

[4] Or, the *system distribution* of the *system spectral dimension* SDSD [11].

- The special case of the *p-distribution* is that *for measuring values of* $\boldsymbol{\alpha} \equiv \boldsymbol{\theta}$, $\boldsymbol{\theta} \in \boldsymbol{\Theta}$,

$$p(\theta|\theta|\boldsymbol{\theta}) = \mathrm{Tr}(\boldsymbol{\theta}\pi_\theta) = \theta \dim \Psi(\theta|\boldsymbol{\theta}), \quad \theta \in \mathbf{S}(\boldsymbol{\theta}) \tag{16}$$

It is the case of *measuring the* $\boldsymbol{\theta}$ *itself*. Thus, by the equation (14), the term ***measuring*** **means just the spectral decomposition of the unit operator 1 within the spectral relation with** $\boldsymbol{\alpha}$ [4, 20]. The act of measuring of the quantity $\boldsymbol{\alpha} \in \mathbf{A}$ in (the system) state $\boldsymbol{\theta}$ gives the value α from the spectrum $\mathbf{S}(\boldsymbol{\alpha})$ with the probability $p(\alpha|\boldsymbol{\alpha}|\boldsymbol{\theta}) = \mathrm{Tr}(\boldsymbol{\theta}\pi_\alpha)$ given by the p-distribution,

$$\sum_{\mathcal{S}(\alpha)} p(\alpha|\boldsymbol{\alpha}|\boldsymbol{\theta}) = \sum_{\mathcal{S}(\alpha)} \mathrm{Tr}(\boldsymbol{\theta}\pi_\alpha) = 1 \tag{17}$$

The ***measured*** **quantity** $\boldsymbol{\alpha}$ **in the state** $\boldsymbol{\theta}$ **is a** ***stochastic*** **quantity with its values [occuring with probabilities** $p(\cdot|\boldsymbol{\alpha}|\boldsymbol{\theta})$**] from its spectrum** $\mathbf{S}(\boldsymbol{\alpha})$. For its *mathematical expectation, medium value* is valid

$$E(\boldsymbol{\alpha}) = \sum_{\alpha \in \mathbf{S}(\boldsymbol{\alpha})} \alpha \,\mathrm{Tr}(\boldsymbol{\theta}\pi_\alpha) = \mathrm{Tr}\left(\boldsymbol{\theta} \sum_{\alpha \in \mathbf{S}(\boldsymbol{\alpha})} \alpha\pi_\alpha\right) = \mathrm{Tr}(\boldsymbol{\theta}\boldsymbol{\alpha}) = (\boldsymbol{\alpha}\psi, \psi) \tag{18}$$

Nevertheless, in the pure state $\boldsymbol{\theta} = \pi\{\psi\}$ the values $i = \alpha \in \mathbf{S}(\boldsymbol{\alpha})$ are measured, $\mathrm{Tr}(\boldsymbol{\theta}_i\boldsymbol{\alpha}) = (\boldsymbol{\alpha}\psi_i, \psi_i) = \alpha_{ii}$.
Let $\boldsymbol{\Psi}$ is an *arbitrary* stationary physical system and $\boldsymbol{\theta} \in \boldsymbol{\Theta} \subset \mathbf{A}$ is its arbitrary state. The *physical entropy* $\mathcal{H}(\boldsymbol{\theta})$ of the system $\boldsymbol{\Psi}$ *in the state* $\boldsymbol{\theta}$ is defined by the equality

$$\mathcal{H}(\boldsymbol{\theta}) \stackrel{\mathrm{Def}}{=} -\mathrm{Tr}(\boldsymbol{\theta} \ln \boldsymbol{\theta}) \tag{19}$$

When $\{\pi_\theta : \theta \in \mathbf{S}(\boldsymbol{\theta})\}$ is the decomposition of **1** spectral equivalent with $\boldsymbol{\theta}$, then it is valid that

$$\boldsymbol{\theta} \ln \boldsymbol{\theta} = \sum_{\theta \in \mathbf{S}(\boldsymbol{\theta})} \theta \ln \theta \, \pi_\theta \text{ and } \mathcal{H}(\boldsymbol{\theta}) = -\sum_{\theta \in \mathbf{S}(\boldsymbol{\theta})} \theta \ln \theta \cdot \dim \Psi(\theta|\boldsymbol{\theta}) \tag{20}$$

3.1.0.3. Theorem:

For a physical system $\boldsymbol{\Psi}$ in any state $\boldsymbol{\theta} \in \boldsymbol{\Theta}$ is valid that

$$\begin{aligned}\mathcal{H}(\boldsymbol{\theta}) &= -\sum_{i=1}^{n} q(i|\boldsymbol{\theta}) \cdot \ln q(i|\boldsymbol{\theta}) = H[q(\cdot|\boldsymbol{\theta})] = \ln n - \sum_{\theta \in \mathbf{S}(\boldsymbol{\theta})} p(\theta|\theta|\boldsymbol{\theta}) \cdot \ln \frac{p(\theta|\theta|\boldsymbol{\theta})}{d(\theta|\boldsymbol{\theta})} \\ &= \ln n - I[p(\cdot|\boldsymbol{\theta}|\boldsymbol{\theta}) \,\|\, d(\cdot|\boldsymbol{\theta})]\end{aligned} \tag{21}$$

where $H(\cdot)$ is the *Shannon entropy*, $I(\cdot\|\cdot)$ is the *information divergence*, $p(\cdot|\boldsymbol{\theta}|\boldsymbol{\theta})$ is the p-distribution for the state $\boldsymbol{\theta}$, $q(\cdot|\boldsymbol{\theta})$ is the q-distribution for the state $\boldsymbol{\theta}$ and $d(\cdot|\boldsymbol{\theta})$ is the d-distribution for the state $\boldsymbol{\theta}$, $n = \dim \Psi$.

3.1.0.4. Proof:

The relations in (21) follows from (20) and from definition (10) of the distribution $q(\cdot|\boldsymbol{\theta})$. From definitions (13) and (16) of the other two distributions follows that [12, 38]

$$\begin{aligned}I[p(\cdot|\boldsymbol{\theta}|\boldsymbol{\theta}) \,\|\, d(\cdot|\boldsymbol{\theta})] &= \sum_{\theta \in \mathbf{S}(\boldsymbol{\theta})} \theta \dim \Psi(\theta|\boldsymbol{\theta}) \cdot \ln \frac{n\theta \cdot \dim \Psi(\theta|\boldsymbol{\theta})}{\dim \Psi(\theta|\boldsymbol{\theta})} \\ &= \sum_{\theta \in \mathbf{S}(\boldsymbol{\theta})} \theta \dim \Psi(\theta|\boldsymbol{\theta}) \cdot \ln n - \mathcal{H}(\boldsymbol{\theta}) = \ln n - \mathcal{H}(\boldsymbol{\theta})\end{aligned} \tag{22}$$

Due to the quantities X, Y, $X|Y$, $Y|X$ describing the information transfer are in our algebraic description denoted as follows, $X \triangleq \boldsymbol{\theta}$, $Y \triangleq \boldsymbol{\alpha}$ or $Y \triangleq (\boldsymbol{\alpha}\|\boldsymbol{\theta})$, $(X|Y) \triangleq (\boldsymbol{\theta}|\boldsymbol{\alpha})$, $(Y|X) \triangleq (\boldsymbol{\alpha}|\boldsymbol{\theta})$. The laws of information transfer are writable in this way too:

$$C = \sup_{\alpha,\theta} I(\boldsymbol{\alpha};\boldsymbol{\theta}), \quad I(\cdot||\cdot) \equiv T(\cdot;\cdot) \tag{23}$$

$$I(\boldsymbol{\theta};\boldsymbol{\alpha}) = I(\boldsymbol{\alpha};\boldsymbol{\theta}) = \mathcal{H}(\boldsymbol{\theta}) - H(\boldsymbol{\theta}|\boldsymbol{\alpha}) = H(\boldsymbol{\alpha}\|\boldsymbol{\theta}) - H(\boldsymbol{\alpha}|\boldsymbol{\theta}) = \mathcal{H}(\boldsymbol{\theta}) + H(\boldsymbol{\alpha}\|\boldsymbol{\theta}) - H(\boldsymbol{\theta},\boldsymbol{\alpha})$$
$$H(\boldsymbol{\theta},\boldsymbol{\alpha}) = H(\boldsymbol{\alpha},\boldsymbol{\theta}) = \mathcal{H}(\boldsymbol{\theta}) + H(\boldsymbol{\alpha}|\boldsymbol{\theta}) = H(\boldsymbol{\alpha}\|\boldsymbol{\theta}) + H(\boldsymbol{\theta}|\boldsymbol{\alpha}) \tag{24}$$

4. Narrow-band quantum transfer channels

Let the symmetric operator ε of energy of quantum particle is considered, the spectrum of which eigenvalues ε_i is $\mathbf{S}(\varepsilon)$. Now the *equidistant* energy levels are supposed. In a pure state $\boldsymbol{\theta}_i$ of the measured (observed) system $\boldsymbol{\Psi}$ the eigenvalue $\varepsilon_i = i{\cdot}\varepsilon$, $\varepsilon > 0$. Further, the *output* quantity $\boldsymbol{\alpha}$ of the *observed* system $\boldsymbol{\Psi}$ is supposed (the system is *cell of the phase space B–E or F–D*) with the spectrum of eigenvalues $\mathbf{S}(\boldsymbol{\alpha}) = \{\alpha_0, \alpha_1, ..., \}$ being measured with probability distribution $\Pr(\cdot) = \{p(0), p(1), p(2), ...\}$

$$p(\alpha_k|\boldsymbol{\alpha}|\boldsymbol{\theta}_i) = \begin{cases} p(k-i) \text{ pro } k \geq i \\ 0 \qquad\qquad k < i \end{cases} \tag{25}$$

Such a situation arises when a particle with energy ε_i is *excited by an impact* from the output environment. The jump of energy level of the impacted particle is from ε_i up to ε_{i+j}, $i + j = k$. The output ε_{i+j} for the excited particle is measured (it is the value on the output of the channel $\mathcal{K} \cong \boldsymbol{\Psi}$. This transition j occurs with the probability distribution

$$\Pr(j),\ j \in \{0, 1, 2, ...\} \tag{26}$$

Let be considered the *narrow–band* systems (with one *constant* level of a particle energy) $\boldsymbol{\Psi}$ of B–E or F–D type [27] (denoted further by $\boldsymbol{\Psi}_{\mathrm{B-E},\varepsilon}$, $\boldsymbol{\Psi}_{\mathrm{F-D},\varepsilon}$).

- In the B–E system, *bosonic*, e.g. the *photonic gas* the B–E distribution is valid

$$\Pr(j) = (1-p)\cdot p^j, \quad j \in \{0, 1, ...\},\ p \in (0,1),\ \ p^{-\frac{\varepsilon}{k\Theta}} \tag{27}$$

- In the F–D system, *fermionic*, e.g. *electron gas* the F–D distribution is valid

$$\Pr(j) = \frac{p^j}{1+p}, \quad j \in \{0,1\},\ p \in (0,1),\ \ p^{-\frac{\varepsilon}{k\Theta}} \tag{28}$$

where parameter p is variable with *absolute temperature* $\Theta > 0$; k is the **Boltzman** constant.

Also a collision with a bundle of j particles with constant energies ε of each and absorbing the energy $j \cdot \varepsilon$ of the bundle is considerable. E.g., by $\boldsymbol{\Psi}$ (e.g. $\boldsymbol{\Psi}_{\mathrm{B-E},\varepsilon}$ is the photonic gas) the monochromatic impulses with amplitudes $i \in \mathcal{S}$ are transferred, nevertheless generated from the environment of the same type but at the temperature T_W, $T_W > T_0$ where T_0 is the temperature of the transfer system $\boldsymbol{\Psi} \cong \mathcal{K}$ (the *noise* temperature).

It is supposed that both pure states $\boldsymbol{\theta}_i$ in the place where the input message is being *coded* - on the *input of the channel* $\boldsymbol{\Psi} \cong \mathcal{K}$ and, also, the measurable values of the quantity $\boldsymbol{\alpha}$ being observed on the place where the *output* message is *decoded* - on the *output of the channel* $\mathbf{\Psi} \cong \mathcal{K}$ are arrangable in such a way that in a given i-th pure state $\boldsymbol{\theta}_i$ of the system $\mathbf{\Psi} \cong \mathcal{K}$ only the values $\alpha_k \in \mathcal{S}(\boldsymbol{\alpha})$, $k = i + j$ are measurable and, that the probability of measuring the k-th value is $\Pr(j) = Pr(k - i)$. This probability distribution describes the *additive noise* in the given channel $\mathbf{\Psi} \cong \mathcal{K}$. Just it is this noise which creates observed values from the output spectrum $\mathcal{S}(\boldsymbol{\alpha}) = \{\alpha_i, \alpha_{i+1}, ...\}$ being the selecting space of the stochastic quantity $\boldsymbol{\alpha}$.

The pure states with energy level $\varepsilon_i = i \cdot \varepsilon$ are achievable by sending i particles with energy ε of each. When the environment, through which these particles are going, generates a bundle of j particles with probability $\Pr(j)$ then, with the same probability the energy $\varepsilon_{i+j} = k{\cdot}\varepsilon$ is decoded on the output.

It is supposed, also, the infinite number of states $\boldsymbol{\theta}_i$, infinite spectrum $\mathbf{S}(\boldsymbol{\alpha})$ of the measured quantity $\boldsymbol{\alpha}$, then $\mathbf{S}(\boldsymbol{\alpha}) = \{0,\ 1,\ 2,\ ...\}$, $\mathbf{S}(\boldsymbol{\alpha}) = D(\boldsymbol{\theta})$; $[\mathbf{S}(\boldsymbol{\alpha}) \stackrel{\triangle}{=} \mathbf{S},\ \alpha_k \stackrel{\triangle}{=} \alpha]$.

The *narrow–band, memory-less (quantum) channel, additive* (with additive noise) operating on the energy level $\varepsilon \in \mathcal{S}(\varepsilon)$ is defined by the tri–partite structure (1),

$$\mathcal{K}_\varepsilon = \{[\mathcal{S}, q(i|\boldsymbol{\theta})],\ p(\alpha|\boldsymbol{\alpha}|\boldsymbol{\theta}_i),\ [\mathcal{S},\ p(\alpha|\boldsymbol{\alpha}|\boldsymbol{\theta})]\}. \tag{29}$$

4.1. Capacity of Bose–Einstein narrow–band channel

Let now $\boldsymbol{\theta}_i \equiv i,\ i = 0, 1,\ ...$ are pure states of a system $\mathbf{\Psi}_{\mathrm{B-E},\varepsilon} \cong \mathcal{K}$ and let $\boldsymbol{\alpha}$ is output quantity taking on in the state $\boldsymbol{\theta}_i$ values $\alpha \in \mathbf{S}$ with probabilities

$$p(\alpha|\boldsymbol{\alpha}|\boldsymbol{\theta}_i) = (1 - p)\, p^{\alpha - i} \tag{30}$$

Thus the distribution $p(\cdot|\boldsymbol{\alpha}|\boldsymbol{\theta}_i)$ is determined by the *forced* (inner–input) state $\boldsymbol{\theta}_i = \pi\{\psi_i\}$ representing the coded input energy at the value $\varepsilon_i = i \cdot \varepsilon \in \mathbf{S}(\varepsilon)$, $\varepsilon = \text{const}$. For the *medium value* $\mathcal{W}$ of the input i is valid

$$\mathcal{W} = \sum_{i=0}^{\infty} i \cdot q(i|\boldsymbol{\theta}) = E(\boldsymbol{\theta}), \quad W = \varepsilon \cdot \mathcal{W} \tag{31}$$

The quantity W is the medium value of the energy coding the input signal i.

For the medium value of the number of particles $j = \alpha - i > 0$ with B–E statistics is valid

$$\sum_{j=0}^{\infty} j \cdot (1 - p)\, p^j = (1 - p) \cdot p \cdot \sum_{j=0}^{\infty} j\, p^{j-1} = (1 - p) \cdot p \cdot \frac{\mathrm{d}}{\mathrm{d}p}\left[\frac{1}{1 - p}\right] = \frac{p}{1 - p}$$

The quantity $E(\boldsymbol{\alpha})$ is the medium value of the output quantity $\boldsymbol{\alpha}$ and

$$E(\boldsymbol{\alpha}) = \sum_{\alpha \in \mathbf{S}(\boldsymbol{\alpha})} \alpha \cdot p(\alpha|\boldsymbol{\alpha}|\boldsymbol{\theta}) \tag{32}$$

where $p(\alpha|\boldsymbol{\alpha}|\boldsymbol{\theta})$ is the probability of measuring the eigenvalue $\alpha = k$ of the output variable $\boldsymbol{\alpha}$. This probability is defined by the state $\boldsymbol{\theta} = \sum_{i=0}^{n} q(i|\boldsymbol{\theta})\, \boldsymbol{\theta}_i$ of the system $\mathbf{\Psi}_{\mathrm{B-E},\varepsilon} \cong \mathcal{K}$

$$p(\alpha|\boldsymbol{\alpha}|\boldsymbol{\theta}) = \sum_{i=0}^{n} q(i|\boldsymbol{\theta}) \cdot p(\alpha|\boldsymbol{\alpha}|\boldsymbol{\theta}_i) = (1 - p) \cdot \sum_{i=0}^{n} q(i|\boldsymbol{\theta}) \cdot p^{\alpha - i} \quad [= \mathrm{Tr}(\boldsymbol{\theta}\pi_\alpha)] \tag{33}$$

From the differential equation with the *condition* for $\alpha = 0$

$$p(\alpha|\boldsymbol{\alpha}|\boldsymbol{\theta}) = p(\alpha-1|\boldsymbol{\alpha}|\boldsymbol{\theta})\cdot p + (1-p)\cdot q(\alpha|\boldsymbol{\theta}),\ \forall\alpha\geq 1;\quad p(0|\boldsymbol{\alpha}|\boldsymbol{\theta}) = (1-p)\cdot q(0|\boldsymbol{\theta}) \tag{34}$$

follows, for the medium value $E(\boldsymbol{\alpha})$ of the output stochastic variable $\boldsymbol{\alpha}$, that

$$\begin{aligned} E(\boldsymbol{\alpha}) &= \sum_{\alpha\geq 1}\alpha\cdot p(\alpha-1|\boldsymbol{\alpha}|\boldsymbol{\theta})\cdot p + \sum_{\alpha\geq 1}\alpha\cdot(1-p)\cdot q(\alpha|\boldsymbol{\theta}) \\ &= \sum_{\alpha\geq 1}\alpha\cdot p(\alpha-1|\boldsymbol{\alpha}|\boldsymbol{\theta})\cdot p + \mathcal{W}\cdot(1-p) = p\cdot E(\boldsymbol{\alpha}) + p\cdot\sum_{\alpha\geq 1}p(\alpha-1|\boldsymbol{\alpha}|\boldsymbol{\theta}) + \mathcal{W}\cdot(1-p)\end{aligned} \tag{35}$$

$$E(\boldsymbol{\alpha})\ \cdot\ (1-p) = p + \mathcal{W}\cdot(1-p) \longrightarrow E(\boldsymbol{\alpha}) = \frac{p}{1+p} + \mathcal{W},\quad \mathcal{W} = E(\boldsymbol{\theta}) > 0,\quad \boldsymbol{\theta}\in\boldsymbol{\Theta}_0 \tag{36}$$

The quantity $H(\boldsymbol{\alpha}\|\boldsymbol{\theta}_i)$ is the p-entropy of measuring $\boldsymbol{\alpha}$ for the input $i\in\mathbf{S}$ being represented by the pure state $\boldsymbol{\theta}_i$ of the system $\boldsymbol{\Psi}_{\mathrm{B-E},\varepsilon}$

$$\begin{aligned} -H(\boldsymbol{\alpha}\|\boldsymbol{\theta}_i) &= \sum_{j\in\mathbf{S}}(1-p)\,p^j\cdot\ln[(1-p)\,p^j] \\ &= (1-p)\cdot\ln(1-p)\cdot\sum_j p^j - (1-p)\cdot p\cdot\ln p\cdot\sum_j j\,p^{j-1} \\ &= \ln(1-p) - (1-p)\cdot p\cdot\ln p\cdot\frac{\mathrm{d}}{\mathrm{d}p}\left[\frac{1}{1-p}\right] = -\frac{h(p)}{1-p},\ \forall i\in\mathbf{S}\end{aligned} \tag{37}$$

where $h(p)\stackrel{\triangle}{=}-(1-p)\cdot\ln(1-p)-p\cdot\ln p$ is the Shannon entropy of Bernoulli distribution $\{p,\ 1-p\}$.

The quantity $H(\boldsymbol{\alpha}|\boldsymbol{\theta})$ is the conditional Shannon entropy of the stochastic quantity $\boldsymbol{\alpha}$ in the state $\boldsymbol{\theta}$ of the system $\boldsymbol{\Psi}_{\mathrm{B-E},\varepsilon}$, not depending on the $\boldsymbol{\theta}$ (the *noise* entropy)

$$H(\boldsymbol{\alpha}|\boldsymbol{\theta}) = \sum_{i=0}^{n}q(i|\boldsymbol{\theta})\cdot H(\boldsymbol{\alpha}\|\boldsymbol{\theta}_i) = \sum_{i=0}^{n}q(i|\boldsymbol{\theta})\cdot\frac{h(p)}{1-p} = \frac{h(p)}{1-p} \tag{38}$$

For capacity $C_{\mathrm{B-E}''}$ of the channel $\mathcal{K}\cong\boldsymbol{\Psi}_{\mathrm{B-E}}$ is, following the capacity definition, valid

$$C_{\mathrm{B-E}''} = \sup_{\boldsymbol{\theta}\in\boldsymbol{\Theta}_0}\ H(\boldsymbol{\alpha}\|\boldsymbol{\theta}) - H(\boldsymbol{\alpha}|\boldsymbol{\theta}) = \sup_{\boldsymbol{\theta}\in\boldsymbol{\Theta}_0}\ H(\boldsymbol{\alpha}\|\boldsymbol{\theta}) - \frac{h(p)}{1-p} \tag{39}$$

where the set $\boldsymbol{\Theta}_0 = \{\boldsymbol{\theta}\in\boldsymbol{\Theta},\ E(\boldsymbol{\theta}) = \mathcal{W}\geq 0\}$ represents the *coding procedure* of the input $i\in\mathbf{S}$. The quantity $H(\boldsymbol{\alpha}\|\boldsymbol{\theta}) = H(p(\cdot|\boldsymbol{\alpha}|\boldsymbol{\theta}))$ is the p-entropy of the output quantity $\boldsymbol{\alpha}$. Its supremum is determined by the **Lagrange** multipliers method:

$$H(\boldsymbol{\alpha}\|\boldsymbol{\theta}) = -\sum_{\alpha\in\mathbf{S}(\boldsymbol{\alpha})}p(\alpha|\boldsymbol{\alpha}|\boldsymbol{\theta})\cdot\ln p(\alpha|\boldsymbol{\alpha}|\boldsymbol{\theta}) = -\sum_{\alpha\in\mathbf{S}(\boldsymbol{\alpha})}p_\alpha\cdot\ln p_\alpha,\quad p_\alpha\stackrel{\triangle}{=}p(\alpha|\boldsymbol{\alpha}|\boldsymbol{\theta}) \tag{40}$$

The conditions for determinating of the *bound* extreme are

$$\sum_{\alpha\in\mathbf{S}(\boldsymbol{\alpha})}p_\alpha = 1,\qquad \sum_{\alpha\in\mathbf{S}(\boldsymbol{\alpha})}\alpha\cdot p_\alpha = E(\boldsymbol{\alpha}) = \text{const.} \tag{41}$$

The Lagrange function

$$L = -\sum_{\alpha} p_\alpha \cdot \ln p_\alpha - \lambda_1 \cdot \sum_{\alpha} p_\alpha + \lambda_1 - \lambda_2 \cdot \sum_{\alpha} \alpha \cdot p_\alpha + \lambda_2 E(\boldsymbol{\alpha}) \tag{42}$$

which gives the condition for the extreme, $\dfrac{\partial L}{\partial p_\alpha} = -\ln p_\alpha - 1 - \lambda_1 - \lambda_2 \cdot \alpha = 0,$ yielding in

$$p_\alpha = e^{-1-\lambda_1} \cdot e^{-\lambda_2 \alpha} = p(\alpha|\boldsymbol{\alpha}|\boldsymbol{\theta}) \quad \text{and further, in } \sum_{\alpha} p_\alpha = \sum_{\alpha} \frac{e^{-1-\lambda_1}}{e^{\lambda_2 \alpha}} = \frac{e^{-1-\lambda_1}}{1-e^{-\lambda_2}} = 1 \tag{43}$$

Then for the medium value $E(\boldsymbol{\alpha})$ the following result is obtained

$$E(\boldsymbol{\alpha}) = \sum_{\alpha} \alpha \cdot p_\alpha = \sum_{\alpha} \alpha \cdot e^{-1-\lambda_1} \cdot e^{-\lambda_2 \alpha} = -e^{-1-\lambda_1} \cdot \frac{\partial}{\partial \lambda_2} \sum_{\alpha} e^{-\lambda_2 \alpha} \tag{44}$$

$$= -e^{-1-\lambda_1} \cdot \frac{\partial}{\partial \lambda_2} \left[\frac{1}{1-e^{-\lambda_2}} \right] = \frac{e^{-1-\lambda_1}}{(1-e^{-\lambda_2})^2} \cdot e^{-\lambda_2} = \frac{e^{-\lambda_2}}{1-e^{-\lambda_2}}$$

By (35), (36) and for the parametr $p = e^{-\frac{\varepsilon}{k\Theta}} = \text{const.}$ ($\varepsilon = \text{const.}$ $\Theta = \text{const.}$) $E(\boldsymbol{\alpha})$ is a function of $\mathcal{W}$ only. For $\lambda_2 = \dfrac{\varepsilon}{kT_W}$ is $e^{-\lambda_2} = p(\mathcal{W})$ and $E(\boldsymbol{\alpha})$ is the medium value for $\boldsymbol{\alpha}$ with the *geometric* probability distribution

$$p(\cdot) = p(\cdot|\boldsymbol{\alpha}|\boldsymbol{\theta}) = [1 - p(\mathcal{W})] \cdot p(\mathcal{W})^\alpha, \quad \alpha \in \mathbf{S}(\boldsymbol{\alpha}) \tag{45}$$

depending only on $\dfrac{\varepsilon}{kT_W}$ or, on the absolute temperature T_W respectively. Thus for $E(\boldsymbol{\alpha})$ is valid that

$$E(\boldsymbol{\alpha}) = \frac{p(\mathcal{W})}{1-p(\mathcal{W})}, \quad p(\mathcal{W}) = e^{-\frac{\varepsilon}{kT_W}} \tag{46}$$

From (35) and (46) is visible that $p(\mathcal{W})$ or T_W respectively is the *only one* root of the equation

$$\frac{p(\mathcal{W})}{1-p(\mathcal{W})} = \frac{p}{1-p} + \mathcal{W}, \text{resp.} \quad \frac{e^{-\frac{\varepsilon}{kT_W}}}{1-e^{-\frac{\varepsilon}{kT_W}}} = \frac{e^{-\frac{\varepsilon}{kT_0}}}{1-e^{-\frac{\varepsilon}{kT_0}}} + \mathcal{W} \tag{47}$$

From (34) and (45) follows that for state $\boldsymbol{\theta} \in \boldsymbol{\Theta}_0$ or, for the q-distribution $q(\cdot|\boldsymbol{\theta})$ respectively, in which the value $H(\boldsymbol{\alpha}\|\boldsymbol{\theta})$ is maximal [that state in which $\boldsymbol{\alpha}$ achieves the distribution $q(\cdot|\boldsymbol{\theta}) = p(\cdot)$], is valid that

$$q(\alpha|\boldsymbol{\theta}) = \frac{1-p(\mathcal{W})}{1-p}, \ \alpha = 0 \ \text{ and } \ q(\alpha|\boldsymbol{\theta}) = \frac{1-p(\mathcal{W})}{1-p} \left[p(\mathcal{W}) - p\right] \cdot p(\mathcal{W})^{\alpha-1}, \ \alpha > 0 \tag{48}$$

For the *effective temperatutre T_W of coding input messages* the distribution (45) supremizes (maximizes) the p-entropy $H(\boldsymbol{\alpha}\|\boldsymbol{\theta})$ of $\boldsymbol{\alpha}$ and, by using (37) with $p(\mathcal{W})$, is gained that

$$\sup_{\theta \in \boldsymbol{\Theta}_0} H(\boldsymbol{\alpha}\|\boldsymbol{\theta}) = \frac{h[p(\mathcal{W})]}{1-p(\mathcal{W})} \tag{49}$$

From (39) and (49) follows [12, 37] the capacity $C_{B-E,\varepsilon}$ of the narrow–band channel $\mathcal{K} \cong \boldsymbol{\Psi}_{B-E}$

$$C_{B-E,\varepsilon} = \frac{h[p(\mathcal{W})]}{1-p(\mathcal{W})} - \frac{h(p)}{1-p} \tag{50}$$

By (35), (36) and (46) the medium value $\mathcal{W}$ of the input message $i \in \mathbf{S}$ is derived,

$$\mathcal{W} = \frac{p(\mathcal{W})}{1-p(\mathcal{W})} - \frac{p}{1-p} \tag{51}$$

By (31) the condition for the *minimal* average energy W_{Krit} needed for coding the input message is

$$\mathcal{W} \geq 0 \quad \text{resp.} \quad W = \varepsilon \cdot \mathcal{W} \geq 0, \quad \varepsilon \in \mathbf{S}(\varepsilon) \tag{52}$$

$$W \geq W_{Krit}, \; W_{Krit} = 0 \tag{53}$$

The relations (35), (36), (46) and (52), (53) yield in

$$E(\boldsymbol{\alpha}) = \frac{p(\mathcal{W})}{1-p(\mathcal{W})} \geq \frac{p}{1-p} \quad \text{and, then} \quad p(\mathcal{W}) \geq p, \; 1-p(\mathcal{W}) \leq 1-p \tag{54}$$

From (47) and (54) follows that for the defined direction of the signal (messaage) transmission at the temperature T_W of its sending and decoding is valid that

$$\frac{p(\mathcal{W})}{1-p(\mathcal{W})} > \frac{p}{1-p}, \; p(\mathcal{W}) > p, \; \mathcal{W} > 0 \tag{55}$$

$$p(\mathcal{W}) = e^{-\frac{\varepsilon}{kT_W}} \geq e^{-\frac{\varepsilon}{kT_0}} = p \quad \text{and thus} \quad T_W \geq T_0 \tag{56}$$

4.2. Capacity of Fermi–Dirac narrow–band channel

Let is now considered, in the same way as it was in the B–E system, the pure states $\theta_i \equiv i$ of the system $\boldsymbol{\Upsilon}$ which are coding the input messages $i = 0, 1, \ldots$ and the output stochastic quantity $\boldsymbol{\alpha}$ having its selecting space $\mathbf{S}$. On the spectrum $\mathbf{S}$ probabilities of realizations $\alpha \in \mathbf{S}$ are defined,

$$p(\alpha|\boldsymbol{\alpha}|\theta_i) = \frac{p^{\alpha-i}}{1+p}, \quad p \in (0,1), \; i = 0, 1, \ldots \tag{57}$$

expressing the additive stochastic transformation of an input i into the output α for wich is valid $\alpha = i$ or $\alpha = i+1$.[5] The uniform energy level $\varepsilon =$ const. of particles is considered.

The quantity W is the mathematical expectation of the energy coding the input signal

$$W = \varepsilon \cdot \mathcal{W}, \quad \mathcal{W} = \sum_{i \in \mathbf{S}} i \cdot q(i|\boldsymbol{\theta}) = E(\boldsymbol{\theta}) \tag{58}$$

The medium value of a stochastic quantity with the F–D statistic is given by

$$\sum_{j \in \{0,1\}} j \cdot \frac{p^j}{1+p} = \frac{p}{1+p} \tag{59}$$

The quantity $E(\boldsymbol{\alpha})$ is the medium value of the output quantity $\boldsymbol{\alpha}$,

$$E(\boldsymbol{\alpha}) = \sum_{\alpha \in \mathbf{S}(\boldsymbol{\alpha})} \alpha \cdot p(\alpha|\boldsymbol{\alpha}|\boldsymbol{\theta}) \tag{60}$$

[5] In accordance with ***Pauli*** *excluding principle* (valid for *fermions*) and a given energetic level $\varepsilon \in \mathbf{S}(\varepsilon)$

where $p(\alpha|\boldsymbol{\alpha}|\boldsymbol{\theta})$ is probability of realization of $\alpha \in \mathbf{S}$ in the state $\boldsymbol{\theta}$ of the system $\boldsymbol{\Psi} \equiv \boldsymbol{\Psi}_{\mathrm{F-D},\varepsilon} \cong \mathcal{K}$,

$$\boldsymbol{\theta} = \sum_{i \in \mathbf{S}} q(i|\boldsymbol{\theta})\, \boldsymbol{\theta}_i, \quad p(\alpha|\boldsymbol{\alpha}|\boldsymbol{\theta}) = \sum_{i=0}^{n} q(i|\boldsymbol{\theta}) \cdot p(\alpha|\boldsymbol{\alpha}|\boldsymbol{\theta}_i) = \frac{1}{1+p} \cdot \sum_{i=0}^{n} q(i|\boldsymbol{\theta}) \cdot p^{\alpha - i} \tag{61}$$

From the differential equation

$$p(\alpha|\boldsymbol{\alpha}|\boldsymbol{\theta}) = \frac{q(\alpha-1|\boldsymbol{\theta}) \cdot p + q(\alpha|\boldsymbol{\theta})}{1+p}, \quad \alpha \geq 1 \tag{62}$$

with the condition for

$$\alpha = 0, \quad p(0|\boldsymbol{\alpha}|\boldsymbol{\theta}) = \frac{1}{1+p} \cdot q(0|\boldsymbol{\theta})$$

follows that for the medium value $E(\boldsymbol{\alpha})$ of the output stochastic variable $\boldsymbol{\alpha}$ is valid that

$$\begin{aligned} E(\boldsymbol{\alpha}) &= \frac{p}{1+p} \cdot \sum_{\alpha \geq 1} \alpha \cdot q(\alpha - 1|\boldsymbol{\theta}) + \frac{1}{1+p} \cdot \sum_{\alpha \geq 1} \alpha \cdot q(\alpha|\boldsymbol{\theta}) \\ &= \frac{p}{1+p} \cdot \sum_{\alpha \geq 1} (\alpha - 1) \cdot q(\alpha-1|\boldsymbol{\theta}) + \frac{p}{1+p} \cdot \sum_{\alpha \geq 1} q(\alpha - 1|\boldsymbol{\theta}) + \frac{1}{1+p} \cdot \mathcal{W} \\ &= \frac{p}{1+p} \cdot \mathcal{W} + \frac{p}{1+p} + \frac{1}{1+p} \cdot \mathcal{W} \longrightarrow \quad E(\boldsymbol{\alpha}) = \frac{p}{1+p} + \mathcal{W}, \ \mathcal{W} = E(\boldsymbol{\theta}), \ \boldsymbol{\theta} \in \boldsymbol{\Theta}_0 \end{aligned} \tag{63}$$

The quantity $H(\boldsymbol{\alpha}\|\boldsymbol{\theta}_i)$ is the p-entropy of measuring $\boldsymbol{\alpha}$ for the input $i \in \mathbf{S}$ being represented by the pure state $\boldsymbol{\theta}_i$ of the system $\boldsymbol{\Psi}_{\mathrm{F-D},\varepsilon}$

$$\begin{aligned} H(\boldsymbol{\alpha}\|\boldsymbol{\theta}_i) &= -\sum_{j=0}^{1} \frac{p^j}{1+p} \cdot \ln \frac{p^j}{1+p} = -\frac{1}{1+p} \cdot \ln \frac{1}{1+p} - \frac{p}{1+p} \cdot \ln \frac{p}{1+p} \\ &= -\left(1 - \frac{p}{1+p}\right) \cdot \ln \left(1 - \frac{p}{1+p}\right) - \frac{p}{1+p} \cdot \ln \frac{p}{1+p} = h\left(\frac{p}{1+p}\right), \ \forall i \in \mathbf{S} \end{aligned} \tag{64}$$

The quantity $H(\boldsymbol{\alpha}|\boldsymbol{\theta})$ is the conditional (the *noise*) Shannon entropy of the stochastic quantity $\boldsymbol{\alpha}$ in the system state $\boldsymbol{\theta}$, but, independent on this $\boldsymbol{\theta}$,

$$H(\boldsymbol{\alpha}|\boldsymbol{\theta}) = \sum_{i=0}^{n} q(i|\boldsymbol{\theta}) \cdot H(\boldsymbol{\alpha}\|\boldsymbol{\theta}_i) = \sum_{i} q(i|\boldsymbol{\theta}) \cdot h\left(\frac{p}{1+p}\right) = h\left(\frac{p}{1+p}\right) \tag{65}$$

For capacity $C_{\mathrm{F-D},\varepsilon}$ of the channel $\mathcal{K} \cong \boldsymbol{\Psi}_{\mathrm{F-D},\varepsilon}$ is, by the capacity definition in (23)-(24), valid that

$$C_{\mathrm{F-D}''} = \sup_{\boldsymbol{\theta} \in \boldsymbol{\Theta}_0} H(\boldsymbol{\alpha}\|\boldsymbol{\theta}) - H(\boldsymbol{\alpha}|\boldsymbol{\theta}) = \sup_{\boldsymbol{\theta} \in \boldsymbol{\Theta}_0} H(\boldsymbol{\alpha}\|\boldsymbol{\theta}) - h\left(\frac{p}{1+p}\right) \tag{66}$$

where the set $\boldsymbol{\Theta}_0 = \{\boldsymbol{\theta} \in \boldsymbol{\Theta} : E(\boldsymbol{\theta}) = \mathcal{W} > 0\}$ represents the coding procedure.

The quantity $H(\boldsymbol{\alpha}\|\boldsymbol{\theta}) = H(p(\cdot|\boldsymbol{\alpha}|\boldsymbol{\theta}))$ is the p-entropy of the stochastic quantity $\boldsymbol{\alpha}$ in the state $\boldsymbol{\theta}$ of the system $\boldsymbol{\Psi}_{\mathrm{F-D},\varepsilon}$. Its supremum is determined by the **Lagrange** multipliers method in the same way as in B–E case and with the same results for the probility distribution $p(\cdot|\boldsymbol{\alpha}|\boldsymbol{\theta})$ (geometric) and the medium value $E(\boldsymbol{\alpha})$

$$p(\cdot) = p(\cdot|\boldsymbol{\alpha}|\boldsymbol{\theta}) = [1 - p(\mathcal{W})] \cdot p(\mathcal{W})^{\alpha}, \ \alpha \in \mathbf{S}(\boldsymbol{\alpha}) \qquad (67)$$
$$E(\boldsymbol{\alpha}) = \frac{p(\mathcal{W})}{1 - p(\mathcal{W})}, \ p(\mathcal{W}) = e^{-\frac{\varepsilon}{kT_W}}$$

Again, the value $E(\boldsymbol{\alpha})$ depends on $\frac{\varepsilon}{kT_W}$, or on absolute temperature T_W respectively, only. By using $E(\boldsymbol{\alpha})$ in (63) it is seen that $p(\mathcal{W})$ or T_W respectively is the only one root of the equation [12, 30]

$$\frac{p(\mathcal{W})}{1 - p(\mathcal{W})} = \frac{p}{1+p} + \mathcal{W}, \ \text{resp.} \ \frac{e^{-\frac{\varepsilon}{kT_W}}}{1 - e^{-\frac{\varepsilon}{kT_W}}} = \frac{e^{-\frac{\varepsilon}{kT_0}}}{1 + e^{-\frac{\varepsilon}{kT_0}}} + \mathcal{W} \qquad (68)$$

For the q-distribution $q(\cdot|\boldsymbol{\theta}) = p(\cdot)$ of states $\boldsymbol{\theta} \in \boldsymbol{\Theta}_0$, for which the relation (62) and (67) is gained, follows that

$$\frac{q(\alpha - 1|\boldsymbol{\theta}) \cdot p + q(\alpha|\boldsymbol{\theta})}{1+p} = [1 - p(\mathcal{W})] \cdot p(\mathcal{W})^{\alpha}, \quad \alpha \in \mathbf{S} \ \text{with conditions} \qquad (69)$$
$$q(0|\boldsymbol{\theta}) = (1+p) \cdot [1 - p(\mathcal{W})] \ \text{and} \ q(1|\boldsymbol{\theta}) = (1+p) \cdot [1 - p(\mathcal{W})] \cdot [p(\mathcal{W}) - p];$$
$$q(\alpha|\boldsymbol{\theta}) = (1+p) \cdot [1 - p(\mathcal{W})] \cdot \left[\left(\sum_{i=0}^{\alpha-1} (-1)^i \cdot p(\mathcal{W})^{\alpha-i} \cdot p^i \right) + (-1)^{\alpha} \cdot p^{\alpha} \right], \ \alpha > 1$$

For the *effective temperatutre T_W of coding the input messages* the distribution (67) supremises (maximizes) the p-entropy $H(\boldsymbol{\alpha}\|\boldsymbol{\theta})$ of $\boldsymbol{\alpha}$ is valid, in the same way as in (49), that

$$\sup_{\boldsymbol{\theta} \in \boldsymbol{\Theta}_0} H(\boldsymbol{\alpha}\|\boldsymbol{\theta}) = \frac{h[p(\mathcal{W})]}{1 - p(\mathcal{W})} \qquad (70)$$

By using (70) in (66) the formula for the $C_{\text{F-D},\varepsilon}$ capacity [12, 37] is gained

$$C_{\text{F-D},\varepsilon} = \frac{h[p(\mathcal{W})]}{1 - p(\mathcal{W})} - h\left(\frac{p}{1+p}\right) \qquad (71)$$

The medium value $\mathcal{W}$ of the input i = 0, 1, 2, ... is limited by a minimal not-zero and positive 'bottom' value $\mathcal{W}_{Krit}$. From (58), (63) and (68) follows

$$E(\boldsymbol{\alpha}) = \frac{p(\mathcal{W})}{1 - p(\mathcal{W})} \geq \frac{p}{1-p}, \ p(\mathcal{W}) = e^{-\frac{\varepsilon}{kT_W}} \geq e^{-\frac{\varepsilon}{kT_0}} = p \ \text{and thus} \ T_W \geq T_0 \qquad (72)$$

$$\mathcal{W} = \frac{p(\mathcal{W})}{1-p(\mathcal{W})} - \frac{p}{1+p} \geq 0, \ \mathcal{W} \geq \frac{2p^2}{1-p^2} = \mathcal{W}_{Krit}, \text{resp.} \ W = \varepsilon \cdot \mathcal{W} \geq \varepsilon \cdot \frac{2e^{-2\frac{\varepsilon}{kT_0}}}{1 - e^{-2\frac{\varepsilon}{kT_0}}} \qquad (73)$$

For the average coding energy W, when the channel $C_{\text{F-D},\varepsilon}$ acts on a uniform energetic level ε, is

$$W \geq W_{Krit} = \frac{2p^2}{1-p^2} \qquad (74)$$

For the F–D channel is then possible speak about the ***effect of the not-zero capacity when the difference between the coding temperatures T_W and the noise temperature T_0* is zero.**[6] This phennomenon is, by necessity, *given by properties of cells of the F–D phase space.*

[6] Not *not-zero capacity* for zero input power as was stated in [37]. The (71) also repares small misprint in [11]

5. Wide–band quantum transfer channels

Till now the narrow–band variant of an information transfer channel $\mathcal{K}_\varepsilon$, $\varepsilon \in \mathbf{S}(\boldsymbol{\varepsilon})$, card $\mathbf{S}(\boldsymbol{\varepsilon}) = 1$ has been dealt. Let is now considered the symmetric operator of energy $\boldsymbol{\varepsilon}$ of a particle, having the spectrum of eigenavalues

$$\mathbf{S}(\boldsymbol{\varepsilon}) = \left\{0, \frac{h}{\tau}, \frac{2h}{\tau}, \ldots, \frac{nh}{\tau}, \ldots\right\} = \left\{\frac{rh}{\tau}\right\}_{r=0,\,1,\,\ldots,\,n}, \text{ card } \mathbf{S}(\boldsymbol{\varepsilon}) = n+1 \tag{75}$$

where $\tau > 0$ denotes the time length of the input signal and h denotes **Planck** constant. The *multi–band physical transfer channel* **K**, *memory-less, with additive noise* is defend by the (arranged) set of narrow–band, *independent* components $\mathcal{K}_\varepsilon$, $\varepsilon \in \mathbf{S}(\boldsymbol{\varepsilon})$,

$$\mathbf{K} = \underset{\varepsilon\in\mathbf{S}(\boldsymbol{\varepsilon})}{\times} \mathcal{K}_\varepsilon = \underset{\varepsilon\in\mathbf{S}(\boldsymbol{\varepsilon})}{\times} \{\boldsymbol{i}_\varepsilon, p(\alpha_\varepsilon|\boldsymbol{\alpha}_\varepsilon|\boldsymbol{\theta}_{\varepsilon,i_\varepsilon}), \boldsymbol{\alpha}_\varepsilon\} = \{\boldsymbol{i}, p(\overline{\alpha}|\boldsymbol{\alpha}|\boldsymbol{\theta}_{\overline{i}}), \boldsymbol{\alpha}\} \tag{76}$$

$$\boldsymbol{i} = \underset{\varepsilon\in\mathbf{S}(\boldsymbol{\varepsilon})}{\times} \boldsymbol{i}_\varepsilon = \underset{\varepsilon\in\mathbf{S}(\boldsymbol{\varepsilon})}{\times} [\mathbf{S}, q_\varepsilon(i_\varepsilon|\boldsymbol{\theta}_\varepsilon)], \; i_\varepsilon \in \mathbf{S} = \{0, 1, 2, \ldots\}$$

$$\boldsymbol{\alpha} = \times_{\varepsilon\in\mathbf{S}(\boldsymbol{\varepsilon})} \boldsymbol{\alpha}_\varepsilon = \underset{\varepsilon\in\mathbf{S}(\boldsymbol{\varepsilon})}{\times} [\mathbf{S}, p(\alpha_\varepsilon|\boldsymbol{\alpha}_\varepsilon|\boldsymbol{\theta}_\varepsilon)], \; \alpha_\varepsilon \in \mathbf{S} = \{0, 1, 2, \ldots\}$$

Due the independency of narrow–band components $\mathcal{K}_\varepsilon$ the vector quantities $\boldsymbol{i}_\varepsilon, \boldsymbol{\alpha}_\varepsilon, \boldsymbol{\theta}_\varepsilon, \boldsymbol{j}_\varepsilon$ are independent stochastic quantities too.

The simultaneous q-*distribution* of the input vector of i_ε and the simultaneous p-*distribution* of measuring the output vector of values α_ε (of the individual narrow–band components $\mathcal{K}_\varepsilon$) are

$$\prod_{\varepsilon\in\mathbf{S}(\boldsymbol{\varepsilon})} q_\varepsilon(i_\varepsilon|\boldsymbol{\theta}_\varepsilon) = q(\overline{i}|\boldsymbol{\theta}), \qquad \boldsymbol{\theta} = \underset{\varepsilon\in\mathbf{S}(\boldsymbol{\varepsilon})}{\times} \boldsymbol{\theta}_\varepsilon, \quad \overline{i} \in \mathbf{S}(\boldsymbol{i}) \tag{77}$$

$$\prod_{\varepsilon\in\mathbf{S}(\boldsymbol{\varepsilon})} p(\alpha_\varepsilon|\boldsymbol{\alpha}_\varepsilon|\boldsymbol{\theta}_\varepsilon) = p(\overline{\alpha}|\boldsymbol{\alpha}|\boldsymbol{\theta}), \quad \boldsymbol{\theta} = \underset{\varepsilon\in\mathbf{S}(\boldsymbol{\varepsilon})}{\times} \boldsymbol{\theta}_\varepsilon, \quad \overline{\alpha} \in \mathbf{S}(\boldsymbol{\alpha}) \tag{78}$$

The system of quantities $\boldsymbol{\theta}_\varepsilon$ (the set of states of the narrow–band components $\mathcal{K}_\varepsilon$) is the state $\boldsymbol{\theta}$ of the multi–band channel **K** in which the (*canonic*) q-*distribution* of the system **K** is defined. Values i', j', α'

$$\alpha' = j' + i'; \;\; \alpha' = \sum_{\varepsilon\in\mathbf{S}(\boldsymbol{\varepsilon})} \alpha_\varepsilon, \; j' = \sum_{\varepsilon\in\mathbf{S}(\boldsymbol{\varepsilon})} j_\varepsilon, \; i' = \sum_{\varepsilon\in\mathbf{S}(\boldsymbol{\varepsilon})} i_\varepsilon; \; j_\varepsilon = \alpha_\varepsilon - i_\varepsilon \geq 0, \; \forall\varepsilon \in \mathbf{S}(\boldsymbol{\varepsilon}), \; \text{card}\mathbf{S}_\varepsilon > 1 \tag{79}$$

are the numbers of the input, output and additive (noise) particles of the *multi–band* channel **K**. In this channel the stochastic transformation of the input i' into the output α' is performed, being determined by additive stochastic transformations of the input i_ε into the output α_ε in individual narrow–band components $\mathcal{K}_\varepsilon$.

Realizations of the stochastic systems $\boldsymbol{i}, \boldsymbol{\alpha}, \boldsymbol{\theta}, \boldsymbol{j}$ are the *vectors (sequences)* $\overline{i}, \overline{\alpha}, \overline{\theta}, \overline{j}$

$$\overline{i} = (i_\varepsilon)_{\varepsilon\in\mathbf{S}(\boldsymbol{\varepsilon})}, \quad \overline{\alpha} = (\alpha_\varepsilon)_{\varepsilon\in\mathbf{S}(\boldsymbol{\varepsilon})}, \quad \overline{\theta} = (\theta_\varepsilon)_{\varepsilon\in\mathbf{S}(\boldsymbol{\varepsilon})}, \quad \overline{j} = (j_\varepsilon)_{\varepsilon\in\mathbf{S}(\boldsymbol{\varepsilon})}; \; \overline{i}, \overline{\alpha}, \overline{j} \in \underset{\varepsilon\in\mathbf{S}(\boldsymbol{\varepsilon})}{\times} \mathbf{S}, \tag{80}$$

$$i_\varepsilon, \alpha_\varepsilon, j_\varepsilon \in \mathbf{S}, \;\; \theta_\varepsilon \in \mathbf{S}(\boldsymbol{\theta}_\varepsilon), \quad \boldsymbol{\theta}_\varepsilon = \sum_{i_\varepsilon\in\mathbf{S}} \theta_\varepsilon \boldsymbol{\theta}_{\varepsilon,i_\varepsilon}, \quad \overline{\theta} \in \mathbf{S}(\boldsymbol{\theta})$$

For the probability of the additive stochastic transformation (77), (78) of input $\bar{i}$ into the output $\bar{\alpha}$ is valid

$$\prod_{\varepsilon\in\mathbf{S}(\varepsilon)} p(\alpha_\varepsilon|\boldsymbol{\alpha}_\varepsilon|\boldsymbol{\theta}_{\varepsilon,i_\varepsilon}) = p(\bar{\alpha}|\boldsymbol{\alpha}|\boldsymbol{\theta}_{\bar{i}}),\ \bar{\alpha}=\bar{j}+\bar{i},\ \bar{i}\in\mathbf{S}(\boldsymbol{i}),\ \bar{j}\in\mathbf{S}(\boldsymbol{j}),\ \bar{\alpha}\in\mathbf{S}(\boldsymbol{\alpha}) \tag{81}$$

The symbol $\boldsymbol{\theta}_{\varepsilon,i_\varepsilon}$ denotes the pure state coding the input $i_\varepsilon \in \mathbf{S}$ of a narrow–band component $\mathcal{K}_\varepsilon$ and the state $\boldsymbol{\theta}_{\bar{i}} = \times_{\varepsilon\in\mathbf{S}(\varepsilon)}\boldsymbol{\theta}_{\varepsilon,i_\varepsilon}$ codes the input $\bar{i}$ for which

$$q(\bar{i}|\boldsymbol{\theta}) = q_{\bar{\theta}} = \prod_{\varepsilon\in\mathbf{S}(\varepsilon)} q_\varepsilon(\theta_\varepsilon) \tag{82}$$

For the multi–band channel **K** the following quantities are defined:[7]

- the *p-entropy of the output* $\boldsymbol{\alpha}$

$$\begin{aligned} H(\boldsymbol{\alpha}\|\boldsymbol{\theta}) &= \sum_{\varepsilon\in\mathbf{S}(\varepsilon)} H(\boldsymbol{\alpha}_\varepsilon\|\boldsymbol{\theta}_\varepsilon) = -\sum_{\varepsilon\in\mathbf{S}(\varepsilon)}\sum_{\alpha_\varepsilon\in\mathbf{S}} p(\alpha_\varepsilon|\boldsymbol{\alpha}_\varepsilon|\boldsymbol{\theta}_\varepsilon)\cdot\ln p(\alpha_\varepsilon|\boldsymbol{\alpha}_\varepsilon|\boldsymbol{\theta}_\varepsilon) \\ &\le \sum_{\varepsilon\in\mathbf{S}(\varepsilon)} \sup_{\theta_\varepsilon} H(\boldsymbol{\alpha}_\varepsilon\|\boldsymbol{\theta}_\varepsilon) \end{aligned} \tag{83}$$

for which, following the output narrow–band B–E and F–D components $\mathcal{K}_\varepsilon \in \mathbf{K}$, is valid that

$$\sup_{\theta\in\overline{\boldsymbol{\Theta}}_0} H(\boldsymbol{\alpha}\|\boldsymbol{\theta}) = \sup_{\theta\in\overline{\boldsymbol{\Theta}}_0}\sum_{\varepsilon\in\mathbf{S}(\varepsilon)} H(\boldsymbol{\alpha}_\varepsilon\|\boldsymbol{\theta}_\varepsilon) = \sum_{\varepsilon\in\mathbf{S}(\varepsilon)}\sup_{\theta_\varepsilon} H(\boldsymbol{\alpha}_\varepsilon\|\boldsymbol{\theta}_\varepsilon) = \sum_{\varepsilon\in\mathbf{S}(\varepsilon)}\frac{h[p_\varepsilon(\mathcal{W})]}{1-p_\varepsilon(\mathcal{W})} \tag{84}$$

- the conditional *noise entropy* (entropy of the *multi–band B–E | F–D noise*)

$$H(\boldsymbol{\alpha}|\boldsymbol{\theta}) = \sum_{\varepsilon\in\mathbf{S}(\varepsilon)} H(\boldsymbol{\alpha}_\varepsilon|\boldsymbol{\theta}_\varepsilon) = \sum_{\varepsilon\in\mathbf{S}(\varepsilon)}\sum_{i\in\mathbf{S}} q(i|\boldsymbol{\theta}_{\varepsilon,i})\cdot H(\boldsymbol{\alpha}_\varepsilon\|\boldsymbol{\theta}_{\varepsilon,i}) = \sum_{\varepsilon\in\mathbf{S}(\varepsilon)}\left[\frac{h(p_\varepsilon)}{1-p_\varepsilon}\middle| h\left(\frac{p_\varepsilon}{1+p_\varepsilon}\right)\right] \tag{85}$$

where $p_\varepsilon(W) = e^{-\frac{\varepsilon}{kT_{0W}}}$, $p_\varepsilon = e^{-\frac{\varepsilon}{kT_0}}$, $T_W \ge T_0 > 0$ and $h(p) = -p\ln p - (1-p)\ln(1-p)$.

- the *transinformation* $T(\boldsymbol{\alpha};\boldsymbol{\theta})$ and the *information capacity* $C(\mathbf{K})$,

$$\begin{aligned} C(\mathbf{K}) &= \sup_{\theta\in\overline{\boldsymbol{\Theta}}_0} T(\boldsymbol{\alpha};\boldsymbol{\theta}) = \sup_{\theta\in\overline{\boldsymbol{\Theta}}_0} H(\boldsymbol{\alpha}\|\boldsymbol{\theta}) - H(\boldsymbol{\alpha}|\boldsymbol{\theta}) \\ &= \sum_{\varepsilon\in\mathbf{S}(\varepsilon)}\frac{h[p_\varepsilon(\mathcal{W})]}{1-p_\varepsilon(\mathcal{W})} - \sum_{\varepsilon\in\mathbf{S}(\varepsilon)}\left[\frac{h(p_\varepsilon)}{1-p_\varepsilon}\middle| h\left(\frac{p_\varepsilon}{1+p_\varepsilon}\right)\right] \end{aligned} \tag{86}$$

The set $\overline{\boldsymbol{\Theta}}_0 = \times_{\varepsilon\in\mathbf{S}(\varepsilon)}\{\boldsymbol{\theta}_\varepsilon \in \Theta_\varepsilon;\ E(\boldsymbol{\theta}_\varepsilon) = W_\varepsilon \ge 0\}$ represents a coding procedure of the input $\bar{i}$ of the **K** into $\boldsymbol{\theta}_{\bar{i}}$, [by transforming each input i_ε into pure state $\boldsymbol{\theta}_{\varepsilon,i_\varepsilon}$, $\forall\varepsilon\in\mathbf{S}(\varepsilon)$].

---

[7] Using the *chain rule* for simultaneous probabilities it is found that for information entropy of an independent stochastic system $\vec{X} = (X_1, X_2, ..., X_n)$ is valid that $H(\vec{X}) = \sum_i H(X_i|X_1,...X_{i-1}) = \sum_i H(X_i)$. Thus the physical entropy $\mathcal{H}(\theta)$ of independent stochastic system, $\boldsymbol{\theta} = \{\theta_\varepsilon\}_\varepsilon$, is the sum of $H_\varepsilon[q(\cdot|\theta_\varepsilon)]$ over $\varepsilon \in \mathbf{S}_\varepsilon$ too.

5.1. Transfer channels with continuous energy spectrum

Let a spectrum of energy with the finite cardinality $n+1$ and a finite time interval $\tau > 0$ are considered

$$\mathbf{S}(\varepsilon) = \{\varepsilon_r\}_{r=0,\,1,\,\ldots,\,n} = \left\{\frac{rh}{\tau}\right\}_{r=0,\,1,\,\ldots,\,n}, \quad \Delta\varepsilon = \frac{h}{\tau},\ \varepsilon_r = \frac{rh}{\tau} = r\cdot\Delta\varepsilon, \tag{87}$$

$$\text{card}\,\mathbf{S}(\varepsilon) = \frac{\varepsilon_n}{\Delta\varepsilon} = n+1,\ n\cdot\Delta\varepsilon = n\cdot\frac{h}{\tau} = \varepsilon_n,\ \frac{\tau}{n} = \frac{h}{\varepsilon_n} = \text{const.}$$

For a transfer channel with the continuous spectrum of energies of particles and with the *band–width* equal to card $\mathbf{S}(\varepsilon) = \frac{\varepsilon_n}{h}$, is valid that

$$\lim_{\tau\to\infty}\varepsilon_r = \lim_{\tau\to\infty}\frac{rh}{\tau} = \lim_{\tau\to\infty} r\cdot\Delta\varepsilon \stackrel{\triangle}{=} r\,\mathrm{d}\varepsilon \text{ resp. } \lim_{\tau\to\infty}\frac{1}{\tau} = \frac{\mathrm{d}\varepsilon}{h},\quad \mathbf{S}(\varepsilon) = \langle 0, \varepsilon_n) \tag{88}$$

But the infinite wide–band and infinite number of particles ($\tau \longrightarrow \infty$, $n \longrightarrow \infty$) will be dealt with. Then

$$\mathbf{S}(\varepsilon) = \{\varepsilon_r\}_{r=0,\,1,\,\ldots} = \left\{\frac{rh}{\tau}\right\}_{r=0,\,1,\,\ldots} = \lim_{\tau\to\infty}\varepsilon_r = \lim_{\tau\to\infty}\frac{rh}{\tau} = \lim_{\varepsilon\to 0} r\,\Delta\varepsilon = r\,\mathrm{d}\varepsilon \tag{89}$$

and thus the *wide–band spectrum* $\mathbf{S}(\varepsilon)$ *of energies* is

$$\lim_{\tau\to\infty}\frac{1}{\tau} = \frac{\mathrm{d}\varepsilon}{h},\quad \mathbf{S}(\varepsilon) = \langle 0, \infty) \tag{90}$$

With the denotation $\alpha_\varepsilon \stackrel{\triangle}{=} \alpha$, $i \stackrel{\triangle}{=} i_\varepsilon$, $j \stackrel{\triangle}{=} j_\varepsilon$, $i_\varepsilon, j_\varepsilon \in \mathbf{S}$, $\alpha_\varepsilon \in \mathbf{S}$, $\varepsilon \in \mathbf{S}(\varepsilon)$ For the p-entropy of the output $\boldsymbol{\alpha}$ of the wide–band transfer channel $\mathbf{K}_{\text{B}-\text{E}|\text{F}-\text{D}}$ is valid that

$$\begin{aligned} H(\boldsymbol{\alpha}\|\boldsymbol{\theta}) &= \lim_{\tau\to\infty}\frac{1}{\tau}\sum_{\varepsilon\in\mathbf{S}(\varepsilon)} H(\boldsymbol{\alpha}_\varepsilon\|\boldsymbol{\theta}_\varepsilon) = \lim_{\tau\to\infty} -\frac{1}{\tau}\sum_{\varepsilon\in\mathbf{S}(\varepsilon)}\sum_{\alpha_\varepsilon\in\mathbf{S}} p(\alpha_\varepsilon|\boldsymbol{\alpha}_\varepsilon|\boldsymbol{\theta}_\varepsilon)\cdot\ln p(\alpha_\varepsilon|\boldsymbol{\alpha}_\varepsilon|\boldsymbol{\theta}_\varepsilon) \qquad (91)\\ &= -\frac{1}{h}\int_0^\infty\left[\sum_{\alpha\in\mathbf{S}} p(\alpha|\boldsymbol{\alpha}_\varepsilon|\boldsymbol{\theta}_\varepsilon)\cdot\ln p(\alpha|\boldsymbol{\alpha}_\varepsilon|\boldsymbol{\theta}_\varepsilon)\right]\mathrm{d}\varepsilon \\ \sup_{\boldsymbol{\theta}} H(\boldsymbol{\alpha}\|\boldsymbol{\theta}) &= \lim_{\tau\to\infty}\frac{1}{\tau}\sum_{\varepsilon\in\mathbf{S}(\varepsilon)}\sup_{\theta_\varepsilon} H(\boldsymbol{\alpha}_\varepsilon\|\boldsymbol{\theta}_\varepsilon) = \frac{1}{h}\int_0^\infty\frac{h[p_\varepsilon(\mathcal{W})]}{1-p_\varepsilon(\mathcal{W})}\,\mathrm{d}\varepsilon \end{aligned}$$

For conditional entropy of the wide–band transfer channel $\mathbf{K}_{\text{B}-\text{E}|\text{F}-\text{D}}$ [entropy of the wide–band noise independent on the system ($\mathbf{K}_{\text{B}-\text{E}|\text{F}-\text{D}}$) state $\boldsymbol{\theta}$] and for its information capacity is valid, by (85) and (86)

$$\begin{aligned} H(\boldsymbol{\alpha}|\boldsymbol{\theta}) &= \lim_{\tau\to\infty}\frac{1}{\tau}\sum_{\varepsilon\in\mathbf{S}(\varepsilon)} H(\boldsymbol{\alpha}_\varepsilon|\boldsymbol{\theta}_\varepsilon) = \lim_{\tau\to\infty}\frac{1}{\tau}\sum_{\varepsilon\in\mathbf{S}(\varepsilon)}\sum_{i\in\mathbf{S}} q(i|\boldsymbol{\theta}_{\varepsilon,i})\cdot H(\boldsymbol{\alpha}_\varepsilon\|\boldsymbol{\theta}_{\varepsilon,i}) \qquad (92)\\ &= \frac{1}{h}\int_0^\infty H(\boldsymbol{\alpha}_\varepsilon|\boldsymbol{\theta}_\varepsilon)\,\mathrm{d}\varepsilon = \frac{1}{h}\int_0^\infty\left[\frac{h(p_\varepsilon)}{1-p_\varepsilon}\middle| h\left(\frac{p_\varepsilon}{1+p_\varepsilon}\right)\right]\mathrm{d}\varepsilon \end{aligned}$$

$$\mathrm{C}(\mathbf{K}_{\text{B}-\text{E}|\text{F}-\text{D}}) = = \frac{1}{h}\int_0^\infty\frac{h[p_\varepsilon(\mathcal{W})]}{1-p_\varepsilon(\mathcal{W})}\mathrm{d}\varepsilon - \frac{1}{h}\int_0^\infty\left[\frac{h(p_\varepsilon)}{1-p_\varepsilon}\middle| h\left(\frac{p_\varepsilon}{1+p_\varepsilon}\right)\right]\mathrm{d}\varepsilon \tag{93}$$

By (52) and (74) the average number of particles on the input of a narrow–band component $\mathcal{K}_\varepsilon$ is

$$\mathcal{W}_\varepsilon = \sum_{i \in \mathbf{S}} i \cdot q(i|\theta_\varepsilon) \geq \left[0 \,\middle|\, \frac{2p_\varepsilon^2}{1-p_\varepsilon^2}\right] \tag{94}$$

and then for the *whole average number of input particles* $\mathcal{W}'$ of the wide–band transfer channel $\mathbf{K}$ is obtained

$$\mathcal{W}' = \frac{1}{h}\int_0^\infty \mathcal{W}_\varepsilon \,\mathrm{d}\varepsilon, \;\; W = \frac{1}{h}\int_0^\infty \varepsilon\, \mathcal{W}_\varepsilon \,\mathrm{d}\varepsilon \tag{95}$$

where W is *whole input energy* and T_W is the *effective coding temperature* being supposingly at the value T_{W_ε}, $T_W = T_{W_\varepsilon}$, $\forall \varepsilon \in \mathbf{S}(\varepsilon)$.

5.2. Bose–Einstein wide–band channel capacity

By derivations (86) and (92), (93) is valid that [12]

$$\mathsf{C}(\mathbf{K}_{\mathrm{B-E}}) = \frac{1}{h}\int_0^\infty \frac{h[p_\varepsilon(\mathcal{W})]}{1-p_\varepsilon(\mathcal{W})}\,\mathrm{d}\varepsilon - \frac{1}{h}\int_0^\infty \frac{h(p_\varepsilon)}{1-p_\varepsilon}\,\mathrm{d}\varepsilon \tag{96}$$

For the first or, for the second integral respectively, obviously is valid

$$\frac{1}{h}\int_0^\infty \frac{h[p_\varepsilon(\mathcal{W})]}{1-p_\varepsilon(\mathcal{W})}\,\mathrm{d}\varepsilon = \frac{\pi^2 \mathrm{k} T_W}{3h} \quad \text{resp.} \quad \frac{1}{h}\int_0^\infty \frac{h(p_\varepsilon)}{1-p_\varepsilon}\,\mathrm{d}\varepsilon = \frac{\pi^2 \mathrm{k} T_0}{3h} \tag{97}$$

Then, for the capacity of the wide–band B–E transfer channel $\mathbf{K}_{\mathrm{B-E}}$ is valid

$$\mathsf{C}(\mathbf{K}_{\mathrm{B-E}}) = \frac{\pi^2 \mathrm{k}}{3h}(T_W - T_0) = \frac{\pi^2 \mathrm{k} T_W}{3h}\cdot\frac{T_W - T_0}{T_W} \triangleq \frac{\pi^2 \mathrm{k} T_W}{3h}\cdot \eta_{\max},\; T_W \geq T_0 \tag{98}$$

and for the *whole average output energy* is valid

$$\lim_{\tau\to\infty}\frac{1}{\tau}\sum_{\varepsilon\in\mathbf{S}(\varepsilon)} \varepsilon \frac{p_\varepsilon(\mathcal{W})}{1-p_\varepsilon(\mathcal{W})} = \frac{1}{h}\int_0^\infty \varepsilon \frac{p_\varepsilon(\mathcal{W})}{1-p_\varepsilon(\mathcal{W})}\,\mathrm{d}\varepsilon = -\frac{\mathrm{k}^2 T_W^2}{h}\int_0^1 \frac{\ln(1-t)}{t}\,\mathrm{d}t = \frac{\pi^2 \mathrm{k}^2 T_W^2}{6h} \tag{99}$$

For the *whole average energy* of the B–E *noise* must be valid

$$\lim_{\tau\to\infty}\frac{1}{\tau}\sum_{\varepsilon\in\mathbf{S}(\varepsilon)} \varepsilon \frac{p_\varepsilon}{1-p_\varepsilon} = \frac{1}{h}\int_0^\infty \varepsilon \frac{p_\varepsilon}{1-p_\varepsilon}\,\mathrm{d}\varepsilon = \frac{\pi^2 \mathrm{k}^2 {T_0}^2}{6h} \tag{100}$$

From the relations (79) among the energies of the output α', of the noise j' and the input i',

$$\frac{\pi^2 \mathrm{k}^2 T_W^2}{6h} = \frac{\pi^2 \mathrm{k}^2 {T_0}^2}{6h} + W \tag{101}$$

the effective coding temperature T_W is derivable, $T_W = T_0 \cdot \sqrt{1 + \frac{6hW}{\pi^2 \mathrm{k}^2 {T_0}^2}}$. Using it in (98) gives

$$\mathsf{C}(\mathbf{K}_{\mathrm{B-E}}) = \frac{\pi^2 \mathrm{k} T_0}{3h}\left(\sqrt{1 + \frac{6hW}{\pi^2 \mathrm{k}^2 {T_0}^2}} - 1\right) \tag{102}$$

For $T_0 \rightarrow 0$ the *quantum aproximation* of $C(\mathbf{K}_{B-E})$, independent on the heat noise energy (deminishes whith temperture's aiming to absolute $0^\circ K$)

$$\lim_{T_0 \to 0} C(\mathbf{K}_{B-E}) = \lim_{T_0 \to 0} \left(\sqrt{\frac{\pi^4 k^2 T_0{}^2}{9h^2} + \pi^2 \frac{2W}{3h}} - \frac{\pi^2 k T_0}{3h} \right) = \pi \sqrt{\frac{2W}{3h}} \tag{103}$$

The *classical approximation* of $C(\mathbf{K}_{B-E})$ is gaind for temperatures $T_0 \gg 0$ ($T_0 \rightarrow \infty$ respectively). It is near to value $\frac{W}{kT_0}$, the Shannon capacity of the wide–band **Gaussian** channel with the whole noise energy kT_0 and with the whole average input energy W. For T_0 from (101), great enough, is gained that[8]

$$C(\mathbf{K}_{B-E}) \doteq \frac{\pi^2 k T_0}{3h} \left(\frac{3hW}{\pi^2 k^2 T_0{}^2} \right) = \frac{W}{kT_0} \tag{104}$$

5.3. Fermi–Dirac wide–band channel capacity

By derivations (86) and (92), (93) is valid that [12]

$$C(\mathbf{K}_{F-D}) = \frac{1}{h} \int_0^\infty \frac{h[p_\varepsilon(W)]}{1 - p_\varepsilon(W)} \, d\varepsilon - \frac{1}{h} \int_0^\infty h\left(\frac{p_\varepsilon}{1 + p_\varepsilon} \right) d\varepsilon = \frac{\pi^2 k T_W}{3h} - \frac{1}{h} \int_0^\infty h\left(\frac{p_\varepsilon}{1 + p_\varepsilon} \right) d\varepsilon \tag{105}$$

For the second integral obviously is valid

$$\frac{1}{h} \int_0^\infty h\left(\frac{p_\varepsilon}{1 + p_\varepsilon} \right) d\varepsilon = \frac{\pi^2 k T_0}{6h} \tag{106}$$

By figuring (105) the capacity of the wide–band F–D channel $\mathbf{K}_{F-D}$ is gained,

$$C(\mathbf{K}_{F-D}) = \frac{\pi^2 k}{3h} \left(T_W - \frac{T_0}{2} \right) \tag{107}$$

and for $T_W > T_0$is writable

$$C(\mathbf{K}_{F-D}) = C(\mathbf{K}_{B-E}) \cdot \frac{2T_W - T_0}{2T_W - 2T_0} \tag{108}$$

For the *whole average output energy* is valid the same as for the B–E case,

$$\frac{1}{h} \int_0^\infty \varepsilon \frac{p_\varepsilon(W)}{1 - p_\varepsilon(W)} \, d\varepsilon = \frac{\pi^2 k^2 T_W^2}{6h} \tag{109}$$

For the *whole average F–D wide–band noise* energy is being derived

$$\lim_{\tau \to \infty} \frac{1}{\tau} \sum_{\varepsilon \in \mathbf{S}(\varepsilon)} \varepsilon \frac{p_\varepsilon}{1 + p_\varepsilon} = \frac{1}{h} \int_0^\infty \varepsilon \frac{e^{-\frac{\varepsilon}{kT_0}}}{1 + e^{-\frac{\varepsilon}{kT_0}}} \, d\varepsilon = \frac{k^2 T_0{}^2}{h} \int_0^\infty x \frac{e^{-x}}{1 + e^{-x}} \, dx \tag{110}$$

$$= -\frac{k^2 T_0{}^2}{h} \int_0^1 \frac{\ln t}{t + 1} \, dt = \frac{\pi^2 k^2 T_0{}^2}{12h}$$

---

[8] For $|x| < 1$, $\sqrt{1+x} = 1 + \frac{1}{2}x - \frac{1}{8}x^2 + \ldots \doteq 1 + \frac{1}{2}x$ where $x = \frac{6hW}{\pi^2 k^2 T_0{}^2} < 1$.

From the relation (79) among the whole output, input, and noise energy,

$$\frac{\pi^2 k^2 T_W^2}{6h} = \frac{\pi^2 k^2 {T_0}^2}{12h} + W \tag{111}$$

follows the effective coding temperature $T_W = T_0 \cdot \sqrt{\frac{1}{2} + \frac{6hW}{\pi^2 k^2 {T_0}^2}}$. Using it in (107) the result is [24]

$$\mathsf{C}(\mathbf{K}_{F-D}) = \frac{\pi^2 k T_0}{3h}\left(\sqrt{\frac{1}{2} + \frac{6hW}{\pi^2 k^2 {T_0}^2}} - \frac{1}{2}\right) \tag{112}$$

For $T_0 \to 0$ the *quantum approximation* capacity $\mathsf{C}(\mathbf{K}_{F-D})$, independent on heat noise energy kT_0 is gaind (the same as in the B–E case (103),

$$\lim_{T_0\to 0} \mathsf{C}(\mathbf{K}_{F-D}) = \lim_{T_0\to 0}\left(\sqrt{\frac{\pi^4 k^2 {T_0}^2}{9h^2}\cdot\frac{1}{2} + \pi^2\frac{2W}{3h}} - \frac{\pi^2 k T_0}{3h}\cdot\frac{1}{2}\right) = \pi\sqrt{\frac{2W}{3h}} \tag{113}$$

The *classical approximation* of the capacity $\mathsf{C}(\mathbf{K}_{F-D})$ is gained for $T_0 \gg 0$[9]

$$\begin{aligned} \mathsf{C}(\mathbf{K}_{F-D}) &= \frac{\pi^2 k T_0}{3h}\left[\frac{1}{\sqrt{2}}\sqrt{1 + \frac{12hW}{\pi^2 k^2 {T_0}^2}} - \frac{1}{2}\right] \doteq \frac{\pi^2 k T_0}{3h}\left[\frac{1}{\sqrt{2}}\left(1 + \frac{6hW}{\pi^2 k^2 {T_0}^2}\right) - \frac{1}{2}\right] \\ &= \frac{\pi^2 k T_0}{6h}\left(\sqrt{2} - 1\right) + \sqrt{2}\frac{W}{kT_0} \quad \left[\underset{T_0\to\infty}{\longrightarrow} \frac{\pi^2 k T_0}{6h}\left(\sqrt{2}-1\right),\ W = \text{const.} \geq W_{crit}\right] \end{aligned} \tag{114}$$

By (74) the condition for the medium value of the input particles of a narrow–band component $\mathcal{K}_\varepsilon$, $\varepsilon \in \mathbf{S}(\varepsilon)$, of the channel $\mathbf{K}_{F-D}$ is valid, $\mathcal{W}_\varepsilon \geq \frac{2p_\varepsilon^2}{1-p_\varepsilon^2}$, from which the condition for the whole input energy of the wide–band channel $\mathbf{K}_{F-D}$ follows. By (95) it is gaind, for $T_W \geq T_0 > 0$, that[10]

$$W \geq \lim_{\tau\to\infty}\frac{1}{\tau}\sum_{\varepsilon\in\mathbf{S}(\varepsilon)} \varepsilon\mathcal{W}_\varepsilon \geq \lim_{\tau\to\infty}\frac{1}{\tau}\sum_{\varepsilon\in\mathbf{S}(\varepsilon)} \varepsilon\frac{2p_\varepsilon^2}{1-p_\varepsilon^2} \doteq \frac{2}{h}\int_0^\infty \varepsilon\frac{e^{-2\frac{\varepsilon}{kT_0}}}{1-e^{-2\frac{\varepsilon}{kT_0}}}\,d\varepsilon = \frac{\pi^2 k^2 {T_0}^2}{12h} = W_{crit} > 0 \tag{115}$$

6. Physical information transfer and thermodynamics

Whether the considered information transfers are narrow–band or wide–band, their algebraic-information description remains the same. So let be considered an arbitrary *stationary physical system* $\mathbf{\Psi}$ of these two band–types as usable for information transfer.

Let a system state $\boldsymbol{\theta}' = \sum_{i=1}^{n} q(i|\boldsymbol{\theta}')\,\pi\{\psi_i'\} \in \boldsymbol{\Theta}$ of the system $\mathbf{\Psi}$ is the *successor (follower, equivocant)* of the system state $\boldsymbol{\theta} = \sum_{i=1}^{n} q(i|\boldsymbol{\theta})\,\pi\{\psi_i\} \in \boldsymbol{\Theta}$, $\boldsymbol{\theta} \longrightarrow \boldsymbol{\theta}'$ is written. The

[9] For $\sqrt{1+x} \doteq 1 + \frac{1}{2}x$ when $|x| < 1$; $x = \frac{12hW}{\pi^2 k^2 {T_0}^2}$.

[10] If, in the special case of F–D channel, it is considered that the value W given by the number of electrons as the average energy of the *modulating current* entering into a wire, over a *time unit*, then it is the average power on the electric resistor $R = 1\Omega$ too.

distribution $q(i|\theta') = \sum_{j=1}^{n} p(i|j)\, q(j|\theta),\ p(i|j) = (\psi_i', \psi_j)^2 = u_{ji}^2$ and $\theta \longrightarrow \theta'$, ensures existence of the *transformation matrix* $[u_{i,j}]$ of a *base* of the space $\Psi = \{\Psi\}_{i=1}^{n}$ into the base $\{\Psi'\}_{i=1}^{n}$.[11]

From the relation $\theta \longrightarrow \theta'$ also is visible that it is *reflexive* and *transitive* relation between states and, thus, it defines (a partial) *arrangment* on the set space $\boldsymbol{\Theta}$. The *terminal, maximal* state for this arrangement is the equilibrial state θ^+ of the system $\boldsymbol{\Psi}$: it is the successor of an arbitrary system state, including itself.

The *statistic, Shannon, information)* entropy $H(\cdot)$ is a generalization of the physical entropy$\mathcal{H}(\theta)$. The quantity I-divergence $I(\cdot\|\cdot)$ is, by (21), a generalization of the physical quantity $I(p\|d) = \mathcal{H}(\theta^+) - \mathcal{H}(\theta)$ where the state

$$\theta^+ = \frac{1}{n}\sum_{i=1}^{n} \theta_i \in \boldsymbol{\Theta} \quad \text{for} \quad \theta_i = \pi\{\psi_i\}, \quad i = 1, 2, ..., n \tag{116}$$

is the *equlibrial state* of the system $\boldsymbol{\Psi}$. The probability distribution into the *canonic components* θ_i of θ^+ is *uniform* and thus

$$\mathcal{H}(\theta^+) = \ln n = \ln \dim(\Psi) \tag{117}$$

Information divergence $I(p\|d) \geq 0$ expresses the *distance* of the two probability distributions $q(\cdot|\theta)$ and $q(\cdot|\theta^+)$ of states (stochastic quantities)

$$\theta = [\mathbf{S}, q(\cdot|\theta)] \quad \text{and} \quad \theta^+ = [\mathbf{S}, q(\cdot|\theta^+)] \tag{118}$$

In the physical sense the divergence $I(p\|d)$ is a *measure of a not-equilibriality* of the state θ of the physical (let say a thermodynamic) system $\boldsymbol{\Psi}$. Is maximized in the initial (starting), not-equilibrium state of the (time) evolution of the $\boldsymbol{\Psi}$. It is clear that $I(p\|d) \equiv T(\alpha; \theta)$

6.0.0.5. $\mathcal{H}$-Theorem, *II. Second Principle of Thermodynamics*:

Let for states $\theta, \theta' \in \boldsymbol{\Theta}$ of the system $\boldsymbol{\Psi}$ is valid that $\theta \rightarrow \theta'$. Then

$$\mathcal{H}(\theta') \geq \mathcal{H}(\theta) \tag{119}$$

and the equality arises for $\theta = \theta'$ only [12, 38].

6.0.0.6. Proof:

(a) For a *strictly convex* function $f(u) = u \cdot \ln u$ the ***Jensen** inequality* is valid [23]

$$f\left[\sum_{j=1}^{n} p(i|j)\, q(j|\theta)\right] \leq \sum_{j=1}^{n} p(i|j)\, f[q(j|\theta)], \quad i = 1, 2, ..., n \tag{120}$$

$$\sum_{i=1}^{n} f\left[\sum_{j=1}^{n} p(i|j)\, q(j|\theta)\right] \leq \sum_{i=1}^{n} p(i|j) \sum_{j=1}^{n} f[q(j|\theta)]$$

$$= \sum_{j=1}^{n} f[q(j|\theta)] = \sum_{j=1}^{n} q(j|\theta) \ln q(j|\theta) = -H[q(\cdot|\theta)] = -\mathcal{H}(\theta) \quad \text{due to} \quad \sum_{i=1}^{n} p(i|j) = 1$$

[11] It is the matrix of the *unitary operator* $\mathbf{u}(t)$ expressing the time evolution of the system $\boldsymbol{\Psi}$.

and for distributions $q(i|\theta')$ is valid $\mathcal{H}(\theta') \geq \mathcal{H}(\theta)$:

$$\sum_{i=1}^{n} f\left(\sum_{j=1}^{n} p(i|j)\, q(j|\theta)\right) = \sum_{i=1}^{n} q(i|\theta') \ln q(i|\theta') = -\mathcal{H}(\theta') \;\; [\leq -\mathcal{H}(\theta)]$$

(b) The equality in (119) arises if and only if the index permutation $[i(1), i(2), ..., i(n)]$ exists that $p(i|j) = \delta[i|i(j)]$, $j = 1, 2, ..., n$; then $q[i(j)|\theta'] = q(j|\theta)$, $j = 1, 2, ...$
Let a fixed j is given. Then, when $0 = p(i|j) = (\psi_i', \psi_j)^2$, $i \neq i(j)$, the orthogonality is valid

$$\Psi(\psi_j|\psi_j) \perp \left[\bigoplus_{i \neq i(j)} \Psi(\psi_i'|\psi_i')\right], \quad \psi_j = \pi\{\psi_j\} = \theta_j, \quad \psi_i' = \pi\{\psi_i'\} = \theta_i' \tag{121}$$

and, consequently, $\psi_j \in \Psi[\psi'_{i(j)}|\psi'_{i(j)}]$, $p[i(j)|j] = (\psi'_{i(j)}, \psi_j)^2 = 1$. It results in $\psi_j = \psi'_{i(j)}$. This prooves that the equality $\mathcal{H}(\theta') = \mathcal{H}(\theta)$ implies the equality $q(j|\theta)\, \pi\{\psi_j\} = q[i(j)|\theta]\, \pi\{\psi'_{i(j)}\}$ and $\theta = \theta'$.

$\mathcal{H}$-theorem says, that a **reversible transition is not possible between any two different states** $\theta \neq \theta'$. From the inequality (119) also follows that any state $\theta \in \Theta$ of the system Ψ is the successor of itself, $\theta \to \theta$ and, that any **reversibility** of the **relation $\theta \to \theta'$ (the transition $\theta' \to \theta$) is not possible within the system only, it is not possible without openning this system Ψ**. The difference

$$\mathcal{H}(\theta^+) - \mathcal{H}(\theta) = \max_{\theta' \in \Theta} \mathcal{H}(\theta') - \mathcal{H}(\theta) = H\left[q(\cdot|\theta^+)\right] - H\left[q(\cdot|\theta)\right] \tag{122}$$

reppresents the information-theoretical expressing of the ***Brillouin** (maximal) entropy defect* ΔH (the Brillouin *negentropic information principle* [2, 30]). For the state θ^+ is valid that $\theta \to \theta^+$, $\forall \theta \in \Theta$. It is also called the *terminal* state or the (*atractor* of the time evolution) of the system Ψ.[12]

6.0.0.7. Gibbs Theorem:

For all θ, $\tilde{\theta} \in \Theta$ of the system Ψ is valid

$$\mathcal{H}(\theta) \leq -\mathrm{Tr}(\theta \ln \tilde{\theta}) \tag{123}$$

and the equality arises only for $\theta = \tilde{\theta}$ [38].

6.0.0.8. Proof:

Let for θ, $\tilde{\theta} \in \Theta$ is valid that $\theta = \sum_{i=1}^{n} q(i|\theta)\, \pi\{\psi_i\}$, $\tilde{\theta} = \sum_{i=1}^{n} q(i|\tilde{\theta})\, \pi\{\psi_i'\}$ and let the operators α, θ are *commuting* $\alpha\theta = \theta\alpha$, $D(\alpha) = \{D_\alpha : \alpha \in \mathbf{S}(\alpha)\}$, $D(\theta) = \{D_\theta : \theta \in \mathbf{S}(\theta)\}$, are their spectral decompositions. Let be the state θ' the successor of θ, $\theta \to \theta'$ and relations $p(\alpha|\alpha|\theta) = \sum_{i \in D_\alpha} q(i|\theta') = p(\alpha|\alpha|\theta')$ and $\mathcal{H}(\theta') \geq \mathcal{H}(\theta)$ are valid. For the matrix (θ_{ij}) of the

[12] In this sense, the physical entropy $\mathcal{H}(\theta)$ (19), (21) determines the direction of the *thermodynamic time arrow* [2], $\frac{\mathcal{H}(\theta') - \mathcal{H}(\theta)}{\Delta t} = \frac{\partial \mathcal{H}}{\partial t} \geq 0$, $\Delta t = t_{\theta'} - t_\theta > 0$. The equality occurs in the *equlibrial (stationary)* state θ^+ of the system Ψ and its environment.

operator $\boldsymbol{\theta}$ in the base $\{\psi'_1, \psi'_2, ..., \psi'_n\}$ is obtained that $\theta_{ij} = \sum_{k=1}^{n} u_{ki}\, u_{kj}\, q(k|\boldsymbol{\theta})$ and thus for operators $\ln \tilde{\boldsymbol{\theta}}$ and $\mathrm{Tr}(\boldsymbol{\theta} \ln \tilde{\boldsymbol{\theta}})$ is valid that

$$\ln \tilde{\boldsymbol{\theta}} = \sum_{i=1}^{n} \ln q(i|\tilde{\boldsymbol{\theta}})\, \pi\{\psi'_i\} \text{ and } -\mathrm{Tr}(\boldsymbol{\theta} \ln \tilde{\boldsymbol{\theta}}) = -\sum_{i=1}^{n} \left[\sum_{k=1}^{n} u_{ki}^2\, q(k|\boldsymbol{\theta}) \right] \ln q(i|\tilde{\boldsymbol{\theta}}) \quad (124)$$
$$= -\sum_{i=1}^{n} q(i|\boldsymbol{\theta}') \ln q(i|\tilde{\boldsymbol{\theta}})$$

For the information divergence of the distributions $q(\cdot|\boldsymbol{\theta}')$, $q(\cdot|\tilde{\boldsymbol{\theta}})$ and the entropy $\mathcal{H}(\boldsymbol{\theta}')$ is valid that

$$I[q(\cdot|\boldsymbol{\theta}')\|q(\cdot|\tilde{\boldsymbol{\theta}})] = \sum_{i=1}^{n} q(i|\boldsymbol{\theta}') \ln \frac{q(i|\boldsymbol{\theta}')}{q(i|\tilde{\boldsymbol{\theta}})} \geq 0, \quad -\mathcal{H}(\boldsymbol{\theta}') \geq \sum_{i=1}^{n} q(i|\boldsymbol{\theta}') \ln q(i|\tilde{\boldsymbol{\theta}}). \quad (125)$$

By (119) for $\boldsymbol{\theta} \to \boldsymbol{\theta}'$ is writable that $\mathcal{H}(\boldsymbol{\theta}) \leq \mathcal{H}(\boldsymbol{\theta}') \leq -\mathrm{Tr}(\boldsymbol{\theta} \ln \tilde{\boldsymbol{\theta}})$. By (123) $-\mathrm{Tr}(\boldsymbol{\theta}' \ln \tilde{\boldsymbol{\theta}}) \geq \mathcal{H}(\boldsymbol{\theta}') \geq \mathcal{H}(\boldsymbol{\theta})$ are valid; the first equality is for

$$I[q(\cdot|\boldsymbol{\theta}')\|q(\cdot|\tilde{\boldsymbol{\theta}})] = 0, \; \mathcal{H}(\boldsymbol{\theta}') = \mathcal{H}(\tilde{\boldsymbol{\theta}}), \; q(i|\boldsymbol{\theta}') = q(i|\tilde{\boldsymbol{\theta}}), \quad i = 1, 2, ..., n, \; \boldsymbol{\theta}' = \tilde{\boldsymbol{\theta}} \quad (126)$$

the second equality is for $\boldsymbol{\theta}' = \boldsymbol{\theta}$. The **Gibbs theorem expresses, in the deductive (matematical-logical) way, the phenomenon of Gibbs paradox.**[13]

From formulas (47), (55), (56) and (68), (72), (73) for the narrow-band B–E and F–D capacities follows that

$$e^{-\frac{\varepsilon}{kT_W}} \cdot e^{\frac{\varepsilon}{kT_0}} \geq 1, \; e^{\frac{\varepsilon}{kT_0}\left(\frac{T_W - T_0}{T_W}\right)} \geq e^0; \; \varepsilon > 0, \, T_0 > 0 \longrightarrow T_W \geq T_0 \longrightarrow \frac{T_W - T_0}{T_W} \triangleq \eta_{\max} \geq 0 \quad (127)$$

and it is seen that the quantity temperature is decisive for studied information transfers. The last relation envokes, inevitably, such an opinion, that these transfers are able be modeled by a *direct* reversible **Carnot** cycle with efficiency $\eta_{max} \in (0, 1\rangle)$. Conditions leading to $C_{[\cdot|\cdot]} < 0$ mean, in such a *direct* thermodynamic model, that its efficiency should be $\eta_{\max} < 0$. This is the contradiction with the *Equivalence Principle of Thermodynamics* [19]; expresses only that the transfer is running in the opposite direction (as for temperatures).

As for B–E channel; for the supposition $W < 0$ the inequalities $T_W < T_0$ and $p(\mathcal{W}) < p$ would be gained which is the *contradiction* with (35), (36) and (47). It would be such a situation with the *information is transferred in a different direction and under a different operation mode*. Our sustaining on the meaning about the original organization of the transfer, for $T_W > T_0$, then leads to the contradiction mentioned above saying only that we are convinced mistakenly about the actual direction of the information transfer. In the case $T_W = T_0$ for the capacity $C_{\text{B-E}''}$ from (50) is valid that $C_{\text{B-E}''} = 0$. Then $W = W_{Krit}$ $[= 0]$ for $p(\mathcal{W}) = p$.

As for F–D channel; for the supposition $\mathcal{W} < \dfrac{2p^2}{1-p^2}$ $T_W < T_0$ and $p(\mathcal{W}) < p$ is gained which is the contradiction with (68). For $T_W = T_0$ is for $C_{\text{F-D}}$ from (71) valid

[13] Derived by the information-thermodynamic way together with the *I.* and *II. Thermodynamic Principle* and with the *Equivalence Principle of Thermodynamics in [16, 17, 19].*

$$C_{F-D} = \frac{h(p)}{1-p} + \frac{p}{1+p} \cdot \ln p - \ln(1+p).^{14}$$

Let be noticed yet the relations between the wide–band B–E and F–D capacities and the model heat efficiency η_{max}. For the B–E capacity (98) is gained that

$$C(\mathbf{K}_{B-E}) = \frac{\pi^2 k T_W}{3h}\left(\frac{T_W - T_0}{T_W}\right) = \frac{\pi^2 k T_W}{3h}\eta_{\max} \underset{\eta_{\max}\to 1}{\longrightarrow} \frac{\pi^2 k T_W}{3h} = C^{\max}(\mathbf{K}_{B-E}) \quad (128)$$

$$C^{\max}(\mathbf{K}_{B-E}) = \sup_{\boldsymbol{\theta}} H(\boldsymbol{\alpha}\|\boldsymbol{\theta}) = H(\boldsymbol{i}) = \mathcal{H}(\boldsymbol{\theta})$$

$$C(\mathbf{K}_{B-E}) > 0,\ T_W > T_0, \quad C(\mathbf{K}_{B-E}) \underset{\substack{T_W\to T_0\\(\eta_{\max}\to 0)}}{\longrightarrow} 0,\ T_W \longrightarrow T_0$$

It is the information capacity for such a direct Carnot cycle where $H(X) = \frac{\pi^2 k T_W}{3h} = C^{\max}(\mathbf{K}_{B-E})$.

For the wide-band F–D capacity from (105) is valid

$$C(\mathbf{K}_{F-D}) = \frac{\pi^2 k T_W}{3h} - \frac{\pi^2 k T_0}{6h},\ T_W \geq T_0 \ \text{ and for } T_W > T_0, \quad (129)$$

$$C(\mathbf{K}_{F-D}) = \frac{\pi^2 k T_W}{3h} \cdot \frac{2T_W - T_0}{2T_W} = \frac{\pi^2 k T_W}{3h} \cdot \frac{2T_W - T_0}{2(T_W - T_0)} \cdot \eta_{\max}$$

$$= C(\mathbf{K}_{B-E}) \cdot \frac{2T_W - T_0}{2(T_W - T_0)}$$

Due to $1 - \eta_{\max} = \frac{T_0}{T_W}$ is valid $T_0 = T_W(1 - \eta_{\max})$ and also $C(\mathbf{K}_{F-D}) = \frac{\pi^2 k T_W}{6h} \cdot (1 + \eta_{\max})$. Then,

$$C\ (\mathbf{K}_{F-D}) = \underset{\eta_{\max}\to 1}{\longrightarrow} \frac{\pi^2 k T_W}{3h}, \quad (130)$$

$$C\ (\mathbf{K}_{F-D}) \underset{\substack{T_W\to T_0\\(\eta_{\max}\to 0)}}{\longrightarrow} \frac{1}{2}H(\boldsymbol{i}) = \frac{1}{2}\mathcal{H}(\boldsymbol{\theta}) = \frac{\pi^2 k T_W}{6h}$$

$$C\ (\mathbf{K}_{F-D}) \in \left\langle \frac{\pi^2 k T_W}{6h}, \frac{\pi^2 k T_W}{3h} \right) = \left\langle \frac{1}{2}\mathcal{H}(\boldsymbol{\theta}), \mathcal{H}(\boldsymbol{\theta}) \right)$$

Again the phenomenon of the not-zero capacity is seen here when the difference between the coding temperature T_W and the noise temperature T_0 is zero. Capacities $C(\mathbf{K}_{F-D}) \geq 0$ are, surely, considerable for $T_W \in \langle \frac{T_0}{2}, T_0 \rangle$ and being given by the property of the F–D phase space cells. Capacities $C(\mathbf{K}_{B-E}) < 0$ and $C(\mathbf{K}_{F-D}) < 0$ are without sense for the given direction of information transfer.

[14] Nevertheless the capacity C_{F-D} for this case $W < W_{krit}$ is set in [12, 13]. *Similar* results as this one and (74) are gained for the **Maxwell–Boltzman** (M–B) system in [13].

Nevertheless, it will be shown that all these processes themselves are not organized cyclically 'by themselves'.

Further the relation between the information transfer in a wide–band B–E (photonic) channel *organized in a cyclical way* and a relevant (reverse) heat cycle will be dealt with. But, firstly, the way in which the capacity formula for an information transfer system of photons was derived in [30] will be reviewed.

6.1. Thermodynamic derivation of wide–band photonic capacity

A transfer channel is now created by the electromagnetic radiation of a system $\mathcal{L} \cong \mathbf{K}_{\mathrm{L-L}}$ of photons being emitted from an *absolute black body* at temperature T_0 and within a frequency bandwidth of $\Delta\nu = R^+$, where ν is the frequency. Then the energy of such radiation is the *energy* of *noise*. A source of *input messages, signals* transmits monochromatic electromagnetic impulses (numbers a_i of photons) into this environment with an average *input energy* W. This source is defined by an alphabet of input messages, signals $\{a_i\}_{i=1}^n$, with a probability distribution $p_i = p(a_i), i = 1, 2, \ldots, n$.[15] The *output (whole, received)* signal is created by *additive* superposition of the input signal and the noise signal. The input signal a_i, within a frequency ν, is represented by the *occupation* number $m = m(\nu)$, which equates to the number of photons of an input field with an energy level $\varepsilon(\nu) = h\nu$. The output signal is represented by the occupation number $l = l(\nu)$. The noise signal, created by the number of photons emitted by absolute black body radiation at temperature T_0, is represented by the occupation number $n = n(\nu)$. The medium values of these quantities (*spectral* densities of the input, noise and output photonic stream) are denoted as $\overline{m}$, $\overline{n}$ and $\overline{l}$. In accordance with the ***Planck** radiation* law, the spectral density $\overline{r}$ of a photonic stream of absolute black body radiation at temperature Θ and within frequency ν, is given by the ***Planck** distribution*,

$$\overline{r}(\nu) = \frac{p(\nu,\Theta)}{1 - p(\nu,\Theta)}, \quad \overline{n}(\nu) = \frac{p(\nu,T_0)}{1 - p(\nu,T_0)}, \quad \overline{l}(\nu) = \frac{p(\nu,T_W)}{1 - p(\nu,T_W)}, \quad p(\nu,\Theta) = e^{-\frac{h\nu}{k\Theta}} \tag{131}$$

Thus, for the average energy P of radiation at temperature Θ within the bandwidth $\Delta\nu = R^+$ is gained that

$$P(\Theta) = \int_0^\infty \overline{\varepsilon}(\nu,\Theta)\mathrm{d}\nu = \frac{\pi^2 \mathrm{k}^2 \Theta^2}{6\hbar} \quad \text{where } \overline{\varepsilon}(\nu,\Theta) = \overline{r}(\nu)h\nu \text{ and } \frac{\mathrm{d}P(\Theta)}{\mathrm{d}\Theta} = \frac{\pi^2 \mathrm{k}^2 \Theta}{3\hbar}. \tag{132}$$

Then, for the average noise energy P_1 at temperature T_0, and for the average output energy P_2 at temperature T_W, both of which occur within the bandwidth $\Delta\nu = R^+$ is valid that

$$P_1(T_0) = \frac{\pi^2 \mathrm{k}^2 {T_0}^2}{6\hbar}, \quad P_2(T_W) = \frac{\pi^2 \mathrm{k}^2 {T_W}^2}{6\hbar}. \tag{133}$$

The entropy H of radiation at temperature ϑ is derived from ***Clausius** definition* of *heat entropy* and thus

$$H = \int_0^\vartheta \frac{1}{\mathrm{k}\Theta} \frac{\mathrm{d}P(\Theta)}{\mathrm{d}\Theta} \mathrm{d}\Theta = \frac{\pi^2 \mathrm{k}\vartheta}{3\hbar} = \frac{2P(\vartheta)}{\mathrm{k}\vartheta} \tag{134}$$

[15] To distinguish between two frequencies mutually deferring at an infinitesimally small $\mathrm{d}\nu$ is needed, in accordance with ***Heisenberg** uncertainty principle*, a time interval spanning the infinite length of time, $\Delta t \longrightarrow \infty$; analog of the *thermodynamic stationarity*.

Thus, for the entropy H_1 of the noise signal and for the entropy H_2 of the ouptut signal on the channel $\mathcal{L}$ is

$$H_1 = \frac{\pi^2 \mathrm{k} T_0}{3\hbar} = \frac{2P_1}{\mathrm{k}T_0}, \quad H_2 = \frac{\pi^2 \mathrm{k} T_W}{3\hbar} = \frac{2P_2}{\mathrm{k}T_W} \tag{135}$$

The information capacity C_{T_W,T_0} of the wide-band photonic transfer channel $\mathcal{L}$ is given by the maximal *entropy defect* [2, 30]) by

$$C_{T_0,T_W} = H_2 - H_1 = \int_{T_0}^{T_W} \frac{1}{\mathrm{k}\Theta} \frac{\mathrm{d}P(\Theta)}{\mathrm{d}\Theta} \mathrm{d}\Theta = \frac{\pi^2 \mathrm{k}}{3\hbar} \int_{T_0}^{T_W} \mathrm{d}\Theta = \frac{\pi^2 \mathrm{k}}{3\hbar} \cdot (T_W - T_0). \tag{136}$$

For $P_2 = P_1 + W$, where W is the average energy of the input signal is then valid that

$$\frac{\pi^2 \mathrm{k}^2 {T_W}^2}{6h} = \frac{\pi^2 \mathrm{k}^2 {T_0}^2}{6h} + W \longrightarrow T_W = T_0 \cdot \sqrt{1 + \frac{6h \cdot W}{\pi^2 \mathrm{k}^2 {T_0}^2}} \tag{137}$$

Then, in accordance with (102), (103), (104) [30]

$$C_{T_0,W}(\mathbf{K}_{\mathrm{L-L}}) = \frac{\pi^2 \mathrm{k} T_0}{3\hbar} \cdot \left(\sqrt{1 + \frac{6h \cdot W}{\pi^2 \mathrm{k}^2 {T_0}^2}} - 1 \right) \tag{138}$$

7. Reverse heat cycle and transfer channel

A *reverse* and reversible Carnot cycle $\mathcal{O}_{rrev}$ starts with the *isothermal expansion* at temperature T_0 (the *diathermic* contact [31] is made between the system $\mathcal{L}$ and the cooler $\mathcal{B}$) when $\mathcal{L}$ is receiving the *pumped out, transferred* heat ΔQ_0 from the $\mathcal{B}$. During the *isothermal compression,* when the temperature of both the system $\mathcal{L}$ and the heater $\mathcal{A}$ is at the same value T_W, $T_W > T_0 > 0$, the *output* heat ΔQ_W is being delivered to the $\mathcal{A}$

$$\Delta Q_W = \Delta Q_0 + \Delta A \tag{139}$$

where ΔA is the *input* mechanical energy (work) delivered into $\mathcal{L}$ during this isothermal compression. It follows from [2, 8, 28] that when an average amount of information ΔI is being *recorded, transmitted, computed,* etc. at temperature Θ, there is a need for the average energy $\Delta W \geq k \cdot \Theta \cdot \Delta I$; at this case $\Delta W \triangleq \Delta A$. Thus $\mathcal{O}_{rrev}$ is considerable as a *thermodynamic model* of information transfer process in the channel $\mathcal{K} \cong \mathcal{L}$ [14]. The following values are *changes* of the information entropies defined on $\mathcal{K}$[16]:

$$H(Y) \triangleq \frac{\Delta Q_W}{\mathrm{k}T_W} \; output \; (\triangleq \Delta I), \quad H(X) \triangleq \frac{\Delta A}{\mathrm{k}T_W} \; input, \quad H(Y|X) \triangleq \frac{\Delta Q_0}{\mathrm{k}T_W} \; noise \tag{140}$$

where k is Boltzman constant. The information transfer in $\mathcal{K} \cong \mathcal{L}$ is *without losses* caused by the *friction, noise heat* ($\Delta Q_{0x} = 0$) and thus $H(X|Y) = 0$.
By assuming that for the changes (140) and $H(X|Y) = 0$ the channel equation (4), (5) and (23) is valid The result is

$$T(X;Y) = \frac{\Delta A}{\mathrm{k}T_W} - 0 = \frac{\Delta Q_W}{\mathrm{k}T_W} \cdot \eta_{max} = H(X) \tag{141}$$

$$T(Y;X) = \frac{\Delta Q_0 + \Delta A}{\mathrm{k}T_W} - \frac{\Delta Q_0}{\mathrm{k}T_W} = \frac{\Delta A}{\mathrm{k}T_W} = H(X).$$

[16] In *information* units *Hartley, nat, bit*; $H(\cdot) = \Delta H(\cdot)$, $H(\cdot|\cdot) = \Delta H(\cdot|\cdot)$.

But the other *information arrangement, description* of a revese Carnot cycle will be used further, given by

$$\Delta Q_0 \sim H(X),\ \Delta Q_W \sim H(Y) \text{ and } \Delta A \sim H(Y|X),\ H(X|Y)=0 \tag{142}$$

In a general (reversible) *discrete* heat cycle $\mathcal{O}$ (with temperatures of its heat reservoires changing in a discrete way) considered as a model of the information transfer process in an transfer channel $\mathbf{K} \cong \mathcal{L}$ [17, 19] is, for the *elementary* changes $H(\Theta_k) \cdot \eta_{[max_k]}$ of information entropies of $\mathcal{L}$, valid that[17]

$$H(\Theta_k) \cdot \eta_{[max_k]} \triangleq \frac{\Delta Q(\Theta_k)}{k\Theta_k} \cdot \eta_{[max_k]},\ k = 1, 2, ..., n \tag{143}$$

where $n \geq 2$ is the maximal number of its *elementary* Carnot cycles $\mathcal{O}_k$.[18] The change of heat of the system $\mathcal{L}$ at temperatures Θ_k is $\Delta Q(\Theta_k)$.

In a general (reversible) *continuous* cycle $\mathcal{O}$ [with temperatures changing continuously, $n \longrightarrow \infty$ in the previous discrete system, at Θ will be $\Delta Q(\Theta)$] considered as an information transfer process in a transfer channel $\mathbf{K} \cong \mathcal{L}$ is valid that

$$dH(\Theta) \triangleq \frac{\delta Q(\Theta)}{k\Theta} = \frac{\frac{\partial Q(\Theta)}{\partial \Theta} d\Theta}{k\Theta} \text{ and} H(\Theta) = \int_0^\Theta \frac{\delta Q(\theta)}{k\theta} d(\theta);\ \Delta Q(\Theta) = \int_0^\Theta \delta Q(\theta) d\theta \tag{144}$$

For the whole cycle $\mathcal{O}_{rrev}$, $T_W > T_0 > 0$, let be $H(X|Y) = 0$ and then

$$H(X) = \frac{S(T_W)}{k} - \frac{S(T_0)}{k} = \oint_{\mathcal{O}_{rrev}} \frac{\delta A(\Theta)}{k\Theta} = \int_{T_0}^{T_W} dH(\Theta) = \frac{2\Delta Q_W}{k{T_W}^2} \cdot (T_W - T_0) \tag{145}$$

$$H(Y) = \frac{S(T_W)}{k} = \int_0^{T_W} \frac{\delta Q_W(\Theta_W)}{k\Theta_W} = \int_0^{T_W} dH(\Theta) = \frac{2\Delta Q_W}{kT_W} \quad [H(X) = H(Y) \cdot \eta_{\max}]$$

$$H(Y|X) = H(Y) - H(X) = \int_0^{T_W} dH(\Theta) - \int_{T_0}^{T_W} dH(\Theta) = \int_0^{T_0} dH(\Theta)$$

$$= \int_0^{T_0} \frac{\partial Q_W(\Theta)}{k\Theta} = \frac{S(T_0)}{k} = \frac{2\Delta Q_W}{k{T_W}^2} \cdot T_0 = \frac{2\Delta Q_W}{kT_W} \cdot \frac{T_0}{T_W} = H(X) \cdot \beta$$

$$T(Y;X) = H(Y) - H(Y|X) = \int_0^{T_W} dH(\Theta) - \int_0^{T_0} dH(\Theta) = \int_{T_0}^{T_W} dH(\Theta)$$

$$= \oint_{\mathcal{O}_{rrev}} \frac{\delta A(\Theta)}{k\Theta} = \frac{2\Delta Q_W}{kT_W} \cdot \eta_{\max} = H(Y) \cdot \eta_{\max} = H(X) = \Delta I$$

Obviously, $T(X;Y) = H(X) - H(X|Y) = T(Y|X)$. Further it is obvious that $T(X;Y)$ is the capacity C_{T_W,T_0} of the channel $\mathbf{K} \cong \mathcal{L}$ too,

$$C_{T_W,T_0} = T(X;Y) = H(X) = \oint_{\mathcal{O}_{rrev}} \frac{\delta A(\Theta)}{k\Theta} = \int_{T_0}^{T_W} \frac{\delta Q_W(\Theta)}{k\Theta} \tag{146}$$

$$= \frac{2\Delta Q_W}{k{T_W}^2} \cdot (T_W - T_0),\ C^{\max}_{T_W,T_0} = H(Y)$$

[17] In reality for the least elementary heat change $\delta Q = \hbar\nu$ is right where $\hbar = \frac{h}{2\pi}$ and h is ***Planck** constant.*

[18] It is provable that the Carnot cycle itself is elenentary, *not dividible* [18].

7.1. Triangular heat cycle

Elementary change $d\Theta$ of temperature Θ of the *environment* of the general continuous cycle $\mathcal{O}$ and thus of its working *medium* $\mathcal{L}$ (both are in the *diathermic contact at* Θ) causes the elementary reversible change of the heat $Q*(\Theta)$ *delivered, (radiated)* into $\mathcal{L}$, just about the value $\delta Q*(\Theta)$,

$$\delta Q*(\Theta) = \frac{\partial Q*(\Theta)}{\partial \Theta} d\Theta, \quad Q*(\Theta_W) = \int_0^{\Theta_W} \frac{\partial Q*(\Theta)}{\partial \Theta} d\Theta \tag{147}$$

The heat $Q*(\Theta_W)$ is the whole heat delivered (reversibly) into $\mathcal{L}$ (at the *end* temperature Θ_W). For the infinitezimal heat $\delta Q*(\Theta)$ delivered (reversibly) into $\mathcal{L}$ at temperature Θ and in accordance with the *Clausius* definition of heat entropy $S*_{\mathcal{L}}$ [22] is valid that

$$\delta Q*(\Theta) = \Theta \cdot dS*_{\mathcal{L}}(\Theta), \quad dS*_{\mathcal{L}}(\Theta) = \frac{\delta Q*(\Theta)}{\Theta} \tag{148}$$

For the whole change of entropy $\Delta S*_{\mathcal{L}}(\Theta_W)$, or for the entropy $S*_{\mathcal{L}}(\Theta_W)$ respectively, delivered into the medium $\mathcal{L}$ by its heating within the temperature interval $(0, \Theta_W\rangle$, is valid that

$$\Delta S*_{\mathcal{L}}(\Theta_W) = \int_0^{\Theta_W} dS*_{\mathcal{L}}(\Theta) = \int_0^{\Theta_W} \frac{\delta Q*(\Theta)}{\Theta} = \int_0^{\Theta_W} \frac{\frac{\partial Q*(\Theta)}{\partial \Theta} d\Theta}{\Theta} = S*_{\mathcal{L}}(\Theta_W) \tag{149}$$

when $S*_{\mathcal{L}}(0) \overset{\text{Def}}{=} 0$ is set down. By (148) for the whole heat $Q*(\Theta_W)$ deliverd into $\mathcal{L}$ within the temperature interval $\Theta \in (0, \Theta_W\rangle$ also is valid that

$$Q*(\Theta_W) = \int_0^{\Theta_W} \Theta dS*_{\mathcal{L}}(\Theta) \tag{150}$$

Then, by *medium value* theorem[19] is valid that $\overline{\Theta_{(0,\Theta_W)}} = \frac{0 + \Theta_W}{2} = \frac{\Theta_W}{2}$ and

$$Q*(\Theta_W) = \int_{S*_{\mathcal{L}}(0)}^{S*_{\mathcal{L}}(\Theta_W)} \Theta dS*_{\mathcal{L}}(\Theta) = [S*_{\mathcal{L}}(\Theta_W) - S*_{\mathcal{L}}(0)] \cdot \overline{\Theta_{(0,\Theta_W)}} \tag{151}$$

For the extremal values T_0 a T_W of the cooler temperature Θ of $\mathcal{O}$ and by (151)

$$Q*_0 \overset{\triangle}{=} Q*(T_0) = \int_0^{T_0} \delta Q*_W(\Theta) \quad \text{and} \quad Q*_W \overset{\triangle}{=} Q*(T_W) = \int_0^{T_W} \delta Q*_W(\Theta) \tag{152}$$

$$Q*_0 = \int_{S*_{\mathcal{L}}(0)}^{S*_{\mathcal{L}}(T_0)} \Theta dS*_{\mathcal{L}}(\Theta) = [S*_{\mathcal{L}}(T_0) - S*_{\mathcal{L}}(0)] \cdot \overline{\Theta_{(0,T_0)}}, \quad \overline{\Theta_{(0,T_0)}} = \frac{T_0}{2}$$

$$Q*_W = \int_{S*_{\mathcal{L}}(0)}^{S*_{\mathcal{L}}(T_W)} \Theta dS*_{\mathcal{L}}(\Theta) = [S*_{\mathcal{L}}(T_W) - S*_{\mathcal{L}}(0)] \cdot \overline{\Theta_{(0,T_W)}}, \quad \overline{\Theta_{(0,T_W)}} = \frac{T_W}{2}$$

With $S*_{\mathcal{L}}(0) = 0$ for the (end) temperatures Θ, T_0, T_W of $\mathcal{L}$ and the relevant heats and their entropies is valid

$$Q*(\Theta) = S*_{\mathcal{L}}(\Theta) \cdot \frac{\Theta}{2} \quad \text{and then} \quad S*_{\mathcal{L}}(\Theta) = \frac{2Q*(\Theta)}{\Theta} \tag{153}$$

$$Q*_W = S*_{\mathcal{L}}(T_W) \cdot \frac{T_W}{2} \quad \text{and then} \quad S*_{\mathcal{L}}(T_W) = \frac{2Q*_W}{T_W}$$

$$Q*_0 = S*_{\mathcal{L}}(T_0) \cdot \frac{T_0}{2} \quad \text{and then} \quad S*_{\mathcal{L}}(T_0) = \frac{2Q*_0}{T_0}$$

[19] Of Integral Calculus.

For the change $\Delta S*_{\mathcal{L}}$ of the thermodynamic entropy $S*_{\mathcal{L}}$ of the system $\mathcal{L}$, at the tememperature Θ running through the interval $\langle T_0, T_W \rangle$, by gaining heat from its environment (the environment of the cycle $\mathcal{O}$), is valid

$$\Delta S*_{\mathcal{L}} = S*_{\mathcal{L}}(T_W) - S*_{\mathcal{L}}(T_0) = 2\left(\frac{Q*_W}{T_W} - \frac{Q*_0}{T_0}\right) = \int_{T_0}^{T_W} \frac{\delta Q*(\Theta)}{\Theta} \tag{154}$$

By (149) for the entropy $S*_{\mathcal{L}}(\Theta_W)$ of $\mathcal{L}$ at variable temperature $\Theta \in \langle 0, \Theta_W \rangle$, $\Theta_W \leq T_W$, is gained that

$$S*_{\mathcal{L}}(\Theta_W) = \int_0^{\Theta_W} \left(\frac{\partial}{\partial\Theta}\left[\frac{S*_{\mathcal{L}}(\Theta)\cdot\Theta}{2}\right]\right)\frac{d\Theta}{\Theta} = 2\cdot\frac{1}{2}\int_0^{\Theta_W} dS*_L(\Theta) \tag{155}$$
$$= \frac{1}{2}\int_0^{\Theta_W} S*'_{\mathcal{L}}(\Theta)d\Theta + \frac{1}{2}\int_0^{\Theta_W} S*_{\mathcal{L}}(\Theta)\frac{d\Theta}{\Theta}$$

and then

$$S*_{\mathcal{L}}(\Theta_W) = \int_0^{\Theta_W} S*_{\mathcal{L}}(\Theta)\frac{d\Theta}{\Theta} = \int_0^{\Theta_W} dS*_{\mathcal{L}}(\Theta) \quad \left[= \frac{2Q*(\Theta_W)}{\Theta_W}\right]$$

and thus
$$S*_{\mathcal{L}}(\Theta)\frac{d\Theta}{\Theta} = dS*_{\mathcal{L}}(\Theta) \tag{156}$$

By the result of derivation (155)-(156) for an arbitrary temperature Θ of medium $\mathcal{L}$ is valid that[20]

$$S*_{\mathcal{L}}(\Theta) = l\cdot\Theta, \quad l = \frac{2Q*(\Theta)}{\Theta^2} \quad\longrightarrow\quad Q*(\Theta) = \lambda\cdot\Theta^2, \quad \lambda = \frac{l}{2} \tag{157}$$

Obviously, from (154) for $\Theta \in \langle T_0, T_W \rangle$ is derivable that

$$\Delta S*_{\mathcal{L}} = l\cdot(T_W - T_0) = l\cdot T_W\cdot(1-\beta), \quad \beta = \frac{T_0}{T_W} \tag{158}$$

Let such a reverse cycle is given that the medium $\mathcal{L}$ of which takes, through the elementary isothermal expansions at temperatures $\Theta \in \langle T_0, T_W \rangle$, the whole heat ΔQ_0

$$\Delta Q_0 = \int_{T_0}^{T_W} \delta Q*(\Theta) = \int_{T_0}^{T_W} l\Theta d\Theta = \frac{l}{2}\cdot\left({T_W}^2 - T_0^2\right) = Q*_W - Q*_0 \tag{159}$$

or, with medium values

$$\Delta Q_0 = \int_{T_0}^{T_W} \delta Q*(\Theta) = \int_{S*_{\mathcal{L}}(T_0)}^{S*_{\mathcal{L}}(T_W)} \Theta dS*_{\mathcal{L}}(\Theta)$$
$$= \overline{\Theta_{W(T_0,T_W)}}\cdot[S*_{\mathcal{L}}(T_W) - S*_{\mathcal{L}}(T_0)] = \frac{T_W + T_0}{2}\cdot 2\left[\frac{Q*_W}{T_W} - \frac{Q*_0}{T_0}\right]$$

and thus equivalently

$$\Delta Q_0 = (T_W + T_0)\cdot\lambda\cdot[T_W - T_0] = \lambda {T_W}^2\cdot(1-\beta^2), \quad \beta = \frac{T_0}{T_W}$$

For a reverse reversible Carnot cycle $\mathcal{O}'_{rrev}$, equivalent with the just considered general continuous heat cycle $\mathcal{O}$, drawing up the same heat ΔQ_0, consumpting the same mechanical

[20] If $\int \frac{f(x)}{x}dx = \int df(x)$, or $\frac{df(x)}{f(x)} = \frac{dx}{x}$, then $\ln|f(x)| = \ln|x| + \ln L$, $L > 0$, $\ln|f(x)| = \ln(L\cdot|x|)$, $f(x) = l\cdot x$, $l \in \mathbb{R}$.

work ΔA and giving, at its higher temperature t_W (the average temperature of the heater of our general cycle), the same heat ΔQ_W, is valid that $\Delta Q_0 = \Delta Q_W \cdot \gamma$ where $\gamma = \frac{\frac{T_0+T_W}{2}}{t_W}$ is the *transform ratio.*

Then

$$\Delta Q_W = \Delta Q_0 \cdot \frac{2t_W}{T_0+T_W} = \frac{l}{2}\cdot\left(T_W{}^2 - T_0{}^2\right)\cdot\frac{2t_W}{T_0+T_W} = l\cdot t_W\cdot(T_W - T_0) \tag{160}$$

$$\Delta A = \Delta Q_W \cdot (1-\gamma) = l\cdot t_W \cdot (T_W - T_0)\cdot\left(1 - \frac{T_0+T_W}{2t_W}\right) \tag{161}$$

$$= \frac{l}{2}\cdot(T_W - T_0)\cdot(2t_W - T_0 - T_W)$$

For the elementary work $\delta A(\cdot,\cdot)$ corresponding with the heat $Q*(\Theta)$ pumped out (reversibly) from $\mathcal{L}$ at the (end, output) temperature Θ of $\mathcal{L}$ and for the entropy $S*_{\mathcal{L}}(\Theta)$ of the whole environment of $\mathcal{O}$ (including $\mathcal{L}$ with $\mathcal{O}$) is valid

$$S*_{\mathcal{L}}(\Theta)\frac{\mathrm{d}\Theta}{\Theta} = \frac{Q*(\Theta)\frac{\mathrm{d}\Theta}{\Theta}}{\Theta} \stackrel{\triangle}{=} \frac{\delta A(\Theta,\mathrm{d}\Theta)}{\Theta} = \mathrm{d}S*_{\mathcal{L}}(\Theta) = l\cdot\mathrm{d}\Theta \tag{162}$$

$$\delta A(\Theta,\mathrm{d}\Theta) = S*_{\mathcal{L}}(\Theta)\mathrm{d}\Theta = l\cdot\Theta\mathrm{d}\Theta \text{ and } \delta A(\mathrm{d}\Theta,\mathrm{d}\Theta) = l\cdot\mathrm{d}\Theta\mathrm{d}\Theta = \mathrm{d}S*_{\mathcal{L}}(\Theta)\mathrm{d}(\Theta) \stackrel{\triangle}{=} \delta A$$

$$\left(\int_{T_0}^{\Theta} l\mathrm{d}\theta\right)\mathrm{d}\Theta = l\cdot(\Theta - T_0)\mathrm{d}\Theta \stackrel{\triangle}{=} \delta A(\Theta,\mathrm{d}\Theta;T_0) \tag{163}$$

For the whole work $\Delta A(\Theta_W;T_0)$ consumpted by the general reverse cycle $\mathcal{O}$ between temperatures T_0 and Θ_W, being coverd by elementary cycles (162), is valid that

$$\Delta A(\Theta_W;T_0) = \int_{T_0}^{\Theta_W}\left[\int_{T_0}^{\Theta}\mathrm{d}S*_{\mathcal{L}}(\theta)\right]\mathrm{d}\Theta = \int_{T_0}^{\Theta_W}\left[\int_{T_0}^{\Theta} l\mathrm{d}\theta\right]\mathrm{d}\Theta = l\cdot\int_{T_0}^{\Theta_W}(\Theta - T_0)\mathrm{d}\Theta \tag{164}$$

$$= \frac{l}{2}\cdot(\Theta_W{}^2 - T_0{}^2) - l\cdot T_0(\Theta_W - T_0) = \frac{l}{2}\cdot\Theta_W{}^2 + \frac{l}{2}\cdot T_0{}^2 - \frac{2l}{2}\cdot T_0\Theta_W = \lambda\cdot(\Theta_W - T_0)^2$$

$$\stackrel{\triangle}{=} \oint_{\mathcal{O}_{(\Theta_W,T_0)}} \delta A = \lambda\cdot T_W{}^2\cdot(1-\beta)^2, \text{ when it is valid that } \Theta_W = T_W,\ \beta = \frac{T_0}{T_W}$$

But, then for the results for ΔA in (160), (161) and (164) follows that

$$t_W = \frac{T_W + \Theta_W}{2} = T_W \text{ and then } t_W = T_W = \text{const.} \tag{165}$$

Thus our general cycle $\mathcal{O}$ is of a triangle shape, $\mathcal{O} \stackrel{\triangle}{=} \mathcal{O}_{rrev\triangle}$ with the *apexes*

$$[lT_0, T_0],\ [lT_W, T_W],\ [lT_0, T_W] \tag{166}$$

and its efficiency is $1 - \gamma = \frac{1-\beta}{2} = \frac{1}{2}\cdot\eta_{\max}$. Thus, the return to the *initial (starting) state* of the medium $\mathcal{L}$ is possible by using the oriented *abscissas* (in the $S - T$ diagram)

$$\overrightarrow{[lT_W, T_W], [lT_0, T_W]} \text{ and } \overrightarrow{[lT_0, T_W], [lT_0, T_0]} \tag{167}$$

For works $\delta A(\Theta, \mathrm{d}\Theta; T_0)$ of elementary Carnot cycles covering cycle $\mathcal{O}_{rrev\triangle}$ (166), the range of their working temperatures is $\mathrm{d}\Theta$ and, for the given heater temperature $\Theta \in \langle T_0, \Theta_W \rangle$ is, by (162)-(163) valid that

$$\begin{aligned} \delta A(\Theta, \mathrm{d}\Theta; T_0) &= \Delta Q_W(\Theta) \cdot \frac{\mathrm{d}\Theta}{\Theta} = l \cdot \Theta \cdot (\Theta - T_0) \frac{\mathrm{d}\Theta}{\Theta} = l \cdot (\Theta - T_0)\mathrm{d}\Theta \\ &= [S*_{\mathcal{L}}(\Theta) - S*_{\mathcal{L}}(T_0)] \cdot \mathrm{d}\Theta \\ \Delta Q_W(\Theta) &= l\Theta\,(\Theta - T_0) \;\; \text{and for } \gamma \text{ used in (160) is gained that} \\ \Delta Q_0(\Theta) &= \Delta Q_W(\Theta) \cdot \gamma(\Theta) = l \cdot \Theta \cdot (\Theta - T_0) \cdot \frac{\Theta + T_0}{2\Theta} = \lambda \cdot \left(\Theta^2 - {T_0}^2\right) \end{aligned} \tag{168}$$

For the whole heats ΔQ_0 a ΔQ_W being changed mutually between the working medium $\mathcal{L}$ of the whole triangular cycle $\mathcal{O}_{rrev\triangle}$ and its environment (166), and for the work ΔA, in its equivalent Carnot cycle $\mathcal{O}'_{rrev}$ with working temperatures $\frac{T_0 + T_W}{2}$ and T_W, will be valid that[21]

$$\Delta Q_0 = \int_{T_0}^{T_W} \left[\int_0^{\Theta} l \mathrm{d}\theta \right] \mathrm{d}\Theta = \frac{l}{2} \cdot {T_W}^2 \cdot (1 - \beta^2) \triangleq W \left[= l \cdot (T_W - T_0) \cdot T_0 + \frac{l}{2} \cdot (T_W - T_0)^2 \right] \tag{169}$$

$$\begin{aligned} \Delta Q_W &= \int_{T_0}^{T_W} \left[\int_0^{T_W} l \mathrm{d}\theta \right] \mathrm{d}\Theta = l \cdot (T_W - T_0) \cdot T_W = l \cdot {T_W}^2 \cdot (1 - \beta) \\ \Delta A &= \frac{1}{2} \int_{T_0}^{T_W} \left[\int_{T_0}^{T_W} l \mathrm{d}\theta \right] \mathrm{d}\Theta = \cdot l {T_W}^2 \cdot (1 - \beta) - l \cdot {T_W}^2 \cdot (1 - \beta) \\ &= l \cdot {T_W}^2 \cdot (1 - \beta) \cdot [1 - \frac{l}{2}(1 + \beta)] = \frac{l}{2} \cdot {T_W}^2 \cdot (1 - \beta)^2 = \oint_{\mathcal{O}_{rrev\triangle}} \delta A \end{aligned} \tag{170}$$

7.2. Capacity corections for wide–band photonic transfer channel

The average *output* energy $P_2(\Theta_W)$ of the message being received within interval $(0, T_W\rangle$ of the temperature Θ of the medium $\mathcal{L} \cong \mathbf{K}_{\mathrm{L-L}}$ from [30], when $0 < \Theta_0 \leq \Theta \leq \Theta_W$ and $\Theta_0 \leq T_0$ and $\Theta_W \leq T_W$ are valid, is given by the sum of the *input* average energy $W(\Theta_W, \Theta_0)$ and the average energy $P_1(\Theta_0)$ of the *additive noise*

$$P_2(\Theta_W) = P_1(\Theta_0) + W(\Theta_W, \Theta_0) \tag{171}$$

The output message bears the *whole average output information* $H_2(\Theta_W)$. By the medium value theorem is possible, for a certain maximal temperature $\Theta_W \leq T_W$ of the temperature $\Theta \in (0, \Theta_W\rangle$, consider that the receiving of the output message is performed at the *average (constant) temperature* $\overline{\Theta_W} = \frac{\Theta_W}{2}$. Then for the whole change of the output information entropy $\Delta H_2 \triangleq H_2(\Theta_W)$ [the thermodynamic entropy $S*_{\mathcal{L}}(\Theta_W)$ in information units] is valid

$$H_2(\Theta_W) = \frac{P_2(\Theta_W)}{\mathrm{k}\overline{\Theta_W}} = \frac{P_1(\Theta_0) + W(\Theta_W, \Theta_0)}{\mathrm{k}\overline{\Theta_W}} \triangleq H_1(\Theta_W, \Theta_0) + H[W(\Theta_W, \Theta_0)] \tag{172}$$

---

[21] Further it willbe layed down $\lambda = \frac{\pi^2 k^2}{6\hbar}$, $l = \frac{\pi^2 k^2}{3\hbar} = 2 \cdot \lambda$.

By (153) is valid that $Q*(\Theta) = \lambda\Theta^2$ and $\delta Q*(\Theta) = l\Theta d\Theta$. Thus for $\Theta \in (0, T_0\rangle$ a $\Theta \in (0, T_W\rangle$ is valid

$$dH_{\Theta_W}(Y) = \frac{\delta Q*(\Theta_W)}{k\Theta_W} = \frac{l}{k}d\Theta_W, \; dH_{\Theta_W}(Y|X) = \frac{2l\Theta d\Theta}{k\Theta_W} \tag{173}$$

With $\Theta_W = T_W$ and $\Theta_0 = T_0$ and with the reducing temperature $\frac{T_W}{2}$ is possible to write

$$P_2 = W + P_1 \triangleq Q*_W = \lambda T_W{}^2 \triangleq Y \tag{174}$$

$$H_2 \triangleq H(Y) = \int_0^{T_W} \frac{l}{k} d\Theta_W = \frac{l}{k} T_W = \frac{P_2}{k\frac{T_W}{2}} = \frac{2(W+P_1)}{kT_W} = \frac{2\lambda T_W{}^2}{kT_W} = \frac{lT_W}{k}$$

$$P_1 \triangleq Q*_0 = \lambda T_0{}^2 \triangleq Y|X$$

$$H_1 \triangleq H(Y|X) = \int_0^{T_0} \frac{2l\Theta d\Theta}{kT_W} = \frac{l}{kT_W} \cdot T_0{}^2 = \frac{P_1}{k\frac{T_W}{2}} = \frac{2\lambda T_0{}^2}{kT_W} = \frac{lT_0{}^2}{kT_W}$$

$$W = Q*_W - Q*_0 \triangleq X$$

$$H[W(T_W, T_0)] \triangleq H(X) = \frac{W}{k\frac{T_W}{2}} = \frac{2\lambda}{kT_W} \cdot (T_W{}^2 - T_0{}^2) \tag{175}$$

By the channel equation (4), (5) and by equations (23)-(24) and also by definitions (174)-(175) and with the loss entropy $H(X|Y) = 0$ it must be valid for the transinformation $T(\cdot;\cdot)$ that

$$T(Y;X) = H(Y) - H(Y|X) = \frac{l}{k} \cdot (T_W - T_0) \cdot (1+\beta) = H(X) \tag{176}$$

$$T(X;Y) = H(X) - H(X|Y) = H(X) = T(X,Y) \text{ and by using } l = \frac{\pi^2 k^2}{3\hbar}, \; \beta = \frac{T_0}{T_W},$$

$$T(Y;X) = \frac{\pi^2 k}{3h} \cdot T_W \cdot (1-\beta^2) = \frac{\pi^2 k}{3h} \cdot T_W \cdot (1-\beta) \cdot (1+\beta) = C_{T_0,W}(\mathbf{K}_{L-L}) \cdot (1+\beta)$$

For the given extremal temperatures T_0, T_W the value $T(X;Y)$ stated this way is the only one, and thus also, it is the information capacity $C'_{T_0,T_W}(W)$ of the channel $\mathbf{K}_{L-L}$ (the *first* correction)

$$C'_{T_0,T_W} = (W) = T(X;Y) = \frac{\pi^2 k T_0}{3h} \cdot \left(\sqrt{1 + \frac{6h \cdot W}{\pi^2 k^2 T_0{}^2}} - 1\right) \cdot (1+\beta) \tag{177}$$

The information capacity correction (177) of the wide–band photonic channel $\mathbf{K}_{L-L}$ [30], stated this way, is $(1+\beta)$-times *higher* than the formulas (102) and (138) say. The reason is in using two different information descriptions of the oriented abscissa $\overrightarrow{[l0,0],[lT_W,T_W]}$ in derivation (138) and (177) which abscissa $\overrightarrow{[l0,0],[lT_W,T_W]}$ is on one line in $S-T$ diagram and is composed from two oriented abscissas,

$$\overrightarrow{[l0,0],[lT_0,T_0]} \text{ and } \overrightarrow{[lT_0,T_0],[lT_W,T_W]} \tag{178}$$

The first abscissa represents the *phase of noise generation* and the second one the *phase of input signal generation*. The whole composed abscissa represents the *phase of whole output signal generation*.

For the **sustaining, in the sense repeatable, cyclical information transfer, the renewal of the initial or starting state of the transfer channel $\mathbf{K}_{L-L} \cong \mathcal{L}$, after any *individual information transfer act* - the *sending input and receiving output message* has been accomplished, is needed.**
Nevertheless, in derivations of the formulas (102), (177) and (138) **this return of the physical medium $\mathcal{L}$**, after accomplishing any individual information transfer act, into the starting state **is either not considered, or, on the contrary, is considered, but by that the whole transfer chain is opened *to cover the energetic needs for this return transition from another, outer resources* than from those ones within the transfer chain itself**. In both these two cases the channel equation is fulfilled. This enables any individual act of information transfer be realized by *external and forced out, repeated starting* of each this individual transfer act.[22] [23]

If for the creation of a cycle the resources of the transfer chain are used only, the need for *another correction*, this time in (177) arises. To express it it will be used the *full cyclical thermodynamic analogy* $\mathbf{K}$ of $\mathbf{K}_{L-L}$ used *cyclically*, $\mathbf{K}'_{L-L}$. The information transfer will be modeled by the cyclical thermodynamic process $\mathcal{O}_{rrev\triangle}$ of reversible changes in the channel $\mathbf{K} \cong \mathbf{K}'_{L-L} \cong \mathcal{L}$ and without opening the transfer chain. (Also $\mathbf{K} \cong \mathbf{K}'_{B-E}$).

7.2.1. Return of transfer medium into initial state, second correction

Now the further correction for capacity formulas (102), (138) and (177) will be dealt with for that case that the *return* of the medium $\mathcal{L}$ into its initial, starting state is performed *within the transfer chain* only. It will be envisiged by a triangular reverse heat cycle $\mathcal{O}_{rrev\triangle}$ created by the oriented abscissas within the apexes in the $S-T$ diagram (166), $[lT_0, T_0]$, $[lT_W, T_W]$, $[lT_0, T_W]$. The abstract experiment from [30] will be now, formally and as an analogy, realized by this *reverse* and reversible heat cycle $\mathcal{O}_{rrev\triangle} \equiv \mathcal{O}'_{rrev}$, described informationaly, and thought as modeling information transfer process in a channel $\mathbf{K} \cong \mathbf{K}'_{L-L|B-E} \cong \mathcal{L}$. Thus the denotation $\mathbf{K} \equiv \mathbf{K}_\triangle$ is usable. By (153)-(157) it will be

$$Q*(\Theta) = \frac{\pi^2 k^2 \Theta^2}{6h} = \frac{l}{2}\cdot\Theta^2,\ Q*_W = \frac{l}{2}\cdot T_W{}^2,\ Q*_0 = \frac{l}{2}\cdot T_0{}^2 \tag{179}$$

The working temperature Θ_0 of cooling and Θ_W of heating are changing by (157),

$$\Theta_0 = \frac{1}{l}\cdot S*_{\mathcal{L}}(\Theta_0) \in \langle T_0, T_W\rangle \text{ and } \Theta_W = T_W = \text{const.} \tag{180}$$

and the heat entropy $S*_{\mathcal{L}}(\Theta)$ of the medium $\mathcal{L}$ is changing by (155)-(156),

$$dS*_{\mathcal{L}} = \frac{\partial Q*(\Theta)}{\partial\Theta}d\Theta\cdot\frac{1}{\Theta} \text{ and then } S*_{\mathcal{L}}(\Theta) = l\cdot\Theta = \frac{2Q*(\Theta)}{\Theta} \tag{181}$$

Using integral (149) it is possible to write that

$$Q*(\Theta) = \int_0^{\Theta}\delta Q*(\theta) = \int_0^{\Theta}\frac{\partial Q*(\theta)}{\partial\theta}d\theta = \int_0^{\Theta} l\theta d\theta \tag{182}$$

---

[22] For these both cases is not possible to construct a construction-relevant heat cycles *described in a proper information way*.

[23] But the modeling by the direct cycle such as in (128) is possible for the *II. Principle of Thermodynamics* is valid in any case and giving the possibility of the cycle description.

For the whole heats ΔQ_0 and ΔQ_W being changed mutually between $\mathcal{L}$ with the cycle $\mathcal{O}_{rrev\triangle}$ and its environment and, for the whole work ΔA for the equivalent Carnot cycle $\mathcal{O}'_{rrev}$ with working temperatures $\frac{T_0+T_W}{2}$ and T_W is valid, by (169)-(170), that

$$W = \Delta Q_0 = \frac{l}{2}\cdot T_W{}^2\cdot(1-\beta^2) \overset{\triangle}{=} X, \quad \Delta Q_W = l\cdot T_W{}^2\cdot(1-\beta) \overset{\triangle}{=} Y \tag{183}$$
$$\Delta Q_W - \Delta Q_0 = \Delta A = \frac{l}{2}\cdot T_W{}^2\cdot(1-\beta)^2 \overset{\triangle}{=} Y|X$$

For the whole work ΔA delivered into the cycle $\mathcal{O}_{rrev\triangle}$, at the temperature T_W, and the entropy $S*_{\mathcal{L}}$ of its working medium $\mathcal{L}$ is valid

$$\frac{\Delta A}{T_W} = \oint_{\mathcal{O}_{rrev\triangle}} \frac{\delta A}{T_W} = \int_{T_0}^{T_W} l(\Theta - T_0)\cdot\frac{\mathrm{d}\Theta}{T_W} \tag{184}$$
$$\oint_{\mathcal{O}_{rrev\triangle}} \frac{\delta A}{T_W} = \frac{1}{2}\int_{T_0}^{T_W}\left[\int_{T_0}^{T_W} \mathrm{d}S*_{\mathcal{L}}(\theta)\right]\frac{\mathrm{d}\Theta}{T_W} = \int_{T_0}^{T_W}\left[S*_{\mathcal{L}}(T_W) - S*_{\mathcal{L}}(T_0)\right]\frac{\mathrm{d}\Theta}{2T_W}$$
$$= \frac{l}{2T_W}\cdot(T_W - T_0)^2 = \frac{l}{2}\cdot T_W\cdot(1-\beta)^2 = \frac{\Delta A}{T_W}$$

Following (4), (5) and (23) and the triangular shape of the cycle $\mathcal{O}_{rrev\triangle}$, the changes of information entropies by expressions (142), (169)-(170) are defined, valid for the equivalent $\mathcal{O}'_{rrev}$[24], see (142),

$$H(X) \overset{\text{Def}}{=} \frac{\Delta Q_0}{\mathrm{k}T_W} = \int_{T_0}^{T_W}\left[\int_0^{\Theta}\frac{\delta Q*(\theta)}{\theta}\right]\frac{\mathrm{d}\Theta}{\mathrm{k}T_W} = \int_{T_0}^{T_W}\left[\int_0^{\Theta} l\mathrm{d}`\right]\frac{1}{\mathrm{k}T_W}\mathrm{d}\Theta \tag{185}$$
$$H(Y) \overset{\text{Def}}{=} \frac{\Delta Q_W}{\mathrm{k}T_W} = \int_{T_0}^{T_W}\left[\int_0^{T_W}\frac{\delta Q*(\theta)}{\theta}\right]\frac{\mathrm{d}\Theta}{\mathrm{k}T_W} = \int_{T_0}^{T_W}\left[\int_0^{T_W} l\mathrm{d}`\right]\frac{1}{\mathrm{k}T_W}\mathrm{d}\Theta$$
$$H(Y|X) \overset{\text{Def}}{=} \frac{\Delta A}{\mathrm{k}T_W} = \frac{1}{2}\int_{T_0}^{T_W}\left[\int_{T_0}^{T_W}\frac{\delta Q*(\theta)}{\theta}\right]\frac{\mathrm{d}\Theta}{\mathrm{k}T_W} = \frac{1}{2}\int_{T_0}^{T_W}\left[\int_{T_0}^{T_W} l\mathrm{d}\theta\right]\frac{1}{\mathrm{k}T_W}\mathrm{d}\Theta$$
$$T(Y;X) = H(Y) - H(Y|X) = \frac{l}{2\mathrm{k}T_W}\int_{T_0}^{T_W}\left[2\int_0^{T_W}\mathrm{d}\theta - \int_{T_0}^{T_W}\mathrm{d}\theta\right]\mathrm{d}\Theta$$

and by figuring these formulas with $l = \frac{\pi^2 k}{3h}$ is gained that

$$H(X) = \frac{l}{2\mathrm{k}}\cdot T_W\cdot(1-\beta^2) = \frac{\pi^2\mathrm{k}T_W}{6h}\cdot(1-\beta^2)$$
$$H(Y) = \frac{l}{\mathrm{k}}\cdot T_W\cdot(1-\beta) = \frac{\pi^2\mathrm{k}T_W}{3h}\cdot(1-\beta)$$
$$H(Y|X) = \frac{l}{2\mathrm{k}}\cdot T_W\cdot(1-\beta)^2 = \frac{\pi^2\mathrm{k}T_W}{6h}\cdot(1-\beta)^2 = \oint_{\mathcal{O}_{rrev}}\frac{\delta A}{\mathrm{k}T_W}$$
$$T(Y;X) = \frac{l}{\mathrm{k}}\cdot T_W\cdot(1-\beta) - \frac{l}{2\mathrm{k}}\cdot T_W\cdot(1-\beta)^2 = \frac{l}{2\mathrm{k}}\cdot T_W\cdot(1-\beta)\cdot(2-1+\beta)$$
$$= \frac{l}{2\mathrm{k}}\cdot T_W\cdot(1-\beta^2) = \frac{\pi^2\mathrm{k}T_W}{6h}\cdot(1-\beta^2) = H(X) = T(X;Y)$$
$$H(X|Y) = 0$$

[24] In accordance with the input energy delivered and the extremal temperatures used in [30].

It is visible that the quantity $H(Y)$ $[= H(X) + H(Y|X)]$ is introduced correctly, for by (185) is valid that

$$H(Y) = \frac{l}{2\mathrm{k}} \cdot T_W \cdot (1-\beta)^2 + \frac{l}{2\mathrm{k}} \cdot T_W \cdot (1-\beta^2) = \frac{l}{2\mathrm{k}} \cdot T_W \cdot (1-\beta) \cdot (1-\beta+1+\beta) \quad (186)$$
$$= \frac{l}{\mathrm{k}} \cdot T_W \cdot (1-\beta)$$

For the transinformation and the information capacity of the transfer organized this way is valid (177),

$$T(X;Y) = \frac{1}{2} C'_{T_0,T_W}(W) \quad (187)$$

With the extremal temperatures T_0 and T_W the information capacity $C*_{T_0,T_W}(W)$ is given by

$$T(X;Y) = C*_{T_0,T_W}(W) \text{ and then } C^{\max} = \lim_{T_0 \longrightarrow T_W} C*_{T_0,T_W}(W) = H(Y) \quad (188)$$

From the difference $\Delta Q_0 \stackrel{\triangle}{=} W = Q*_0 - Q*_W$ (in $\mathcal{L}$) follows that the temperature $T_W = T_0 \cdot \sqrt{1 + \frac{6h \cdot W}{\pi^2 \mathrm{k}^2 T_0{}^2}}$.
Then, for the transinformation, in the same way as in (187), is now valid

$$T(X;Y) = \frac{\pi^2 k}{6h} \cdot \left(1-\beta^2\right) = \frac{\pi^2 k}{3h} \cdot (T_W - T_0) \cdot \frac{1+\beta}{2} = \frac{\pi^2 \mathrm{k} T_0}{3h} \cdot \left(\sqrt{1 + \frac{6h \cdot W}{\pi^2 \mathrm{k}^2 T_0{}^2}} - 1\right) \cdot \frac{1+\beta}{2} \quad (189)$$

The transformation $T(X;Y)$ is the capacity $C(\mathbf{K}_\triangle)$ and it is possible to write

$$T(X;Y) = C(\mathbf{K}_\triangle) = C*_{T_0,T_W}(W) = \frac{\pi^2 \mathrm{k} T_0}{6hT_W} \cdot \left(\sqrt{1 + \frac{6h \cdot W}{\pi^2 \mathrm{k}^2 T_0{}^2}} - 1\right) \cdot (T_0 + T_W) \quad (190)$$
$$= C*_{T_W}(W) = \frac{W}{\mathrm{k}T_W} = C*_{T_0}(W) = \frac{W}{\mathrm{k}T_0 \cdot \sqrt{1 + \frac{6h \cdot W}{\pi^2 \mathrm{k}^2 T_0{}^2}}} \stackrel{\triangle}{=} C(\mathbf{K}'_{\mathrm{L-L|B-E}})$$

which value is $2\times$ *less* than (177) and $\frac{2}{1+\beta}\times$ *less* than (138).

For $T_0 \longrightarrow 0$ the *quantum approximation* $C(W)$ of the capacity $C*_{T_0,T_W}(W)$ is obtained, independent on the noise energy (the noise power deminishes near the abslute 0° K)

$$C(W) = \lim_{T_0 \to 0} \left(\sqrt{\frac{\pi^4 \mathrm{k}^2 T_0{}^2}{6^2 \hbar^2} + \pi^2 \frac{W}{6h}} - \frac{\pi^2 \mathrm{k} T_0}{6h}\right) \cdot (1+\beta) = \pi \cdot \sqrt{\frac{W}{6h}} \quad (191)$$

The *classical aproximation* $C_{T_0}(W)$ of $C*_{T_0,T_W}(W)$ is gained for $T_0 \gg 0$. This value is near Shannon capacity of the wide–band *Gaussian* channel with noise energy $\mathrm{k}T_0$ and with the whole average input energy (energy) W; in the same way as in (104) is now gained

$$C_{T_0}(W) \doteq \frac{\pi^2 \mathrm{k} T_0}{6h} \left(\frac{3h \cdot W}{\pi^2 \mathrm{k}^2 T_0{}^2}\right) \cdot (1+\beta) = \frac{W}{2\mathrm{k}T_0} \cdot (1+\beta) \longrightarrow \frac{W}{\mathrm{k}T_0} \quad (192)$$

The mutual difference of results (189) and and (102), (138) [12, 30] is given by **the necessity of the returning the transfer medium, the channel $\mathbf{K}_\triangle \cong \mathbf{K}'_{\mathrm{L-L}|B-E} \cong \mathcal{L}$ into its initial state after each individual information transfer act has been accomplished and, by the relevant temperatutre reducing of the heat ΔQ_0** [by T_W in (183)-(189)]. Thus, our thermodynamic cyclical model $\mathbf{K}_\triangle \cong \mathcal{O}_{rrev\triangle}$ for the repeatible information transfer through the channel $\mathbf{K}'_{\mathrm{L-L|B-E}}$ is of the information capacity (189), while in [12, 30] the information capacity of the *one-act* information transfer is stated.[25] By (189) the *whole energy costs* for the cyclical information transfer considered is countable.[26]

8. Conclusion

After each completed 'transmission of an input message and receipt of an output message' ('one-act' transfer) the transferring system must be reverted to its starting state, otherwise the constant (in the sense repeatable) flow of information could not exist. The author believes that either the opening of the chain was presupposed in the original derivation in [30], or that the return of transferring system to its starting state was not considered at all, it was not counted-in. In our derivations this needed state transition is considered be powered within the transfer chain itself, without its openning. Although our derivation of the information capacity for a cyclical case (using the cyclic thermodynamic model) results in a lower value than the original one it seems to be more exact and its result as more precise from the *theoretic point of view*, extending and not ceasing the previous, *original* result [12, 30] which *remains* of its *technology-drawing value*. Also it forces us in being aware and respecting of the *global costs* for (any) communication and its evaluation and, as such, it is of a *gnoseologic character*.

Author details

Bohdan Hejna
Institute of Chemical Technology Prague, Department of Mathematics, Studentská 6, 166 28 Prague 6, Czech Republic

9. References

[1] Bell, D. A. *Teorie informace*; SNTL: Praha, 1961.
[2] Brillouin, L. *Science and Information Theory*; Academia Press: New York, 1963.
[3] Cholevo, A. S. On the Capacity of Quantum Communication Channel. *Problems of Information Transmission* 1979, *15* (4), 3–11.
[4] Cholevo, A. S. *Verojatnostnyje i statističeskije aspekty kvantovoj teorii*; Nauka: Moskva, 1980.
[5] Cover, T. M.; Thomas, J. B. *Elements of Information Theory*; Wiley: New York, 1991.
[6] Davydov, A. S. *Kvantová mechanika*; SPN: Praha, 1978.
[7] Emch, G. G. *Algebraičeskije metody v statističeskoj mechanike i kvantovoj těorii polja*; Mir: Moskva, 1976.

[25] For one-act information transfer the choose between two information descriptions is possible which result in capacity (138) or (177).

[26] For the energy $T(X;Y)\cdot T_W$ on the output the energy $\frac{2}{1+\beta}\cdot T_W \times$ *greater* is needed on the input of the transfer channel which is in accordance with the *II. Principle of thermodynamics* for $\frac{2}{1+\beta} > 1$ is valid.

[8] Gershenfeld, N. Signal entropy and the thermodynamics of computation. *IBM Systems Journal* 1996, *35* (3/4), 577–586. DOI 10.1147/sj.353.0577.

[9] Halmos, P. R. *Konečnomernyje vektornyje prostranstva*; Nauka: Moskva, 1963.

[10] Hašek, O.; Nožička, J. *Technická mechanika pro elektrotechnické obory II.*; SNTL: Praha, 1968.

[11] Hejna, B.; Vajda, I. Information transmission in stationary stochastic systems. *AIP Conf. Proc.* 1999, *465* (1), 405–418. DOI: 10.1063/1.58272.

[12] Hejna, B. Informační kapacita stacionárních fyzikálních systému. Ph.D. Dissertation, ÚTIA AV ČR, Praha, FJFI ČVUT, Praha, 2000.

[13] Hejna, B. Generalized Formula of Physical Channel Capacities. *International Journal of Control, Automation, and Systems* 2003, 15.

[14] Hejna, B. Thermodynamic Model of Noise Information Transfer. In *AIP Conference Proceedings*, Computing Anticipatory Systems: CASYS'07 – Eighth International Conference; Dubois, D., Ed.; American Institute of Physics: Melville, New York, 2008; pp 67–75. ISBN 978-0-7354-0579-0. ISSN 0094-243X.

[15] Hejna, B. Proposed Correction to Capacity Formula for a Wide-Band Photonic Transfer Channel. In *Proceedings of International Conference Automatics and Informatics'08*; Atanasoff, J., Ed.; Society of Automatics and Informatics: Sofia, Bulgaria, 2008; pp VII-1–VIII-4.

[16] Hejna, B. Gibbs Paradox as Property of Observation, Proof of II. Principle of Thermodynamics. In *AIP Conf. Proc.*, Computing Anticipatory Systems: CASYS'09: Ninth International Conference on Computing, Anticipatory Systems, 3–8 August 2009; Dubois, D., Ed.; American Institute of Physics: Melville, New York, 2010; pp 131–140. ISBN 978-0-7354-0858-6. ISSN 0094-243X.

[17] Hejna, B. *Informační termodynamika I.: Rovnovážná termodynamika přenosu informace*; VŠCHT Praha: Praha, 2010. ISBN 978-80-7080-747-7.

[18] Hejna, B. *Informační termodynamika II.: Fyzikální systémy přenosu informace*; VŠCHT Praha: Praha, 2011. ISBN 978-80-7080-774-3.

[19] Hejna, B. Information Thermodynamics, *Thermodynamics - Physical Chemistry of Aqueous Systems*, Juan Carlos Moreno-Pirajaán (Ed.), ISBN: 978-953-307-979-0, InTech, 2011 Available from: http://www.intechopen.com/articles/show/title/information-thermodynamics

[20] Helstroem, C. W. *Quantum Detection and Estimation Theory*; Pergamon Press: London, 1976.

[21] Horák, Z.; Krupka, F. *Technická fyzika*; SNTL/ALFA: Praha, 1976.

[22] Horák, Z.; Krupka, F. *Technická fysika*; SNTL: Praha, 1961.

[23] Jaglom, A. M.; Jaglom, I. M. *Pravděpodobnost a teorie informace*; Academia: Praha, 1964.

[24] Chodasevič, M. A.; Sinitsin, G. V.; Jasjukevič, A. S. *Ideal fermionic communication channel in the number-state model*;Division for Optical Problems in Information technologies, Belarus Academy of Sciences, 2000.

[25] Kalčík, J.; Sýkora, K. *Technická termomechanika*; Academia: Praha, 1973.

[26] Karpuško, F. V.; Chodasevič, M. A. *Sravnitel'noje issledovanije skorostnych charakteristik bozonnych i fermionnych kommunikacionnych kanalov*; Vesšč NAN B, The National Academy of Sciences of Belarus, 1998.

[27] Landau, L. D.; Lifschitz, E. M. *Statistical Physics*, 2nd ed.; Pergamon Press: Oxford, 1969.

[28] Landauer, M. Irreversibility and Heat Generation in the Computing Process. *IBM J. Res. Dev.* 2000, *44* (1/2), 261.

[29] Lavenda, B. H. *Statistical Physics*; Wiley: New York, 1991.

[30] Lebedev, D.; Levitin, L. B. Information Transmission by Electromagnetic Field. *Information and Control* 1966, *9*, 1–22.
[31] Maršák, Z. *Termodynamika a statistická fyzika*; ČVUT: Praha, 1995.
[32] Marx, G. *Úvod do kvantové mechaniky*; SNTL: Praha, 1965.
[33] Moore, W. J. *Fyzikální chemie*; SNTL: Praha, 1981.
[34] Prchal, J. *Signály a soustavy*; SNTL / ALFA: Praha, 1987.
[35] Shannon, C. E. A Mathematical Theory of Communication. *The Bell Systems Technical Journal* 1948, *27*, 379–423, 623–656.
[36] Sinitsin, G. V.; Chodasevič, M. A.; Jasjukevič, A. S. *Elektronnyj kommunikacionnyj kanal v modeli svobodnogo vyroždennogo fermi-gaza*; Belarus Academy of Sciences, 1999.
[37] Vajda, I. *Teória informácie a štatistického rozhodovania*; Alfa: Bratislava, 1982.
[38] Watanabe, S. *Knowing and Guessing*; Wiley: New York, 1969.

Section 3

Property Prediction and Thermodynamics

Thermodynamic Properties and Applications of Modified van-der-Waals Equations of State

Ronald J. Bakker

Additional information is available at the end of the chapter

1. Introduction

Physical and chemical properties of natural fluids are used to understand geological processes in crustal and mantel rock. The fluid phase plays an important role in processes in diagenesis, metamorphism, deformation, magmatism, and ore formation. The environment of these processes reaches depths of maximally 5 km in oceanic crusts, and 65 km in continental crusts, e.g. [1, 2], which corresponds to pressures and temperatures up to 2 GPa and 1000 °C, respectively. Although in deep environments the low porosity in solid rock does not allow the presence of large amounts of fluid phases, fluids may be entrapped in crystals as fluid inclusions, i.e. nm to μm sized cavities, e.g. [3], and fluid components may be present within the crystal lattice, e.g. [4]. The properties of the fluid phase can be approximated with equations of state (Eq. 1), which are mathematical formula that describe the relation between intensive properties of the fluid phase, such as pressure (p), temperature (T), composition (x), and molar volume (V_m).

$$p\left(T, V_m, x\right) \tag{1}$$

This pressure equation can be transformed according to thermodynamic principles [5], to calculate a variety of extensive properties, such as entropy, internal energy, enthalpy, Helmholtz energy, Gibbs energy, et al., as well as liquid-vapour equilibria and homogenization conditions of fluid inclusions, i.e. dew point curve, bubble point curve, and critical points, e.g. [6]. The partial derivative of Eq. 1 with respect to temperature is used to calculate total entropy change (dS in Eq. 2) and total internal energy change (dU in Eq. 3), according to the Maxwell's relations [5].

$$dS = \left(\frac{\partial p}{\partial T}\right)_{V, n_T} dV \tag{2}$$

$$dU = \left[T \cdot \left(\frac{\partial p}{\partial T} \right)_{V,n_T} - p \right] dV \tag{3}$$

where n_T is the total amount of substance in the system. The enthalpy (H) can be directly obtained from the internal energy and the product of pressure and volume according to Eq. 4.

$$H = U + p \cdot V \tag{4}$$

The Helmholtz energy (A) can be calculated by combining the internal energy and entropy (Eq. 5), or by a direct integration of pressure (Eq. 1) in terms of total volume (Eq. 6).

$$A = U - TS \tag{5}$$

$$dA = -pdV \tag{6}$$

The Gibbs energy (G) is calculated in a similar procedure according to its definition in Eq. 7.

$$G = U + p \cdot V - T \cdot S \tag{7}$$

The chemical potential (μ_i) of a specific fluid component (i) in a gas mixture or pure gas (Eq. 8) is obtained from the partial derivative of the Helmholtz energy (Eq. 5) with respect to the amount of substance of this component (n_i).

$$\mu_i = \left(\frac{\partial A}{\partial n_i} \right)_{T,V,n_j} \tag{8}$$

The fugacity (f) can be directly obtained from chemical potentials (Eq. 9) and from the definition of the fugacity coefficient (φ_i) with independent variables V and T (Eq. 10).

$$RT \ln \left(\frac{f_i}{f_i^0} \right) = \mu_i - \mu_i^0 \tag{9}$$

where μ_i^0 and f_i^0 are the chemical potential and fugacity, respectively, of component i at standard conditions (0.1 MPa).

$$RT \ln \varphi_i = \int_V^\infty \left[\left(\frac{\partial p}{\partial n_i} \right)_{T,V,n_j} - \frac{RT}{V} \right] dV - RT \ln z\,. \tag{10}$$

where φ_i and z (compressibility factor) are defined according to Eq. 11 and 12, respectively.

$$\varphi_i = \frac{f_i}{x_i \cdot p} \tag{11}$$

$$z = \frac{pV}{n_T RT} \tag{12}$$

2. Two-constant cubic equation of state

The general formulation that summarizes two-constant cubic equations of state according to van der Waals [7], Redlich and Kwong [8], Soave [9], and Peng and Robinson [10] is illustrated in Eq. 13 and 14, see also [11]. In the following paragraphs, these equations are abbreviated with *Weos, RKeos, Seos,* and *PReos*.

$$p = \frac{RT}{V_m - \zeta_1} - \frac{\zeta_2}{V_m \cdot (V_m + \zeta_3) + \zeta_4 \cdot (V_m - \zeta_4)} \tag{13}$$

$$p = \frac{n_T RT}{V - n_T \zeta_1} - \frac{n_T^2 \zeta_2}{V^2 + n_T \zeta_3 V + n_T \zeta_4 V - n_T^2 \zeta_4^2} \tag{14}$$

where p is pressure (in MPa), T is temperature (in Kelvin), R is the gas constant (8.3144621 J·mol^{-1}K^{-1}), V is volume (in cm^3), V_m is molar volume (in cm^3·mol^{-1}), n_T is the total amount of substance (in mol). The parameters ζ_1, ζ_2, ζ_3, and ζ_4 are defined according to the specific equations of state (Table 1), and are assigned specific values of the two constants a and b, as originally designed by Waals [7]. The a parameter reflects attractive forces between molecules, whereas the b parameter reflects the volume of molecules.

	W	RK	S	PR
ζ_1	b	b	b	b
ζ_2	a	$a \cdot T^{-0.5}$	a	a
ζ_3	-	b	b	b
ζ_4	-	-	-	b

Table 1. Definitions of ζ_1, ζ_2, ζ_3, and ζ_4 according to van der Waals (W), Redlich and Kwong (RK), Soave (S) and Peng and Robinson (PR).

This type of equation of state can be transformed in the form of a cubic equation to define volume (Eq. 15) and compressibility factor (Eq. 16).

$$a_0 V^3 + a_1 V^2 + a_2 V + a_3 = 0 \tag{15}$$

$$b_0 z^3 + b_1 z^2 + b_2 z + b_3 = 0 \tag{16}$$

where a_0, a_1, a_2, and a_3 are defined in Eq. 17, 18, 19, and 20, respectively; b_0, b_1, b_2, and b_3 are defined in Eq. 21, 22, 23, and 24, respectively.

$$a_0 = p \tag{17}$$

$$a_1 = n_T p \cdot (\zeta_3 + \zeta_4 - \zeta_1) - n_T RT \tag{18}$$

$$a_2 = -n_T^2 p \cdot \left(\zeta_4^2 + \zeta_1\zeta_3 + \zeta_1\zeta_4\right) - n_T^2 RT \cdot \left(\zeta_3 + \zeta_4\right) + n_T^2 \zeta_2 \tag{19}$$

$$a_3 = n_T^3 p \cdot \zeta_1 \zeta_4^2 + n_T^3 RT \cdot \zeta_4^2 - n_T^3 \cdot \zeta_1 \zeta_2 \tag{20}$$

$$b_0 = \left(\frac{RT}{p}\right)^3 \tag{21}$$

$$b_1 = \left(\frac{RT}{p}\right)^2 \cdot \left(\zeta_3 + \zeta_4 - \zeta_1 - \frac{RT}{p}\right) \tag{22}$$

$$b_2 = \left(\frac{RT}{p}\right) \cdot \left(-\zeta_4^2 - \left(\zeta_3 + \zeta_4\right) \cdot \left(\zeta_1 + \frac{RT}{p}\right) + \frac{\zeta_2}{p}\right) \tag{23}$$

$$b_3 = \left(\zeta_1 + \frac{RT}{p}\right) \cdot \zeta_4^2 - \frac{\zeta_1 \zeta_2}{p} \tag{24}$$

The advantage of a cubic equation is the possibility to have multiple solutions (maximally three) for volume at specific temperature and pressure conditions, which may reflect coexisting liquid and vapour phases. Liquid-vapour equilibria can only be calculated from the same equation of state if multiple solution of volume can be calculated at the same temperature and pressure. The calculation of thermodynamic properties with this type of equation of state is based on splitting Eq. 14 in two parts (Eq. 25), i.e. an ideal pressure (from the ideal gas law) and a departure (or residual) pressure, see also [6].

$$p = p_{ideal} + p_{residual} \tag{25}$$

where

$$p_{ideal} = \frac{n_T RT}{V} \tag{26}$$

The residual pressure ($p_{residual}$) can be defined as the difference (Δp, Eq. 27) between ideal pressure and reel pressure as expressed in Eq. 14 .

$$\Delta p = p_{residual} = -\frac{n_T RT}{V} + \frac{n_T RT}{V - n_T \zeta_1} - \frac{n_T^2 \zeta_2}{V^2 + n_T \zeta_3 V + n_T \zeta_4 V - n_T^2 \zeta_4^2} \tag{27}$$

The partial derivative of pressure with respect to temperature (Eq. 28) is the main equation to estimate the thermodynamic properties of fluids (see Eqs. 2 and 3).

$$\frac{\partial p}{\partial T} = \frac{\partial p_{ideal}}{\partial T} + \frac{\partial \Delta p}{\partial T} \tag{28}$$

where

$$\begin{aligned}\frac{\partial \Delta p}{\partial T} &= -\frac{n_T R}{V} + \frac{n_T R}{V - n_T\zeta_1} + \frac{n_T RT}{\left(V - n_T\zeta_1\right)^2}\cdot\frac{\partial(n_T\zeta_1)}{\partial T} \\ &\quad - \frac{1}{V^2 + n_T\zeta_3 V + n_T\zeta_4 V - n_T^{\,2}\zeta_4^{\,2}}\cdot\frac{\partial(n_T^{\,2}\zeta_2)}{\partial T} \\ &\quad + \frac{n_T^{\,2}\zeta_2}{\left(V^2 + n_T\zeta_3 V + n_T\zeta_4 V - n_T^{\,2}\zeta_4^{\,2}\right)^2}\cdot\frac{\partial\left(n_T\zeta_3 V + n_T\zeta_4 V - n_T^{\,2}\zeta_4^{\,2}\right)}{\partial T}\end{aligned} \tag{29}$$

The parameters ζ_1, ζ_3, and ζ_4 are usually independent of temperature, compare with the b parameter (Table 1). This reduces Eq. 29 to Eq. 30.

$$\frac{\partial \Delta p}{\partial T} = -\frac{n_T R}{V} + \frac{n_T R}{V - n_T\zeta_1} - \frac{1}{V^2 + n_T\zeta_3 V + n_T\zeta_4 V - n_T^{\,2}\zeta_4^{\,2}}\cdot\frac{\partial(n_T^{\,2}\zeta_2)}{\partial T} \tag{30}$$

Other important equations to calculate thermodynamic properties of fluids are partial derivatives of pressure with respect to volume (Eq. 31 and 32).

$$\frac{\partial p}{\partial V} = -\frac{n_T RT}{\left(V - n_T\zeta_1\right)^2} + \frac{n_T^{\,2}\zeta_2}{\left(V^2 + n_T\zeta_3 V + n_T\zeta_4 V - n_T^{\,2}\zeta_4^{\,2}\right)^2}\cdot\left(2V + n_T\zeta_3 + n_T\zeta_4\right) \tag{31}$$

$$\begin{aligned}\frac{\partial^2 p}{\partial V^2} &= \frac{2n_T RT}{\left(V - n_T\zeta_1\right)^3} - \frac{2n_T^{\,2}\zeta_2}{\left(V^2 + n_T\zeta_3 V + n_T\zeta_4 V - n_T^{\,2}\zeta_4^{\,2}\right)^3}\cdot\left(2V + n_T\zeta_3 + n_T\zeta_4\right)^2 \\ &\quad + \frac{2n_T^{\,2}\zeta_2}{\left(V^2 + n_T\zeta_3 V + n_T\zeta_4 V - n_T^{\,2}\zeta_4^{\,2}\right)^2}\end{aligned} \tag{32}$$

Eqs. 31 and 32 already include the assumption that the parameters ζ_1, ζ_2, ζ_3, and ζ_4 are independent of volume. Finally, the partial derivative of pressure in respect to the amount of substance of a specific component in the fluid mixture (n_i) is also used to characterize thermodynamic properties of fluid mixtures (Eq. 33).

$$\begin{aligned}\frac{\partial p}{\partial n_i} &= \frac{RT}{V - n_T\zeta_1} + \frac{n_T RT}{\left(V - n_T\zeta_1\right)^2}\cdot\frac{\partial(n_T\zeta_1)}{\partial n_i} \\ &\quad - \frac{1}{V^2 + n_T\zeta_3 V + n_T\zeta_4 V - n_T^{\,2}\zeta_4^{\,2}}\cdot\frac{\partial(n_T^{\,2}\zeta_2)}{\partial n_i} \\ &\quad + \frac{n_T^{\,2}\zeta_2}{\left(V^2 + n_T\zeta_3 V + n_T\zeta_4 V - n_T^{\,2}\zeta_4^{\,2}\right)^2}\cdot\left[\left(\frac{\partial(n_T\zeta_3)}{\partial n_i} + \frac{\partial(n_T\zeta_4)}{\partial n_i}\right)\cdot V - 2n_T\zeta_4\frac{\partial(n_T\zeta_4)}{\partial n_i}\right]\end{aligned} \tag{33}$$

3. Thermodynamic parameters

The entropy (S) is obtained from the integration defined in Eq. 2 at constant temperature (Eqs. 34 and 35).

$$\int_{S_0}^{S_1} dS = \int_{V_0}^{V_1} \left(\frac{\partial p}{\partial T}\right)_{V,n_T} dV \tag{34}$$

$$S_1 = S_0 + \int_{V_0}^{V_1} \left(\frac{\partial p_{ideal}}{\partial T} + \frac{\partial \Delta p}{\partial T}\right) dV \tag{35}$$

The limits of integration are defined as a reference ideal gas at S_0 and V_0, and a real gas at S_1 and V_1. This integration can be split into two parts, according to the ideal pressure and residual pressure definition (Eqs. 25, 26, and 27). The integral has different solutions dependent on the values of ζ_3 and ζ_4: Eq. 36 for $\zeta_3 = 0$ and $\zeta_4 = 0$, and Eqs. 37 and 38 for $\zeta_3 > 0$.

$$S_1 = S_0 + n_T R \ln\left(\frac{V_1}{V_0}\right) + n_T R \ln\left(\frac{V_1 - n_T\zeta_1}{V_0 - n_T\zeta_1} \cdot \frac{V_0}{V_1}\right) + \left(\frac{1}{V_1} - \frac{1}{V_0}\right) \cdot \frac{\partial (n_T{}^2 \zeta_2)}{\partial T} \tag{36}$$

$$\begin{aligned} S_1 = {} & S_0 + n_T R \ln\left(\frac{V_1}{V_0}\right) + n_T R \ln\left(\frac{V_1 - n_T\zeta_1}{V_0 - n_T\zeta_1} \cdot \frac{V_0}{V_1}\right) \\ & - \frac{1}{q} \cdot \frac{\partial (n_T{}^2 \zeta_2)}{\partial T} \cdot \ln\left(\frac{2V_1 + n_T(\zeta_3 + \zeta_4) - q}{2V_1 + n_T(\zeta_3 + \zeta_4) + q}\right) + \frac{1}{q} \cdot \frac{\partial (n_T{}^2 \zeta_2)}{\partial T} \cdot \ln\left(\frac{2V_0 + n_T(\zeta_3 + \zeta_4) - q}{2V_0 + n_T(\zeta_3 + \zeta_4) + q}\right) \end{aligned} \tag{37}$$

where

$$q = n_T \sqrt{4\zeta_4{}^2 + (\zeta_3 + \zeta_4)^2} \tag{38}$$

The *RKeos* and *Seos* define q as $n_T b$, whereas in the *PReos* q is equal to $n_T b\sqrt{8}$, according to the values for ζ_3 and ζ_4 listed in Table 1. Eqs. 36 and 37 can be simplified by assuming that the lower limit of the integration corresponds to a large number of V_0. As a consequence, part of the natural logarithms in Eqs. 36 and 37 can be replaced by the unit value 1 or 0 (Eqs. 39, 40, and 41).

$$\lim_{V_0 \to \infty} \left(\frac{V_0}{V_0 - n_T\zeta_1}\right) = 1 \tag{39}$$

$$\lim_{V_0 \to \infty} \left(\frac{1}{V_0}\right) = 0 \tag{40}$$

$$\lim_{V_0 \to \infty} \left(\frac{2V_0 + n_T(\zeta_3 + \zeta_4) - q}{2V_0 + n_T(\zeta_0 + \zeta_1) + q}\right) = 1 \tag{41}$$

The entropy change that is caused by a volume change of ideal gases corresponds to the second term on the right-hand side of Eqs. 36 and 37. This term can be used to express the behaviour of an ideal mixture of perfected gases. Each individual gas in a mixture expands from their partial volume (v_i) to the total volume at a pressure of 0.1 MPa, which results in a new expression for this term (Eq. 42)

$$n_T R \ln\left(\frac{V_1}{V_0}\right)_{ideal.mix} = \sum_i \left[n_i R \ln\left(\frac{V_1}{v_i}\right)\right] \tag{42}$$

where n_i is the amount of substance of component i in the fluid mixture. In addition, the partial volume of an ideal gas is related to the standard pressure p_0 (0.1 MPa) according to the ideal gas law (Eq. 43, compare with Eq. 26).

$$v_i = \frac{n_i RT}{p_0} \tag{43}$$

Finally, the entropy of fluid phases containing gas mixtures at any temperature and total volume according to the two-constant cubic equation of state is given by Eq. 44 for $\zeta_3 = 0$ and $\zeta_4 = 0$, and Eq. 45 for $\zeta_3 > 0$.

$$S = S_0 + \sum_i \left[n_i R \ln\left(\frac{p_0 V}{n_i RT}\right)\right] + n_T R \ln\left(\frac{V - n_T\zeta_1}{V}\right) + \frac{1}{V}\cdot\frac{\partial(n_T{}^2\zeta_2)}{\partial T} \tag{44}$$

$$S = S_0 + \sum_i \left[n_i R \ln\left(\frac{p_0 V}{n_i RT}\right)\right] + n_T R \ln\left(\frac{V - n_T\zeta_1}{V}\right) - \frac{1}{q}\cdot\frac{\partial(n_T{}^2\zeta_2)}{\partial T}\cdot \ln\left(\frac{2V + n_T(\zeta_3+\zeta_4) - q}{2V + n_T(\zeta_3+\zeta_4) + q}\right) \tag{45}$$

The subscripts "1" for the upper limit of integration is eliminated to present a pronounced equation. The standard state entropy (S_0) of a mixture of ideal gases is defined according to the arithmetic average principle (Eq. 46).

$$S_0 = \sum_i n_i \cdot s_i^0 \tag{46}$$

where s_i^0 is the molar entropy of a pure component i in an ideal gas mixture at temperature T.

The internal energy (U, see Eq. 3) is obtained from the pressure equation (Eq. 14) and its partial derivative with respect to temperature (Eqs. 28 and 30):

$$\int_{U_0}^{U_1} dU = \int_{V_0}^{V_1} \left(T\cdot\frac{\partial p}{\partial T} - p\right) dV \tag{47}$$

$$U_1 = U_0 + \int_{V_0}^{V_1} \left(\frac{1}{V^2 + n_T\zeta_3 V + n_T\zeta_4 V - n_T{}^2\zeta_4{}^2}\cdot\left(n_T{}^2\zeta_2 - T\frac{\partial(n_T{}^2\zeta_2)}{\partial T}\right)\right) dV \tag{48}$$

Similar to the integral in the entropy definition (see Eqs. 44 and 45), Eq. 48 has different solutions dependent on the values of ζ_3 and ζ_4: Eq. 49 for $\zeta_3 = 0$ and $\zeta_4 = 0$, and Eq. 50 for $\zeta_3 > 0$.

$$U = U_0 - \frac{1}{V} \cdot \left(n_T^2 \zeta_2 - T \frac{\partial (n_T^2 \zeta_2)}{\partial T} \right) \tag{49}$$

$$U = U_0 + \frac{1}{q} \cdot \left(n_T^2 \zeta_2 - T \frac{\partial (n_T^2 \zeta_2)}{\partial T} \right) \cdot \ln \left(\frac{2V + n_T(\zeta_3 + \zeta_4) - q}{2V + n_T(\zeta_3 + \zeta_4) + q} \right) \tag{50}$$

The definition of q is given in Eq. 38. The standard state internal energy (U_0) of a mixture of ideal gases is defined according to the arithmetic average principle (Eq. 51).

$$U_0 = \sum_i n_i \cdot u_i^0 \tag{51}$$

where u_i^0 is the molar internal energy of a pure component i in an ideal gas mixture at temperature T.

Enthalpy (Eq. 52 for $\zeta_3 = 0$ and $\zeta_4 = 0$, and Eq. 53 for $\zeta_3 > 0$), Helmholtz energy (Eq. 55 for $\zeta_3 = 0$ and $\zeta_4 = 0$, and Eq. 56 for $\zeta_3 > 0$), and Gibbs energy (Eq. 58 for $\zeta_3 = 0$ and $\zeta_4 = 0$, and Eq. 59 for $\zeta_3 > 0$) can be obtained from the definitions of pressure, entropy and internal energy according to standard thermodynamic relations, as illustrated in Eq. 4, 5, and 7. Standard state enthalpy (H_0), standard state Helmholtz energy (A_0), and standard state Gibbs energy (G_0) of an ideal gas mixture at 0.1 MPa and temperature T are defined in Eqs. 54, 57, and 60, respectively.

$$H = U_0 + \frac{n_T RTV}{V - n_T \zeta_1} - \frac{1}{V} \cdot \left(2n_T^2 \zeta_2 - T \frac{\partial (n_T^2 \zeta_2)}{\partial T} \right) \tag{52}$$

$$\begin{aligned} H = U_0 &+ \frac{n_T RTV}{V - n_T \zeta_1} - \frac{n_T^2 \zeta_2 V}{V^2 + n_T \zeta_3 V + n_T \zeta_4 V - n_T^2 \zeta_4^2} \\ &+ \frac{1}{q} \cdot \left(n_T^2 \zeta_2 - T \frac{\partial (n_T^2 \zeta_2)}{\partial T} \right) \cdot \ln \left(\frac{2V + n_T(\zeta_3 + \zeta_4) - q}{2V + n_T(\zeta_3 + \zeta_4) + q} \right) \end{aligned} \tag{53}$$

$$H_0 = U_0 + n_T RT \tag{54}$$

$$A = U_0 - TS_0 - \sum_i \left[n_i RT \ln \left(\frac{p_0 V}{n_i RT} \right) \right] - n_T RT \ln \left(\frac{V - n_T \zeta_1}{V} \right) - \frac{n_T^2 \zeta_2}{V} \tag{55}$$

$$\begin{aligned} A = U_0 &- TS_0 - \sum_i \left[n_i RT \ln \left(\frac{p_0 V}{n_i RT} \right) \right] - n_T RT \ln \left(\frac{V - n_T \zeta_1}{V} \right) \\ &+ \frac{n_T^2 \zeta_2}{q} \cdot \ln \left(\frac{2V + n_T(\zeta_3 + \zeta_4) - q}{2V + n_T(\zeta_3 + \zeta_4) + q} \right) \end{aligned} \tag{56}$$

$$A_0 = U_0 - TS_0 \tag{57}$$

$$G = U_0 - TS_0 - RT\sum_i\left[n_i \ln\left(\frac{p_0 V}{n_i RT}\right)\right] - n_T RT \ln\left(\frac{V - n_T\zeta_1}{V}\right) + \frac{n_T RTV}{V - n_T\zeta_1} - \frac{2n_T^{\ 2}\zeta_2}{V} \tag{58}$$

$$\begin{aligned} G = U_0 &- TS_0 - RT\sum_i\left[n_i \ln\left(\frac{p_0 V}{n_i RT}\right)\right] - n_T RT \ln\left(\frac{V - n_T\zeta_1}{V}\right) + \frac{n_T RTV}{V - n_T\zeta_1} \\ &+ \frac{n_T^{\ 2}\zeta_2}{q}\cdot\ln\left(\frac{2V + n_T(\zeta_3+\zeta_4) - q}{2V + n_T(\zeta_3+\zeta_4) + q}\right) - \frac{n_T^{\ 2}\zeta_2 V}{V^2 + n_T\zeta_3 V + n_T\zeta_4 V - n_T^{\ 2}\zeta_4^{\ 2}} \end{aligned} \tag{59}$$

$$G_0 = U_0 - TS_0 + n_T RT \tag{60}$$

The Helmholtz energy equation (Eqs. 55, 56, and 57) is used for the definition of chemical potential (μ_i) of a component in either vapour or liquid phase gas mixtures (compare with Eq. 8), Eq. 61 for $\zeta_3 = 0$ and $\zeta_4 = 0$, and Eq. 62 for $\zeta_3 > 0$, calculated with two-constant cubic equations of state.

$$\begin{aligned} \mu_i = &\frac{\partial U_0}{\partial n_i} - T\frac{\partial S_0}{\partial n_i} - RT\ln\left(\frac{p_0 V}{n_i RT}\right) + RT - RT\ln\left(\frac{V - n_T\zeta_1}{V}\right) \\ &+ \frac{n_T RT}{V - n_T\zeta_1}\cdot\frac{\partial(n_T\zeta_1)}{\partial n_i} - \frac{1}{V}\cdot\frac{\partial n_T^{\ 2}\zeta_2}{\partial n_i} \end{aligned} \tag{61}$$

$$\begin{aligned} \mu_i = &\frac{\partial U_0}{\partial n_i} - T\frac{\partial S_0}{\partial n_i} - RT\ln\left(\frac{p_0 V}{n_i RT}\right) + RT - RT\ln\left(\frac{V - n_T\zeta_1}{V}\right) + \frac{n_T RT}{V - n_T\zeta_1}\cdot\frac{\partial(n_T\zeta_1)}{\partial n_i} \\ &+ \left(\frac{\partial n_T^{\ 2}\zeta_2}{\partial n_i}\cdot\frac{1}{q} - \frac{n_T^{\ 2}\zeta_2}{q^2}\cdot\frac{\partial q}{\partial n_i}\right)\cdot\ln\left(\frac{2V + n_T(\zeta_3+\zeta_4) - q}{2V + n_T(\zeta_3+\zeta_4) + q}\right) \\ &+ \frac{n_T^{\ 2}\zeta_2}{q}\cdot\frac{1}{2V + n_T(\zeta_3+\zeta_4) - q}\cdot\left(\frac{\partial n_T\zeta_3}{\partial n_i} + \frac{\partial n_T\zeta_4}{\partial n_i} - \frac{\partial q}{\partial n_i}\right) \\ &- \frac{n_T^{\ 2}\zeta_2}{q}\cdot\frac{1}{2V + n_T(\zeta_3+\zeta_4) + q}\cdot\left(\frac{\partial n_T\zeta_3}{\partial n_i} + \frac{\partial n_T\zeta_4}{\partial n_i} + \frac{\partial q}{\partial n_i}\right) \end{aligned} \tag{62}$$

where

$$\frac{\partial q}{\partial n_i} = \frac{1}{q}\cdot\left[4n_T\zeta_4\frac{\partial n_T\zeta_4}{\partial n_i} + n_T(\zeta_3+\zeta_4)\cdot\left(\frac{\partial n_T\zeta_3}{\partial n_i} + \frac{\partial n_T\zeta_4}{\partial n_i}\right)\right] \tag{63}$$

The definitions of the partial derivative of q in respect to amount of substance (Eq. 63) according to $\zeta_3 = b$ and $\zeta_4 = 0$ [8, 9] is illustrated in Eq. 64, and $\zeta_3 = b$ and $\zeta_4 = b$ [10] in Eq. 65.

$$\frac{\partial q}{\partial n_i} = \frac{\partial (n_T b)}{\partial n_i} \tag{64}$$

$$\frac{\partial q}{\partial n_i} = \sqrt{8} \cdot \frac{\partial (n_T b)}{\partial n_i} \tag{65}$$

The fugacity coefficient (φ_i) is defined according to Eqs. 9 and 10 from the difference between the chemical potential of a real gas mixture and an ideal gas mixture at standard conditions (0.1 MPa), see Eq. 66 for $\zeta_3 = 0$ and $\zeta_4 = 0$, and Eq. 67 for $\zeta_3 > 0$. Fugacity coefficient defined in Eq. 66 is applied to *Weos* and Eq. 67 is applied to *RKeos, Seos,* and *PReos.*

$$RT\ln(\varphi_i) = -RT\ln\left(\frac{pV}{n_T RT}\right) - RT\ln\left(\frac{V - n_T\zeta_1}{V}\right) + \frac{n_T RT}{V - n_T\zeta_1} \cdot \frac{\partial (n_T\zeta_1)}{\partial n_i} - \frac{1}{V} \cdot \frac{\partial n_T^2\zeta_2}{\partial n_i} \tag{66}$$

$$\begin{aligned} RT\ln(\varphi_i) = & -RT\ln\left(\frac{pV}{n_T RT}\right) - RT\ln\left(\frac{V - n_T\zeta_1}{V}\right) + \frac{n_T RT}{V - n_T\zeta_1} \cdot \frac{\partial (n_T\zeta_1)}{\partial n_i} \\ & + \left(\frac{\partial n_T^2\zeta_2}{\partial n_i} \cdot \frac{1}{q} - \frac{n_T^2\zeta_2}{q^2} \cdot \frac{\partial q}{\partial n_i}\right) \cdot \ln\left(\frac{2V + n_T(\zeta_3 + \zeta_4) - q}{2V + n_T(\zeta_3 + \zeta_4) + q}\right) \\ & + \frac{n_T^2\zeta_2}{q} \cdot \frac{1}{2V + n_T(\zeta_3 + \zeta_4) - q} \cdot \left(\frac{\partial n_T\zeta_3}{\partial n_i} + \frac{\partial n_T\zeta_4}{\partial n_i} - \frac{\partial q}{\partial n_i}\right) \\ & - \frac{n_T^2\zeta_2}{q} \cdot \frac{1}{2V + n_T(\zeta_3 + \zeta_4) + q} \cdot \left(\frac{\partial n_T\zeta_3}{\partial n_i} + \frac{\partial n_T\zeta_4}{\partial n_i} + \frac{\partial q}{\partial n_i}\right) \end{aligned} \tag{67}$$

4. Spinodal

The stability limit of a fluid mixture can be calculated with two-constant cubic equations of state, e.g. see [6]. This limit is defined by the spinodal line, i.e. the locus of points on the surface of the Helmholtz energy or Gibbs energy functions that are inflection points, e.g. see [12] and references therein. The stability limit occurs at conditions where phase separation into a liquid and vapour phase should take place, which is defined by the binodal. Metastability is directly related to spinodal conditions, for example, nucleation of a vapour bubble in a cooling liquid phase within small constant volume cavities, such as fluid inclusions in minerals (< 100 μm diameter) occurs at conditions well below homogenization conditions of these phases in a heating experiment. The maximum temperature difference of nucleation and homogenization is defined by the spinodal. In multi-component fluid systems, the partial derivatives of the Helmholtz energy with respect to volume and amount of substance of each component can be arranged in a matrix that has a determinant (D_{spin}) equal to zero (Eq. 68) at spinodal conditions.

$$D_{spin} = \begin{vmatrix} A_{VV} & A_{n_1 V} & A_{n_2 V} & \cdots \\ A_{V n_1} & A_{n_1 n_1} & A_{n_2 n_1} & \cdots \\ A_{V n_2} & A_{n_1 n_2} & A_{n_2 n_2} & \cdots \\ \vdots & \vdots & \vdots & \ddots \end{vmatrix} = 0 \tag{68}$$

This matrix is square and contains a specific number of columns that is defined by the number of differentiation variables, i.e. volume and number of components in the fluid mixture minus 1. The individual components of this matrix are defined according to Eqs. 69, 70, 71, 72, 73, and 74. The exact definition of these components according to two-constant cubic equations of state can be obtained from the web site http://fluids.unileoben.ac.at (see also [6]).

$$A_{VV} = \left(\frac{\partial^2 A}{\partial V^2} \right)_{n_1, n_2, \cdots} \tag{69}$$

$$A_{n_1 n_1} = \left(\frac{\partial^2 A}{\partial {n_1}^2} \right)_{n_2, V, \cdots} \tag{70}$$

$$A_{n_2 n_2} = \left(\frac{\partial^2 A}{\partial {n_2}^2} \right)_{n_1, V, \cdots} \tag{71}$$

$$A_{n_1 V} = \left(\frac{\partial^2 A}{\partial n_1 \partial V} \right)_{n_2, \cdots} = A_{V n_1} \tag{72}$$

$$A_{n_2 V} = \left(\frac{\partial^2 A}{\partial n_2 \partial V} \right)_{n_1, \cdots} = A_{V n_2} \tag{73}$$

$$A_{n_1 n_2} = \left(\frac{\partial^2 A}{\partial n_1 \partial n_2} \right)_{V, \cdots} = A_{n_2 n_1} \tag{74}$$

The determinant in Eq. 68 is calculated with the Laplacian expansion that contains "*minors*" and "*cofactors*", e.g. see [13]. The mathematical computation time increases exponential with increasing number of components. Therefore, the *LU* decomposition [14] can be applied in computer programming to reduce this time.

The spinodal curve, binodal curve and critical point of a binary CO_2-CH_4 mixture with $x(CO_2) = 0.9$ are illustrated in Figure 1, which are calculated with the *PReos* [10]. The spinodal has a small loop near the critical point, and may reach negative pressures at lower temperatures. The binodal remains within the positive pressure part at all temperatures. The binodal is obtained from equality of fugacity (Eq. 66 and 67) of each component in both liquid and vapour phase, and marks the boundary between a homogeneous fluid mixture and fluid immiscibility [6, 15].

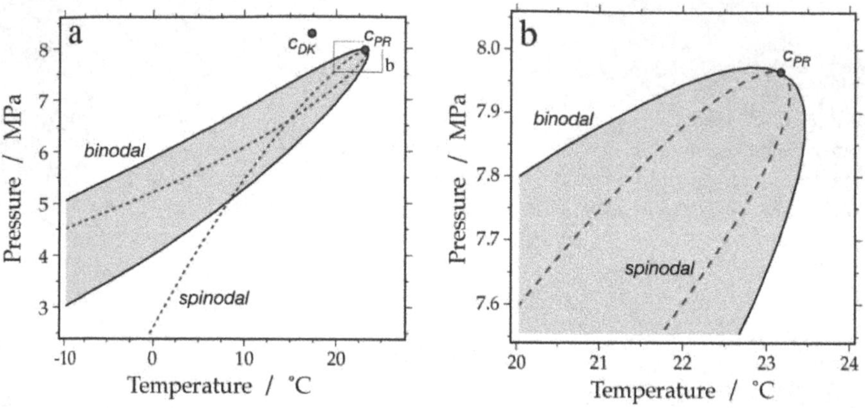

Figure 1. (**a**) Temperature-pressure diagram of a binary CO_2-CH_4 fluid mixture, with $x(CO_2) = 0.9$. The shaded area illustrates *T*-*p* condition of immiscibility of a CO_2-rich liquid phase and a CH_4-rich vapour phase (the binodal). The red dashed line is the spinodal. All lines are calculated with the equation of state according to *PReos* [10]. The calculated critical point is indicated with C_{PR}. C_{DK} is the interpolated critical point from experimental data [16]. (**b**) enlargement of (**a**) indicated with the square in thin lines.

5. Pseudo critical point

The pseudo critical point is defined according to the first and second partial derivatives of pressure with respect to volume (Eqs. 31 and 32). This point is defined in a *p*-*V* diagram where the inflection point and extremum coincide at a specific temperature, i.e. Eqs. 31 and 32 are equal to 0. The pseudo critical point is equal to the critical point for pure gas fluids, however, the critical point in mixtures cannot be obtained from Eqs. 31 and 32. The pseudo critical point estimation is used to define the two-constants (*a* and *b*) for pure gas fluids in cubic equations of state according to the following procedure. The molar volume of the pseudo critical point that is derived from Eqs 31 and 32 is presented in the form of a cubic equation (Eq. 75).

$$0 = V_m^3 - 3\zeta_1 \cdot V_m^2 + \left[3\zeta_4^2 - 3\zeta_1 \cdot (\zeta_3 + \zeta_4)\right] \cdot V_m - \zeta_1 \cdot \left[\zeta_4^2 + (\zeta_3 + \zeta_4)^2\right] + \zeta_4^2 \cdot (\zeta_3 + \zeta_4) \quad (75)$$

The solution of this cubic equation can be obtained from its reduced form, see page 9 in [15]:

$$x^3 + f \cdot x + g = 0 \quad (76)$$

where

$$f = 3 \cdot \left[\zeta_4^2 - \zeta_1 \cdot (\zeta_1 + \zeta_3 + \zeta_4)\right] \quad (77)$$

$$g = -2\zeta_1^3 + \zeta_1 \cdot \left[2\zeta_4^2 - (\zeta_3 + \zeta_4)^2\right] + (\zeta_3 + \zeta_4) \cdot \left[\zeta_4^2 - 3\zeta_1^2\right] \quad (78)$$

$$V_m = x + \zeta_1 \quad (79)$$

The values of f and g in terms of the b parameters for the individual two-constant cubic equations of state are given in Table 2. The molar volume at pseudo critical conditions is directly related to the b parameter in each equation of state: *Weos* Eq. 80; *RKeos* Eq. 81; *Seos* also Eq. 81; and *PReos* Eq. 82.

$$Weos: \quad V_m^{pc} = 3 \cdot b \quad (80)$$

$$RKeos: \quad V_m^{pc} = \frac{b}{\sqrt[3]{2} - 1} \approx 3.847322 \cdot b \quad (81)$$

$$PReos: \quad V_m^{pc} = \left[1 + Q\right] \cdot b \approx 3.951373 \cdot b \quad (82)$$

where Q is defined according to Eq. 83, the superscript "*pc*" is the abbreviation for "*pseudo critical*".

$$Q = \left(4 + \sqrt{8}\right)^{1/3} + \left(4 - \sqrt{8}\right)^{1/3} \quad (83)$$

Equation of state	*f*	*g*	*b in Eqs. 80-82*	*b in Eqs. 94-96*	*difference*
van der Waals [7]	$-3b^2$	$-2b^3$	31.3727	42.8453	37 %
Redlich and Kwong [8]	$-6b^2$	$-6b^3$	24.4633	29.6971	21 %
Soave [9]	$-6b^2$	$-6b^3$	24.4633	29.6971	21 %
Peng and Robinson [10]	$-6b^2$	$-8b^3$	23.8191	26.6656	12 %

Table 2. Definitions of f and g according to Eq. 77 and 78, respectively. The values of b are calculated for the critical conditions of pure CO_2: $V_{m,C} = 94.118\ cm^3 \cdot mol^{-1}$, $T_C = 304.128$ K and $p_C = 7.3773$ MPa [18]. The last column gives the percentage of difference between the values of b (Eqs. 80-82 and 94-96).

The temperature at pseudo critical conditions is obtained from the combination of Eqs. 80-82 and the first partial derivative of pressure with respect to volume (Eq. 31).

$$Weos: \quad T^{pc} = \frac{8}{27} \cdot \frac{\zeta_2}{bR} \approx 0.29629630 \cdot \frac{\zeta_2}{bR} \quad (84)$$

$$RKeos: \quad T^{pc} = \left(\sqrt[3]{2} - 1\right)^2 \cdot \left(\frac{3\zeta_2}{bR}\right) \approx 0.20267686 \cdot \frac{\zeta_2}{bR} \quad (85)$$

$$PReos: \quad T^{pc} = \frac{2Q + 4}{\left(Q + 4 + 2/Q\right)^2} \cdot \left(\frac{3\zeta_2}{bR}\right) \approx 0.17014442 \cdot \frac{\zeta_2}{bR} \quad (86)$$

where Q is defined according to Eq. 83. The order of equations (84, 85, 86) is according to the order of equations of state in Eq. 80, 81,and 82. The parameter ζ_2 is used in Eqs. 84, 85 and, 86 instead of the constant a (see Table 1). Eq. 87 illustrates the transformation of Eq. 85 for the *RKeos* [8] by substitution of ζ_2 according to its value given in Table 1.

$$T^{pc} = \left(\sqrt[3]{2} - 1\right)^{4/3} \cdot \left(\frac{3a}{bR}\right)^{2/3} \tag{87}$$

Any temperature dependency of the a constant has an effect on the definition of the pseudo critical temperature. The pressure at pseudo critical condition (Eqs. 88-90) is obtained from a combination of the pressure equation (Eq.14), pseudo critical temperature (Eqs. 84-87) and pseudo critical molar volume (Eqs. 80-82).

$$Weos: \quad p^{pc} = \frac{1}{27} \cdot \frac{\zeta_2}{b^2} \approx 0.03703704 \cdot \frac{\zeta_2}{b^2} \tag{88}$$

$$RKeos: \quad p^{pc} = \left(\sqrt[3]{2} - 1\right)^3 \cdot \frac{\zeta_2}{b^2} \approx 0.01755999 \cdot \frac{\zeta_2}{b^2} \tag{89}$$

$$PReos: \quad p^{pc} = \frac{Q^2 - 2}{\left(Q^2 + 4Q + 2\right)^2} \cdot \frac{\zeta_2}{b^2} \approx 0.01227198 \cdot \frac{\zeta_2}{b^2} \tag{90}$$

where Q is defined according to Eq. 83. The order of equations (88, 89, and 90) is according to the order of equations of state in Eqs. 80, 81, and 82. These equations define the relation between the a and b constant in two-constant cubic equations of state and critical conditions, i.e. temperature, pressure, and molar volume of pure gas fluids. Therefore, knowledge of these conditions from experimental data can be used to determine the values of a (or ζ_2) and b, which can be defined as a function of only temperature and pressure (Eqs. 91-93, and 94-96, respectively).

$$Weos: \quad \zeta_2 = \frac{27}{64} \cdot \frac{R^2 {T_C}^2}{p_C} = 0.421875 \cdot \frac{R^2 {T_C}^2}{p_C} \tag{91}$$

$$RKeos: \quad \zeta_2 = \frac{1}{9 \cdot \left(\sqrt[3]{2} - 1\right)} \cdot \frac{R^2 {T_C}^2}{p_C} \approx 0.42748024 \cdot \frac{R^2 {T_C}^2}{p_C} \tag{92}$$

$$PReos: \quad \zeta_2 = \frac{\left(Q^2 + 4Q + 2\right)^2 \cdot \left(Q^2 - 2\right)}{4Q^2 \cdot \left(Q + 2\right)^2} \cdot \frac{R^2 {T_C}^2}{p_C} \approx 0.45723553 \cdot \frac{R^2 {T_C}^2}{p_C} \tag{93}$$

$$Weos: \quad b = \frac{1}{8} \cdot \frac{RT_C}{p_C} = 0.125 \cdot \frac{RT_C}{p_C} \tag{94}$$

$$RKeos: \quad b = \frac{\left(\sqrt[3]{2}-1\right)}{3} \cdot \frac{RT_C}{p_C} \approx 0.08664035 \cdot \frac{RT_C}{p_C} \tag{95}$$

$$PReos: \quad b = \frac{\left(Q^2-2\right)}{2Q^2 \cdot \left(Q+2\right)} \cdot \frac{RT_C}{p_C} \approx 0.07779607 \cdot \frac{RT_C}{p_C} \tag{96}$$

where T_C and p_C are the critical temperature and critical pressure, and Q is defined according to Eq. 83. The order of equations (91-93, and 94-96) is according to the order of equations of state in Eqs. 80-82. Comparison of the value of b calculated with experimental critical volume (Eqs. 80, 81 and 82) and critical temperature and pressure (Eqs. 94, 95, and 96) is illustrated in Table 2. The difference indicates the ability of a specific equation of state to reproduce fluid properties of pure gases. A large difference indicates that the geometry or morphology of the selected equation of state in the p-V-T-x parameter space is not exactly reproducing fluid properties of pure gases. The empirical modifications of the van-der-Waals equation of state according to Peng and Robinson [10] result in the most accurate equation in Table 2 (11% for pure CO_2).

6. Critical point and curve

The critical point is the highest temperature and pressure in a pure gas system where boiling may occur, i.e. where a distinction can be made between a liquid and vapour phase at constant temperature and pressure. At temperatures and pressures higher than the critical point the pure fluid is in a homogeneous supercritical state. The critical point of pure gases and multi-component fluid mixtures can be calculated exactly with the Helmholtz energy equation (Eqs. 55-57) that is obtained from two-constant cubic equations of state, e.g. see [17, 18], and it marks that part of the surface described with a Helmholtz energy function where two inflection points of the spinodal coincide. Therefore, the conditions of the spinodal are also applied to the critical point. In addition, the critical curve is defined by the determinant (D_{crit}) of the matrix illustrated in Eq. 97, see also [6].

$$D_{crit} = \begin{vmatrix} A_{VV} & A_{n_1 V} & A_{n_2 V} & \cdots \\ A_{V n_1} & A_{n_1 n_1} & A_{n_2 n_1} & \cdots \\ \vdots & \vdots & \vdots & \vdots \\ D_V & D_{n_1} & D_{n_2} & \ddots \end{vmatrix} = 0 \tag{97}$$

The number of rows in Eq.97 is defined by the differentiation variables volume and number of components minus 2. The last row is reserved for the partial derivatives of the determinant D_{spin} from Eq. 68:

$$D_V = \frac{\partial D_{spin}}{\partial V} \tag{98}$$

$$D_{n_1} = \frac{\partial D_{spin}}{\partial n_1} \tag{99}$$

$$D_{n_2} = \frac{\partial D_{spin}}{\partial n_2} \tag{100}$$

The derivatives of the spinodal determinant (Eqs. 98-100) are calculated from the sum of the element-by-element products of the matrix of "cofactors" (or adjoint matrix) of the spinodal (Eq. 101) and the matrix of the third derivatives of the Helmholtz energy function (Eq. 102).

$$\begin{vmatrix} C_{VV} & C_{n_1 V} & C_{n_2 V} & \cdots \\ C_{V n_1} & C_{n_1 n_1} & C_{n_2 n_1} & \cdots \\ C_{V n_2} & C_{n_1 n_2} & C_{n_2 n_2} & \cdots \\ \vdots & \vdots & \vdots & \ddots \end{vmatrix} \tag{101}$$

$$\begin{vmatrix} A_{VVK} & A_{n_1 VK} & A_{n_2 VK} & \cdots \\ A_{V n_1 K} & A_{n_1 n_1 K} & A_{n_2 n_1 K} & \cdots \\ A_{V n_2 K} & A_{n_1 n_2 K} & A_{n_2 n_2 K} & \cdots \\ \vdots & \vdots & \vdots & \ddots \end{vmatrix} \tag{102}$$

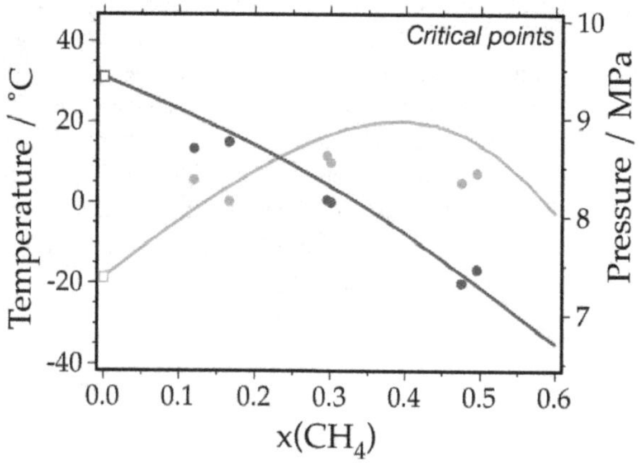

Figure 2. Calculated critical points of binary CO_2-CH_4 fluid mixtures in terms of temperature (red line) and pressure (green line), obtained from the *PReos* [10]. Solid circles are experimental data [16, 19]. The open squares are the critical point of pure CO_2 [20].

where C_{xy} are the individual elements in the matrix of "cofactors", as obtained from the Laplacian expansion. The subscript K refers to the variable that is used in the third differentiation (volume, amount of substance of the components 1 and 2. To reduce computation time in software that uses this calculation method, the *LU* decomposition has been used to calculate the determinant in Eq. 97. The determinants in Eqs. 68 and 97 are both used to calculate exactly the critical point of any fluid mixture and pure gases, based on two-constant cubic equations of state that define the Helmholtz energy function.

An example of a calculated critical curve, i.e. critical points for a variety of compositions in a binary fluid system, is illustrated in Figure 2. The prediction of critical temperatures of fluid mixtures corresponds to experimental data [16, 19], whereas calculated critical pressures are slightly overestimated at higher fraction of CH_4. This example illustrates that the *PReos* [10] is a favourable modification that can be used to calculate sub-critical conditions of CO_2-CH_4 fluid mixtures.

7. Mixing rules and definitions of ζ_1 and ζ_2

All modifications of the van-der-Waals two-constant cubic equation of state [7] have an empirical character. The main modifications are defined by Redlich and Kwong, Soave and Peng and Robinson (see Table 1), and all modification can by summarized by specific adaptations of the values of ζ_1, ζ_2, ζ_3, *and* ζ_4 to fit experimental data. The original definition [7] of ζ_1 *(b)* and ζ_2 *(a)* for pure gases is obtained from the pseudo critical conditions (Eqs. 91-93, and 94-96). This principle is adapted in most modifications of the van-der-Waals equation of state, e.g. *RKeos* [8]. Soave [9] and Peng and Robinson [10] adjusted the definition of ζ_2 with a temperature dependent correction parameter α (Eqs. 103-105).

$$\zeta_2 = a_C \cdot \alpha \tag{103}$$

$$\alpha = \left[1 + m\left(1 - \sqrt{\frac{T}{T_C}}\right)\right]^2 \tag{104}$$

$$m = \sum_{i=0,1,2} m_i \cdot \omega^i \tag{105}$$

where a_C is defined by the pseudo critical conditions (Eqs. 91-93), and ω is the acentric factor. The summation in Eq. 105 does not exceed $i = 2$ for Soave [9] and Peng and Robinson [10]. The definition of the acentric factor is arbitrary and chosen for convenience [5] and is a purely empirical modification. These two equations of state have different definitions of pseudo critical conditions (see Eqs. 91-93 and 94-96), therefore, the values of m_i must be different for each equation (Table 3).

	Soave [9]	*Peng and Robinson [10]*
m_0	0.480	0.37464
m_1	1.574	1.54266
m_2	-0.176	-0.26992

Table 3. Values of the constant m_i in Eq. 105.

The two-constant cubic equation of state can be applied to determine the properties of fluid mixtures by using "*mixing rules*" for the parameters ζ_1 and ζ_2 which are defined for individual pure gases according to pseudo critical conditions. These mixing rules are based on simplified molecular behaviour of each component (i and j) in mixtures [21, 22] that describe the interaction between two molecules:

$$\zeta_1^{mix} = \sum_i \sum_j x_i \cdot \zeta_1(i) \tag{106}$$

$$\zeta_2^{mix} = \sum_i \sum_j x_i x_j \cdot \zeta_2(i,j) \tag{107}$$

where

$$\zeta_2(i,j) = \sqrt{\zeta_2(i) \cdot \zeta_2(j)} \tag{108}$$

These mixing rules have been subject to a variety of modifications, in order to predict fluid properties of newly available experimental data of mixtures. Soave [9] and Peng and Robinson [10] modified Eq. 108 by adding an extra correction factor (Eq. 109).

$$\zeta_2(i,j) = \left(1 - \delta_{ij}\right) \cdot \sqrt{\zeta_2(i) \cdot \zeta_2(j)} \tag{109}$$

where δ_{ij} has a constant value dependent on the nature of component i and j.

8. Experimental data

As mentioned before, modifications of two-constant cubic equation of state was mainly performed to obtain a better fit with experimental data for a multitude of possible gas mixtures and pure gases. Two types of experimental data of fluid properties were used: 1. homogeneous fluid mixtures at supercritical conditions; and 2. immiscible two-fluid systems at subcritical conditions (mainly in petroleum fluid research). The experimental data consist mainly of pressure, temperature, density (or molar volume) and compositional data, but can also include less parameters. Figure 3 gives an example of the misfit between the first type of experimental data for binary CO_2-CH_4 mixtures [19] and calculated fluid properties with *RKeos* [8] at a constant temperature (15 °C). The *RKeos* uses the pseudo critical defined parameters ζ_1 and ζ_2 (Eqs. 92 and 95) and mixing rules according to Eqs. 106-108 and is only approximately reproducing the fluid properties of CO_2-CH_4 mixtures at subcritical conditions

Experimental data of homogeneous supercritical gas mixtures in the ternary CO_2-CH_4-N_2 system [23] are compared with the two-constant cubic equations of state in Table 4. The *Weos* [7] clearly overestimates (up to 14.1 %) experimentally determined molar volumes at 100 MPa and 200 °C. The *Seos* [9] is the most accurate model in Table 4, but still reach deviations of up to 2.3 % for CO_2-rich gas mixtures. The *PReos* [10] gives highly underestimated molar volumes at these conditions.

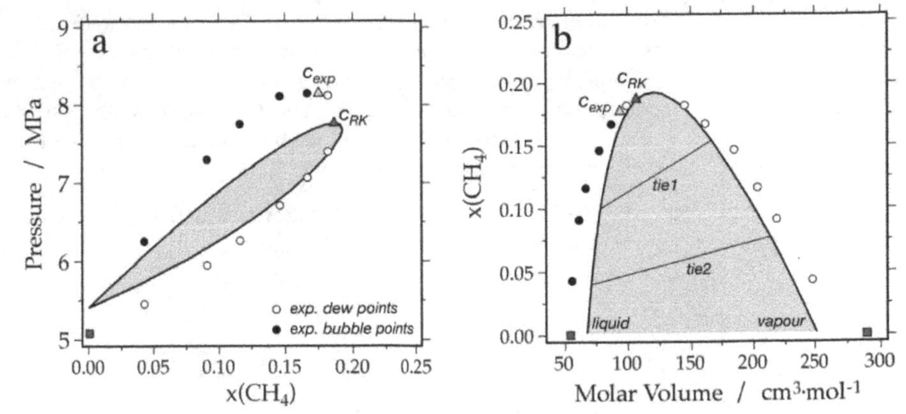

Figure 3. Modelled immiscibility of binary CO_2-CH_4 gas mixtures (shaded areas) in a pressure - amount CH_4 fraction diagram (**a**) and amount CH_4 fraction - molar volume diagram (**b**) at 15 °C. The solid and open circles are experimental data [16]. The red squares are the properties of pure CO_2 [20]. The yellow triangle (C_{exp}) is the interpolated critical point for experimental data, and the green triangle (C_{RK}) is the calculated critical point [8]. *tie1* and *tie 2* in (**b**) are calculated tie-lines between two phases at constant pressures 6.891 and 6.036 MPa, respectively.

composition			*V_m(exp)*	*W*	*RK*	*S*	*PR*
CO_2	*CH_4*	*N_2*	*$cm^3 \cdot mol^{-1}$*				
0.8	0.1	0.1	56.64	64.61 (14.1%)	54.90 (-3.1%)	57.94 (2.3%)	53.59 (-5.4%)
0.8	0.2	0.2	58.92	65.81 (11.7%)	56.61 (-3.9%)	59.61 (1.2%)	56.93 (-6.1%)
0.4	0.3	0.3	61.08	67.08 (9.6%)	58.27 (-4.6%)	61.12 (0.1%)	56.93 (-6.8%)
0.2	0.4	0.4	62.90	68.28 (8.6%)	59.83 (-4.9%)	62.42 (-0.8%)	58.28 (-7.3%)

Table 4. Comparison of supercritical experimental molar volumes [23] at 100 MPa and 200 °C with two-constant cubic equations of state (abbreviations see Table 1). The percentage of deviation from experimentally obtained molar volumes is indicated in brackets.

Figure 3 and Table 4 illustrate that these modified two-constant cubic equations of state still need to be modified again to obtain a better model to reproduce fluid properties at sub- and supercritical conditions.

9. Modifications of modified equations of state

The number of publications that have modified the previously mentioned two-constant cubic equations of state are numerous, see also [11], and they developed highly complex, but purely empirical equations to define the parameters ζ_1 and ζ_2. A few examples are illustrated in the following paragraphs.

9.1. Chueh and Prausnitz [24]

The constant values in the definition of ζ_1 and ζ_2 (Eqs. 92 and 95) are modified for individual gases by Chueh and Prausnitz [24]. This equation is an arbitrary modification of the *RKeos*

[8]. Consequently, the calculation of the value of ζ_1 and ζ_2 is not any more defined by pseudo critical conditions, which give exact mathematical definition of these constants. Although the prediction of fluid properties of a variety of gas mixtures was improved by these modifications, the morphology of the Helmholtz energy equation in the *p-V-T-x* parameter space is not any more related to observed fluid properties. The theory of pseudo critical conditions is violated according to these modifications.

The mixing rules in Eqs. 106-108 were further refined by arbitrary definitions of critical temperature, pressure, volume and compressibility for fluid mixtures.

$$\zeta_2(i,j) = \frac{\Omega_i + \Omega_j}{2} \cdot \frac{R^2 T_{Cij}^2}{p_{Cij}} \tag{110}$$

$$a_{ij} = \frac{\Omega_i + \Omega_j}{2} \cdot \frac{R^2 T_{Cij}^{2.5}}{p_{Cij}} \tag{111}$$

where Ω_i and Ω_j are the newly defined constant values of component *i* and *j*, and T_{Cij} and p_{Cij} are defined according to complex mixing rules [see 24]. The values of T_{Cij} and p_{Cij} are not related to true critical temperatures and pressures of specific binary gas mixtures.

The prediction of the properties of homogeneous fluids at supercritical conditions (Table 5) is only slightly improved compared to *RKeos* [10], but it is not exceeding the accuracy of the *Seos* [11]. At sub-critical condition (Figure 4), the Chueh-Prausnitz equation is less accurate than the Redlich-Kwong equation (compare Figure 3) in the binary CO_2-CH_4 fluid mixture at 15 °C.

composition			$V_m(exp)$	*CP*	*H*	*B1*	*B2*
CO_2	CH_4	N_2	$cm^3 \cdot mol^{-1}$				
0.8	0.1	0.1	56.64	56.42 (-0.4%)	55.96 (-0.6%)	56.84 (0.4%)	56.53 (-0.2%)
0.8	0.2	0.2	58.92	57.85 (-1.8%)	57.68 (-2.1%)	59.43 (0.9%)	58.81 (-0.2%)
0.4	0.3	0.3	61.08	59.21 (-3.1%)	59.17 (-3.1%)	61.67 (1.0%)	60.79 (-0.5%)
0.2	0.4	0.4	62.90	60.44 (-3.9%)	60.38 (-4.0%)	63.45 (0.9%)	62.40 (-0.8%)

Table 5. The same experimental molar volumes as in Table 4 compared with two-constant equations of state according to Chueh and Prausnitz [24] (CP), Holloway [25, 26] (H), Bakker [27] [B1], and Bakker [28] (B2). The percentage of deviation from experimentally obtained molar volumes is indicated in brackets.

9.2. Holloway [25, 26] and Bakker [27]

The equation of Holloway [25, 26] is another modification of the *RKeos* [8]. The modification is mainly based on the improvement of predictions of homogenous fluid properties of H_2O and CO_2 mixtures, using calculated experimental data [29]. The value for ζ_1 and ζ_3 (both *b*) of H_2O is arbitrarily selected at 14.6 $cm^3 \cdot mol^{-1}$, whereas other pure gases are defined according to pseudo critical conditions. The definition of ζ_2 (i.e. *a*) for H_2O as a function of temperature was subjected to a variety of best-fit procedures [25, 26]. The fitting was improved from four experimental data points [25] to six [26] (Figure 5), but was restricted to temperatures above 350 °C. Bakker [27] improved the best-fit equation by including the entire data set [29], down to 50 °C (Eq. 112).

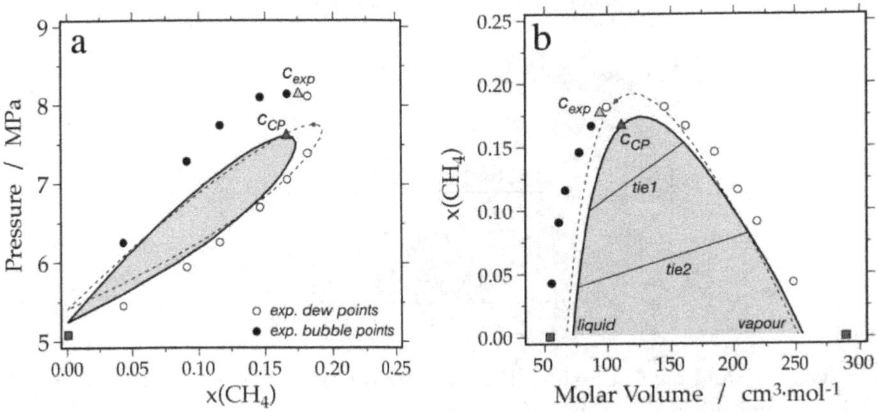

Figure 4. See Figure 3 for details. The *RKeos* is indicated by dashed lines in (**a**) and (**b**). The shaded areas are immiscibility conditions calculated with the Chueh-Prausnitz equation. *tie1* and *tie 2* in (**b**) are calculated tie-lines between two phases at constant pressures 6.944 and 5.984 MPa, respectively.

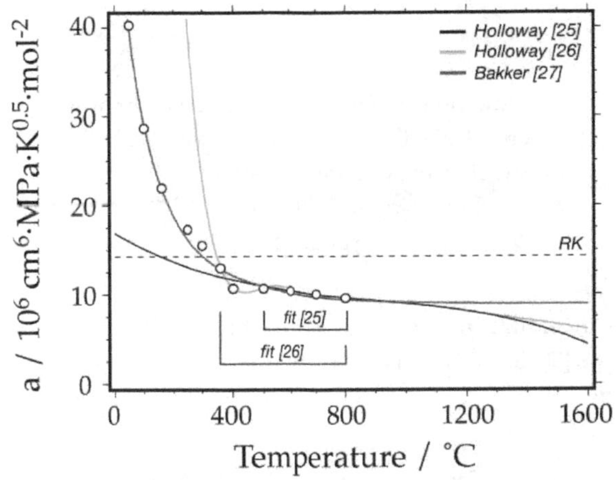

Figure 5. Temperature dependence of the *a* constant for pure H_2O in the modified cubic equation of state [25, 26]. The open circles are calculated experimental data [29]. *fit [25]* is the range of fitting in the definition of Holloway [25], and *fit [26]* of Holloway [26]. *RK* illustrates the constant value calculated from pseudo critical condition [8].

$$a_{H_2O} = \left(9.4654 - \frac{2.0246 \cdot 10^3}{T} + \frac{1.4928 \cdot 10^6}{T^2} + \frac{7.57 \cdot 10^8}{T^3}\right) \cdot 10^6 \quad (112)$$

where T is temperature in Kelvin, and the dimension of a is $cm^6 \cdot MPa \cdot K^{0.5} \cdot mol^{-2}$. The properties of homogeneous pure CO_2, CH_4 and N_2 fluids [27] were also used to obtain a temperature dependent a constant (Eqs. 113, 114, and 115, respectively).

$$a_{CO_2} = \left(-1.2887 + \frac{5.9363 \cdot 10^3}{T} - \frac{1.4124 \cdot 10^6}{T^2} + \frac{1.1767 \cdot 10^8}{T^3}\right) \cdot 10^6 \quad (113)$$

$$a_{CH_4} = \left(-1.1764 + \frac{3.5216 \cdot 10^3}{T} - \frac{1.155 \cdot 10^6}{T^2} + \frac{1.1767 \cdot 10^8}{T^3}\right) \cdot 10^6 \quad (114)$$

$$a_{N_2} = \left(0.060191 - \frac{0.20059 \cdot 10^3}{T} + \frac{0.15386 \cdot 10^6}{T^2}\right) \cdot 10^6 \quad (115)$$

The a_{ij} value of fluid mixtures with a H_2O and CO_2 component (as in Eqs. 106-108 and 110-111) is not defined by the value of pure H_2O and CO_2 (Eqs. 112 and 113), but from a temperature independent constant value (Eqs. 116 and 117, respectively). In addition, a correction factor is used only for binary H_2O-CO_2 mixtures, see [25, 29].

$$a_0(H_2O) = 3.5464 \cdot 10^6 \quad MPa \cdot cm^6 \cdot K^{0.5} \cdot mol^2 \quad (116)$$

$$a_0(CO_2) = 4.661 \cdot 10^6 \quad MPa \cdot cm^6 \cdot K^{0.5} \cdot mol^2 \quad (117)$$

Table 5 illustrates that the equation of Holloway [25] is not improving the accuracy of predicted properties of supercritical CO_2-CH_4-N_2 fluids, compared to Chueh-Prausnitz [24] or *Seos* [9], and it is only a small improvement compared to the *RKeos* [8]. The accuracy of this equation is highly improved by using the definitions of a constants according to Bakker [27] (see Eqs. 112-115), and result in a maximum deviation of only 1% from experimental data in Table 5.

Experimental data, including molar volumes of binary H_2O-CO_2 fluid mixtures at supercritical conditions [30, 31, 32] are used to estimate fugacities of H_2O and CO_2 according to Eq. 118 (compare Eq. 10).

$$RT \ln \varphi_i = \int_0^p \left[V_{m,i} - V_m^{ideal}\right] dp \quad (118)$$

where $V_{m,i} - V_m^{ideal}$ is the difference between the partial molar volume of component i and the molar volume of an ideal gas (see also Eq. 13). The difference between Eqs. 118 and 10 is the mathematical formulation and the use of different independent variables, which are temperature and pressure in Eq. 118. The integration to calculate the fugacity coefficient can be graphically obtained by measuring the surface of a diagram of the difference between the ideal molar volume and the partial molar volume (i.e. $V_{m,i}$ - V_m^{ideal}) as a function of pressure (Figure 6). The surface obtained from experimental data can be directly compared to calculated curves from equations of state, according to Eq. 10 (Table 6).

The dashed line in Figure 6 is calculated with another type of equation of state: a modification of the Lee-Kesler equation of state [33] that is not treated in this manuscript because it is not a two-constant cubic equation of state. Fugacity estimations of H_2O are similar according to both equations, and reveal only a minor improvement for the two-constant cubic equation of state [27]. The experimental data to determine fugacity of CO_2 in this fluid mixture is inconsistent at relative low pressures (< 100 MPa). The calculated fugacity [27] is approximately compatible with the experimental data from [31, 32].

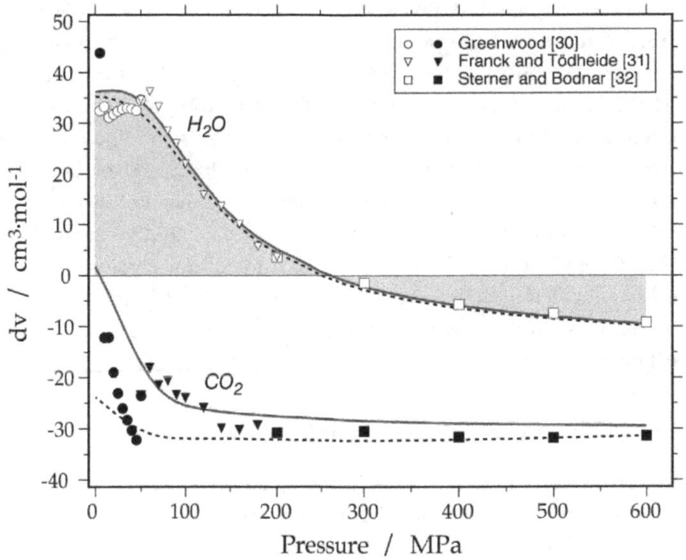

Figure 6. Fugacity estimation in a pressure - dv diagram at 873 K and a composition of $x(CO_2) = 0.3$ in the binary H_2O-CO_2 system, where dv is the molar volume difference of an ideal gas and the partial molar volume of either H_2O or CO_2 in binary mixtures. Experimental data are illustrated with circles, triangles and squares (solid for CO_2 and open for H_2O. The red lines are calculated with Bakker [27], and the shaded area is a measure for the fugacity coefficient of H_2O (Eq. 118).

Pressure (MPa)	*Exp. fugacity (MPa)*	*B1 fugacity (MPa)*
10	6.692	6.659 (-0.5%)
50	27.962	27.3061 (-2.3%)
100	45.341	44.6971 (-1.4%)
200	77.278	75.0515 (-2.9%)
300	114.221	111.072 (-2.8%)
400	160.105	157.145 (-1.8%)
500	219.252	216.817 (-1.1%)
600	295.350	294.216 (-04%)

Table 6. Fugacities of H_2O in H_2O-CO_2 fluid mixtures, $x(CO_2) = 0.3$, at 873.15 K and variable pressures. B1 fugacity is calculated with Bakker [27]. The deviation (in %) is illustrated in brackets.

9.3. Bowers and Helgeson [34] and Bakker [28]

Most natural occurring fluid phases in rock contain variable amounts of NaCl, which have an important influence on the fluid properties. Bowers and Helgeson [34] modified the *RKeos* [8] to be able to reproduce the properties of homogeneous supercritical fluids in the H_2O-CO_2-NaCl system, but only up to 35 mass% NaCl. The model is originally restricted between 350 and 600 °C and pressures above 50 MPa, according to the experimental data [35] that was used to design this equation. This model was modified by Bakker [28] including CH_4, N_2, and additionally any gas with *a* (ζ_2) and b (ζ_1) constants defined by the pseudo critical conditions (Eqs. 91-93 and 94-96). Experimental data in this multi-component fluid system with NaCl can be accurately reproduced up to 1000 MPa and 1300 K. Table 5 illustrates that this modification results in the best estimated molar volumes in the ternary CO_2-CH_4-N_2 fluid system at 100 MPa and 673 K. Similar to all modifications of the *RKeos* [8], this model cannot be used in and near the immiscibility conditions and critical points (i.e. sub-critical conditions).

10. Application to fluid inclusion research

Knowledge of the properties of fluid phases is of major importance in geological sciences. The interaction between rock and a fluid phase plays a role in many geological processes, such as development of magma [36], metamorphic reactions [37] and ore formation processes [38]. The fluid that is involved in these processes can be entrapped within single crystal of many minerals (e.g. quartz), which may be preserved over millions of years. The information obtained from fluid inclusions includes 1. fluid composition; 2. fluid density; 3. temperature and pressure condition of entrapment; and 4. a temporal evolution of the rock can be reconstructed from presence of various generation of fluid inclusions. An equation of state of fluid phases is the major tool to obtain this information. Microthermometry [39] is an analytical technique that directly uses equations of state to obtain fluid composition and density of fluid inclusions. For example, cooling and heating experiment may reveal fluid phase changes at specific temperatures, such as dissolution and homogenization, which can be transformed in composition and density by using the proper equations of state.

The calculation method of fluid properties is extensive and is susceptible to errors, which is obvious from the mathematics presented in the previous paragraphs. The computer package FLUIDS [6, 40, 41] was developed to facilitate calculations of fluid properties in fluid inclusions, and fluids in general. This package includes the group "Loners" that handles a large variety of equations of state according to individual publications. This group allows researchers to perform mathematical experiments with equations of state and to test the accuracy by comparison with experimental data.

The equations of state handled in this study can be downloaded from the web site **http://fluids.unileoben.ac.at** and include 1. "*LonerW*" [7]; 2. "*LonerRK*" [8]; 3. "*LonerS*" [9]; 4. "*LonerPR*" [10]; 5. "*LonerCP*" [24]; 6. "*LonerH*" [25, 26, 27]; and 7. "*LonerB*" [28, 34]. Each program has to possibility to calculate a variety of fluid properties, including pressure, temperature, molar volume, fugacity, activity, liquid-vapour equilibria, homogenization conditions, spinodal, critical point, entropy, internal energy, enthalpy, Helmholtz energy, Gibbs energy, chemical potentials of pure gases and fluid mixtures. In addition, isochores can be calculated and exported in a text file. The diagrams and tables presented in this study are all calculated with these programs.

Author details

Ronald J. Bakker
Department of Applied Geosciences and Geophysics, Resource Mineralogy, Montanuniversitaet, Leoben, Austria

11. References

[1] Dziewonski AM, Anderson DL (1981) Preliminary reference Earth model. Phys. Earth Planet. In. 25: 297-356.
[2] Press F, Siever R (1999) Understanding Earth. Freeman, New York, 679 p.
[3] Roedder E (1984) Fluid inclusions, Reviews in Mineralogy 12, Mineralogical Association of America, 646 p.
[4] Bakker RJ (2009) Reequilibration of fluid inclusions: Bulk diffusion. Lithos 112: 277-288.
[5] Prausnitz JM, Lichtenthaler RN, Gomes de Azevedo E (1986) Molecular thermodynamics of fluid-phase equilibria. Prentice-Hall, Englewood Cliffs, NJ, 600 p.
[6] Bakker RJ (2009) Package FLUIDS. Part 3: correlations between equations of state, thermodynamics and fluid inclusions. Geofluids 9: 63-74.

[7] Waals JD van der (1873) De continuiteit van den gas- en vloeistof-toestand. PhD Thesis, University Leiden, 134 p.

[8] Redlich OR, Kwong JNS (1949) On the thermodynamics of solutions, V: An equation of state, fugacities of gaseous solutions. Chem Rev. 44: 233-244.

[9] Soave G (1972) Equilibrium constants from a modified Redlich-Kwong equation of state. Chem. Eng. Sci. 27: 1197-1203.

[10] Peng DY, Robinson DB (1976) A new two constant equation of state. Ind. Eng. Chem. Fundam. 15: 59-64.

[11] Reid RC, Prausnitz JM, Poling BE (1989) The properties of gases and liquids. McGraw-Hill Book Company, NJ, 741 p.

[12] Levelt-Sengers J (2002) How fluids unmix, discoveries by the school of van der Waals and Kamerlingh Onnes. Koninklijke Nederlandse Akademie van Wetenschappen, Amsterdam, 302 p.

[13] Beyer WH (1991) CRC Standard mathematical tables and formulae. CRC Press, Boca Raton, Fl, 609 p.

[14] Horn RA, Johnson CR (1985) Matrix analysis. Cambridge University Press, Cambridge, 561 p.

[15] Prausnitz JM, Anderson TF, Grens EA, Eckert CA, Hsieh R, O'Connell JP (1980) Computer calculations for the multicomponent vapor-liquid and liquid-liquid equilibria. Prentice-Hall, Englewood Cliffs, NJ, 353 p.

[16] Donnelly HG, Katz DL (1954) Phase equilibria in the carbon dioxide - methane system. Ind. Eng. Chem. 46: 511-517.

[17] Baker LE, Luks KD (1980) Critical point and saturation pressure calculations for multipoint systems. Soc. Petrol. Eng. J. 20: 15-4.

[18] Konynenburg PH van, Scott RL (1980) Critical lines and phase equilibria in binary van der Waals mixtures. Philos. T. Roy. Soc. 298: 495-540.

[19] Arai Y, Kaminishi GI, Saito S (1971) The experimental determination of the P-V-T-X relations for carbon dioxide-nitrogen and carbon dioxide-methane systems. J. Chem. Eng. Japan 4: 113-122.

[20] Span R, Wagner W (1996) A new equation of state for carbon dioxide covering the fluid region from the triple point temperature to 1100 K at pressures up to 800 MPa. J. Phys. Chem. Ref. Data 25:1509-1596.

[21] Lorentz HA (1881) Über die Anwendung des Satzes vom Virial in den kinetischen Theorie der Gase. Ann. Phys. 12: 127-136.

[22] Waals JD van der (1890) Molekulartheorie eines Körpers, der aus zwei verschiedenen Stoffen besteht. Z. Ph. Chem. 5: 133-173

[23] Seitz JC, Blencoe JG, Joyce DB, Bodnar RJ (1994) Volumetric properties of CO_2-CH_4-N_2 fluids at 200 °C and 1000 bars: a comparison of equations of state and experimental data. Geochim. Cosmochim. Acta 58: 1065-1071.

[24] Chueh PL, Prausnitz JM (1967) Vapor-liquid equilibria at high pressures. Vapor-phase fugacity coefficients in non-polar and quantum-gas mixtures. Ind. Eng. Chem. Fundam. 6: 492-498.
[25] Holloway JR (1977) Fugacity and activity of molecular species in supercritical fluids. In: Fraser DG, editor. Thermodynamics in geology, pp 161-182.
[26] Holloway JR (1981) Composition and volumes of supercritical fluids in the earth's crust. In Hollister LS, Crawford MI, editors. Short course in fluid inclusions: Applications to petrology, pp. 13-38.
[27] Bakker RJ (1999a) Optimal interpretation of microthermometrical data from fluid inclusions: thermodynamic modelling and computer programming. Habilitation Thesis, University Heidelberg, 50 p.
[28] Bakker RJ (1999b) Adaptation of the Bowers and Helgeson (1983) equation of state to the H_2O-CO_2-CH_4-N_2-NaCl system. Chem. Geol. 154: 225-236.
[29] Santis R de, Breedveld GJF, Prausnitz JM (1974) Thermodynamic properties of aqueous gas mixtures at advanced pressures. Ind. Eng. Chem. Process, Dess, Develop. 13: 374-377.
[30] Greenwood HJ (1969) The compressibility of gaseous mixtures of carbon dioxide and water between 0 and 500 bars pressure and 450 and 800 °Centigrade. Am. J.Sci. 267A: 191-208.
[31] Franck EU, Tödheide K (1959) Thermische Eigenschaften überkritischer Mischungen von Kohlendioxyd und Wasser bis zu 750 °C und 2000 Atm. Z. Phys. Chem. Neue Fol. 22: 232-245.
[32] Sterner SM, Bodnar RJ (1991) Synthetic fluid inclusions X. Experimental determinations of the P-V-T-X properties in the CO_2-H_2O system to 6 kb and 700 °C. Am. J. Sci. 291: 1-54.
[33] Duan Z, Møller N, Weare JH (1996) A general equation of state for supercritical fluid mixtures and molecular simulation of mixtures PVTX properties. Geochim. Cosmochim. Acta 60: 1209-1216.
[34] Bowers TS, Helgeson HC (1983) Calculation of the thermodynamic and geochemical consequences of non-ideal mixing in the system H_2O-CO_2-NaCl on phase relations in geological systems: equation of state for H_2O-CO_2-NaCl fluids at high pressures and temperatures. Geochim. Cosmochim. Acta 47: 1247-1275.
[35] Gehrig M (1980) Phasengleichgewichte und pVT-daten ternärer Mischungen aus Wasser, Kohlendioxide und Natriumchlorid bis 3 kbar und 550 °C. University Karlsruhe, PhD-thesis, Hochschul Verlag, Freiburg, 109 p.
[36] Thompson JFH (1995) Magmas, fluids, and ore deposits. Short course 23, Mineralogical Association of Canada.
[37] Spear FS (1995) Metamorphic phase equilibria and pressure-temperature-time paths. Mineralogical Society of America, Monograph, 799 p.
[38] Wilkinson JJ (2001) Fluid inclusions in hydrothermal ore deposits. Lithos 55: 229-272.

[39] Shepherd TJ, Rankin AH, Alderton DHM (1985) A practical guide to fluid inclusion studies. Blackie, Glasgow, 239 p.
[40] Bakker RJ (2003) Package *FLUIDS* 1. Computer programs for analysis of fluid inclusion data and for modelling bulk fluid properties. Chem. Geol. 194: 3-23.
[41] Bakker RJ, Brown PE (2003) Computer modelling in fluid inclusion research. In: Samson I, Anderson A, Marshall D, editors. Short course 32, Mineralogical Association of Canada, pp. 175-212.

Chapter 7

Thermodynamics Simulations Applied to Gas-Solid Materials Fabrication Processes

Elisabeth Blanquet and Ioana Nuta

Additional information is available at the end of the chapter

1. Introduction

The development and the design of materials and/or the processes of their fabrication are generally very time consumer and with expensive operations. Various methods of development can be conceived. Often, "empirical" approaches are adopted: the choice of the experimental parameters is established either on technological or commercial criteria, the optimization being the results of a "trial and error" approach, or on the results of design of experiments (DOE) approach targeted at a property of a material or a parameter of a very particular process. Another approach is to use process modeling: to simulate the process by a more or less simplified model. The modeling of gas-solid materials fabrication processes brings together several physical and chemical fields with variable complexity, starting from thermodynamics and\or kinetics studies up to the mass and heat transport coupled with databases and with thermodynamic and/or kinetics transport properties.

The objective of this chapter is to illustrate the interest areas computer-aided materials design and of processes optimization based on the thermodynamic simulation and giving some interesting examples in different domains. Databases as well as their necessary tools for the implementation of the thermodynamic calculations will be described.

The thermodynamic simulations of multicomponent systems contribute at two important points: the selection of the material and the optimization of the conditions of fabrication. In order to obtain a finely targeted product which meet specific functionalities, it is necessary to answer the following questions:

- what type of composition, quantity, and microstructure of the material allow to obtain such properties?
- it is possible to elaborate the material? By what process, with which reagent/ species and which operating conditions?
- is stable this material during a treatment in temperature, and under a given atmosphere?

- does this material react with its environment (substrate, oven, atmosphere)?

These questions are connected because the properties of the material, essentially conditioned by the microstructure of the final material, are going to depend on its chemical composition, to the process, to the operating conditions and on dimensions of the fabrication equipment.

The answers to these questions can be provided and shown from calculation of phase diagrams, evaluation of chemical reactions, calculation of equilibrium pressures, and from reaction diagrams.

2. Description of the methodologies

To evaluate equilibrium state, two possible approaches exist. The first one is to choose *a priori* a limited number of species and the simple chemical reactions which are susceptible to represent the studied process, and to estimate one or several most favorable reactions. It can be reminded the very classic use of the diagrams of Ellingham for the synthesis of metals from their oxides.

The second approach, more complex, is based on the analysis and the consideration of all the species belonging to the chemical system in the studied process.

The optimization procedure must have the following stages:

a. The analysis of the system with the inventory of all the species reasonably susceptible to be present during the reactions taking place in the process.
b. The construction of a consistent set of thermodynamic data for these species.
c. The thermodynamic calculations at equilibrium of complex system
d. The best representation of the results for the users

Thus, the thermodynamic calculations often give satisfactory results for processes which use high temperatures and residence time or reaction but for processes at low temperature, the kinetic factors must be not neglected. That is why the recent developments of thermodynamic softwares tend to include descriptions of phenomena of diffusion and reaction kinetics.

The thermodynamic approach gives the superior limit of possibilities of process (considering the reaction rates as infinite). It can be the only way of modeling for a complex system where the mechanisms of reaction and the kinetic data are badly known.

To include some dynamic aspects (mass transport) in the modeling, an approach which takes into account the evaluation of flows will be presented. It concerns applications where the total pressure is low (<10 Pa).

2.1. The analysis of the system

The analysis of the system consists in listing the following points: the range of temperature T, range of pressure P or of volume V, the duration of the process, the list of the reagent species, the inventory of the components of the reactor and the nature of the atmosphere. For these last points, it means elaborating a list of all the compounds, the gaseous species, the elements and the solid solutions which result from the combination of the basic elements of the system.

This list is automatically generated thanks to interfaces with databases. However, it is advisable to make sure that the used database is very complete. As an example, the list corresponding to the chemical system Si-C-H-Ar (proceeded CVD (Chemical Vapor Deposition) contains about sixty species – excluding the hydrocarbons C_xH_y where x>3 [1].

2.2. Calculation of a thermochemical equilibrium

In a process reactor, at constant pressure, the balance is reached when the total free Gibbs energy function of the system is minimal (equation 1). To determine the nature and the proportion of the present phases at equilibrium, it is necessary to have the description of the energies of Gibbs of all these phases.

$$\frac{\Delta G}{RT} = \sum_{j=1}^{Ne} qj \cdot \frac{\Delta Gj}{RT} \qquad (1)$$

where qj the number of moles of the species j, Gj the molar free energy of Gibbs of the species j, Ne total number of species.

The Gibbs energy can be described from the enthalpy (H) and the entropy (S):

$$G(T) = H(T) - T * S(T) \qquad (2)$$

with

$$H(T) = \Delta H(298K) + \int_{298K}^{T} c_P(T)dT \qquad (3)$$

$$S(T) = S(298K) + \int_{298K}^{T} \frac{c_P(T)}{T} dT \qquad (4)$$

The necessary data are thus: Cp(T), ΔH(298K), S(298K) and the data of possible phases transitions Ttrans, ΔHtrans(T).

Various formalisms are adopted for the analytical expression of the function Cp (T). Among them, the formalisms of the SGTE (Scientific Group Thermodata Europe) [2] (equation 5) and of the NASA [3](equation 6) are :

$$c_P(T) = a + bT + cT^2 + \frac{d}{T^2} \qquad (5)$$

$$c_P(T) = a + bT + cT^2 + dT^3 + eT^4 \qquad (6)$$

where a, b, c, d, e are adjustable parameters.

So, it can be described analytically the Gibbs energy G for a stoichiometric compound (equation 7), for a gas (equation 8) and a solution phase (equation 9):

$$G(T) = A + BT + CT\ln T + DT^2 + ET^3 + \frac{F}{T} \tag{7}$$

$$G(T,P) = G(T) + RT\ln\frac{\overline{P}}{P_0} \tag{8}$$

$$\left.\begin{array}{l} G(T,x) = \sum_i G_i^{ref}(T) + G^{id}(T,x_i) + G^{excès}(T,x_i) \\ G^{id}(T,x_i) = RT\sum_i (x_i \ln x_i) \\ G^{excès}(T,x_i) \quad \textit{described from a model} \end{array}\right\} \tag{9}$$

Besides, as neither the enthalpy nor the entropy can be described in an absolute way, a reference state must be used for these two functions of state. For the entropy, the adopted convention consists in taking a zero value at 0 K. In the case of the enthalpy, the most common convention is to choose the stable structure of the element at T = 298K, as standard reference state (e.g. Al cfc, Ti hcp, O_2 gas ...). For the reference state, ΔH(298K)=0 and S(0K)=0.

As the reliability of the results of the thermodynamics simulation depends widely on the quality and on the consistence of the necessary data, it is advisable to attach an importance to the consistence of the available information: thermodynamics measurements, theoretical calculations, characterizations (X-ray diffraction, Environmental Scanning Microscopy), balance of phases (diagrams). In the Table 1 are given some experimental and theoretical techniques usually used to obtain the thermodynamic data.

The thermodynamic information are accessible in compilations of binary phases diagrams (for example Hansen [5], Elliot [6], Massalski [7]), ternary (Ternary Alloys [8]), or specialized journals (CALPHAD, Journal of Phase Equilibria, Intermetallics...), or tables (JANAF Thermochemical Tables [9], Barin [10], Gurvich [11]).

Today, most of data are available in international electronic databases. In Europe, the economic interest group "Scientific Group Thermodata Europe [12]" proposes common data bases for compounds, pure substances and for solutions. Also let us quote the "Coach" data bank (more than 5000 listed species) proposed by Thermodata [1], well adapted to simulate gas/solid processes, the FACT bank (oxides/salts) proposed by the company GTT [13] and the Research Center in Calculation Thermodynamics [14], base TCRAS [15], bases NASA combustion [16], NIST [17].

	Experimental	Theory
ΔH(T)	Calorimetry (dissolution)	Ab initio calculations Estimations : Miedema [4], analogy
Cp(298K), S(298K)	Temperature measurement at low temperature	
Cp(T)		Estimations : Neumann-Kopp law(ΔCp=0 for a condensed compound)
Cp, S(T)	Differential Scanning Calorimetry (DSC)	volume-specific heat capacity calculations
G(T)	Electromotive force, Mass Spectrometry (activity data, partial pressures at equilibrium)	
T^{trans}, H^{trans} (T)	Differential Thermal Analysis (DTA) Thermogravimetric analysis (TGA)	

Table 1. The classically used techniques to obtain the thermodynamic data

2.3. Calculations of complex equilibrium

The software of complex equilibrium is based on the minimization at constant temperature T of the Gibbs energy and constant pressure P (equation 1) or Helmholtz energy (equation 10), at constant volume:

$$\frac{\Delta F}{RT} = \sum_{j=1}^{Ne} qj \frac{\Delta Gj}{RT} - \frac{P.V}{RT} \tag{10}$$

q_j the number of moles of the species j .

The constraints of mass equilibrium of each present element in the chemical system expressed according to the number of atoms on the pure element i (C elements) are translated by the equation (11):

$$n_i = \sum_{e=1}^{Ne} q_e T_i^e \text{ with i } = 1^{\circ}\ldots\text{C} \tag{11}$$

where T_i^e represents the stoichiometry of the species e for the element i.

These C equations can be translated under the matrix shape (equation 12):

$$[n] = [T]*[q] \text{ with Ne} > \text{C} \tag{12}$$

There are multiple algorithms allowing this minimization. Various classifications were given, the most exhaustive having been supplied by Smith and Missen [18]. In a simple way, two groups of algorithms can be distinguished: on one hand the methods of direct minimization, about zero order for the calculation of the function G, on the other hand, the methods of the first order based on the equality of the chemical potential which require the calculation of the function derivatives. These last ones also include the methods of second order, using among others the algorithm of Newton-Ralphson which is based on the second derivatives. It is necessary to note that the methods of the first order must be perfectly controlled because they can lead to a maximum instead of a minimum and consequently to a wrong result.

A method of the first group is described below: the matrix T is decomposed into a regular square matrix Tp of dimension C and a matrix Td of dimension (C, N_e-C) such as:

$$[n]=[T_p][q_p]+[T_d][q_d] \tag{13}$$

The C species which constitute the matrix column q_p are called the "main species" because they are chosen among the most important species and have by definition a linear independent stoichiometry. The N_e-C remaining species of the matrix T_d is called "derived species" although they are chosen as variables from the minimization. So the N_e-C values q_d are given by the procedure of minimization, C values q_p is calculated by resolving the linear system:

$$[q_p]=[T_p]^{-1}[n]-[T_p]^{-1}[T_d][q_d] \tag{14}$$

An iteration of this method is divided into two steps. Firstly, the phase of exploration, every variable is modified by a value + or - h.

If X_{n-1} is the vector representing the variables after n-1 iterations: the species i having a step h_i and G_i the value of the function

$$G_i^+ = \Delta G(x_1, x_2, ..., x_{i-1}, x_i + h_i, x_{i+1}, ..., x_v) \tag{15}$$

if $G_i^+ < G_i$ either,

$$G_i^{++} = \Delta G(x_1, x_2, ..., x_{i-1}, x_i + 2h_i, x_{i+1}, ..., x_v) \tag{16}$$

or,

$$G_i^- = \Delta G(x_1, x_2, ..., x_{i-1}, x_i - h_i, x_{i+1}, ..., x_v) \tag{17}$$

When the exploration phase is ended, the next step is to move to the second algorithm phase where from the values of G^+ and X_n^+ issued from the exploration phase, we calculate $X^{++} = X_{n-1} + (X_n^+ - X_{n-1})$ as well as the corresponding value G^{++} to obtain the optimal set [19]. To proceed the mimimization procedure, a certain number of more and more friendly softwares are available commercially. It can be listed as example:

- « Gemini1/Gemini2 » [1]
- « FactSage » [14]

- « MTDATA » [20]
- « Thermocalc » [21]

3. Applications

Thomas [22], Bernard [23] and Pons [24] present few examples on CVD processes to illustrate the use of an a priori thermodynamic analysis. In the following paragraphs, it was chosen to show other few examples which evidence the help of thermodynamic modeling in industrial bottlenecks:

- Thermal stability of Metal-Organic Precursors used in CVD and ALD processes.
- Stability of SiC in H_2 atmosphere
- HfO_2 plasma etching
- SiO_2 PVD evaporation-condensation deposition process.

In the last two examples which correspond to processes operating at low pressure (<1 Pa), in addition to pure thermodynamic approach, a dynamic approach was presented which includes calculations of the major species flows.

3.1. Microelectronics: Thermal stability of metal-organic precursors used in CVD and ALD processes

In the pursuit of smaller and faster devices manufacture, microelectronics industry scales down feature sizes and thus has to develop new materials and processes. Nowadays, organometallic precursors are widely used in ALD (Atomic Layer Deposition) and CVD (Chemical Vapor Deposition) deposition processes due to low deposition temperature (generally below 523 K). The objective of computational modeling for gaseous phase processes like ALD or CVD is to correlate the as-grown material quality (uniformity, growth rate, cristallinity, composition, etc) to general parameters such as growth conditions, reactor geometry, as well as local parameters that are actual flow, thermal fields and chemical kinetics at the solid/gas interface.

The gaseous precursors compounds used for the transport of the elements to be deposited by these processes have to meet several physicochemical properties requirements including relatively high volatility, convenient decomposition behavior and thermal stability. The tantalum organometallic precursor pentakis dimethylamino tantalum (PDMAT), remains an attractive solution for tantalum nitride films deposition. Unfortunately, information on physical and chemical behavior of this kind of precursor is scarce and namely species that are formed during vaporization and transported to the deposition chamber remain generally unknown. Thus, the knowledge of thermodynamics of these gaseous compounds could help in the understanding of the transport and growth mechanisms. Indeed, thanks to thermodynamics, it is possible to evaluate what evolves at equilibrium in the precursor source, in the input lines and in the deposition chamber where deposition reactions occur. To control, optimize and understand any ALD or CVD processes, thermodynamic simulations are very useful and therefore data should be primarily assessed.

3.1.1. Assessment of PDMAT thermodynamic data by mass spectrometry

In order to deposit thin layers of TaN using PDMAT, ALD experiments evidenced a cracking of this precursor in the ALD reactor [25].

Cracking reactions of PDMAT can be complex and occur at the same time. A quantitative interpretation of cracking reactions can be deduced from observed molecules by mass spectrometry [26, 27] with the condition that all products and reactants of the reaction are observed and measured by the mass spectrometer at the same time. Without additional hydrogen contribution, the two following cracking reactions of Ta $[N(CH_3)_2]_5(g)$ could occur:

i. either the Ta- $[N(CH_3)_2]$ bond breaks with the additional break of H-CH_2 to produce $HN(CH_3)_2$ (the so-called β substitution):

$$Ta\left[N(CH_3)_2\right]_5(g) \rightarrow Ta\left[N(CH_3)_2\right]_3\left[NCH_3CH_2\right](g) + HN(CH_3)_2(g), \tag{18}$$

ii. or the only bond break of Ta- $[N(CH_3)_2]$,

$$Ta\left[N(CH_3)_2\right]_5(g) \rightarrow Ta\left[N(CH_3)_2\right]_4(g) + N(CH_3)_2(g). \tag{19}$$

As no $N(CH_3)_2$ (g) radical was detected, the observed $HN(CH_3)_2$ (g) molecule could be formed by the following complete and rapid reaction,

$$N(CH_3)_2(g) + ½\,H_2(g)\ (\text{ou H }(g)) \rightarrow HN(CH_3)_2(g). \tag{20}$$

Consequently, it could be assumed that the measured $HN(CH_3)_2$ (g) amount is the same as the initial produced amount of $N(CH_3)_2$ (g): this radical spontaneously reacts totally according to the reaction (20) after being produced by reaction (19). This mechanism could explain why $N(CH_3)_2$ (g) was not detected. So, in this study, the total cracking reaction was finally considered:

$$Ta\left(N(CH_3)_2\right)_5(g) + ½\,H_2(g) \rightarrow Ta\left(N(CH_3)_2\right)_4(g) + HN(CH_3)_2(g) \tag{21}$$

Another cracking reaction could be noticed: $OTaN_4C_8H_{24}$ (g) molecule broke down into $OTaN_3C_6H_{18}$ (g) according to the following reaction,

$$OTaN_4C_8H_{24}(g) + ½\,H_2(g) \rightarrow OTaN_3C_6H_{18}(g) + HN(CH_3)_2(g) \tag{22}$$

because the energy of the Ta-N bond is lower than the Ta-O bond.

The experimental study of these two reactions (21) and (22) requires to know or measure $H_2(g)$ pressure. Hydrogen could come from either equilibrium with cracking cell deposits or either molecules losing one or more hydrogen atoms. In this last case, a new molecule should be present. In this study, as no hydrogen was detected or introduced intentionally in the cell, it was assumed that the amino radical is totally consumed and produces $HN(CH_3)_2$ (g) with just a sufficient hydrogen amount. So, it can be assumed that the partial pressure of

$p(HN(CH_3)_2)$ is quite equal to the partial pressure of $p(N(CH_3)_2)$. That allows us to calculate the equilibrium constant of reaction (19):

$$K_p(T) = \frac{p(N(CH_3)_2) \cdot p(Ta[N(CH_3)_2]_4}{p(Ta[N(CH_3)_2]_5} \tag{23}$$

Pressure measurements of these three molecules by mass spectrometry lead to the evaluation of standard enthalpy at 298 K from the third law of thermodynamics:

$$\Delta_r H^0_{298K} = -RT \ln K_p(T) - T \cdot \Delta_r fef^0_T \tag{24}$$

Measured partial pressures of Ta $[N(CH_3)_2]_5$ (g), Ta $[N(CH_3)_2]_4$ (g) and $HN(CH_3)_2$ (g) are elsewhere reported [26, 27].

From this, the average value of $\Delta_r H°_{298\,K}$ was evaluated to be equal to (85±5) kJ/mol.

3.1.2. Thermodynamic simulation of PDMAT (thermal cracking)

Thermodynamic simulations, based on the Gibbs free energy minimization of the Ta-C-N-H-(O)-(Ar) system were performed using GEMINI software [1] to provide the nature of the species that should be present at equilibrium under experimental conditions. The sets of thermodynamic data which have been used come from SGTE 2007 database [28] and from the mass spectrometry study for Ta $[N(CH_3)_2]_5$(g), Ta $[N(CH_3)_2]_4$(g),and NC_2H_6 (g) gaseous species [26]. Without any available literature data or any estimates, it cannot be considered any thermodynamic description of $OTaN_xC_yH_z$ (g) gaseous species and intermediate $TaN_xC_yH_z$ (g) species such as $TaN_3C_6H_{16}$ (g), even though these species are expected to appear as observed in mass spectrometric measurements and to play a role in PDMAT cracking and in Ta containing solid formation [26]. Two kinds of simulations have been performed within a temperature range from 400 to 750 K and at 10 Pa, which is our typical mass spectrometric total pressure in the cracking cell. First, homogeneous equilibrium was investigated - no solid phase is allowed to be formed - which corresponds to no deposition i.e. transport in gas lines held at temperature above the saturated one (Figure 1).

Second, a heterogeneous equilibrium - the solid phase is allowed to be formed - has also been simulated, which corresponds to the deposition process occurring in the ALD reactor and in the cracking cell.

In all these thermodynamic simulations, it appeared that Ta $[N(CH_3)_2]_5$(g), (PDMAT) is not stable. In Figure 1 the homogeneous equilibrium calculation show that Ta $[N(CH_3)_2]_4$(g) is stable but disappears after 450 K and Ta(g) is the only one main tantalum containing species after 415 K - but this species will soon be condensed due to large over saturation-. Added to Ta(g), a lot of cracking gaseous species such as N_2 (g), CH_4 (g), H_2 (g) originate from the complete amine decomposition and indeed among these species, NC_2H_7 (g) and NC_2H_6 (g) do not appear. The heterogeneous equilibrium calculations shows the formation of C solid that corresponds to the amine decomposition and this amount of free carbon increases with increasing temperature. Also, the formation of solid TaN was observed within the whole investigated temperature range and no gaseous tantalum containing species pertained contrary to mass spectrometric experiments.

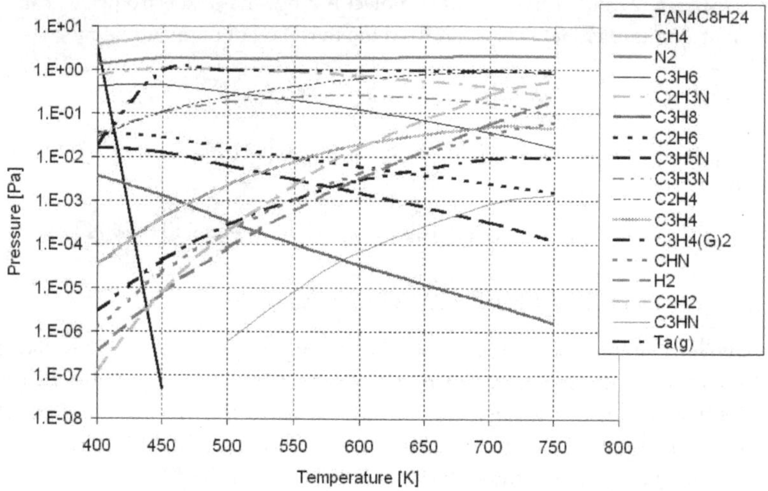

Figure 1. Homogeneous thermodynamic simulations performed starting from 1 mole of the compound Ta $[N(CH_3)_2]_5(g)$ and 0.001 mol of Ar and for an applied total pressure equal to 10 Pa in order to be compared with the mass spectrometric experiments.

3.1.3. *Conclusions*

It is to be concluded that discrepancies exist between thermodynamic simulations and mass spectrometric experiments. Indeed, thermodynamics predicted total cracking of both Ta $[N(CH_3)_2]_5(g)$ and amine molecules in the whole investigated temperature ranges, while in mass spectrometric experiments, Ta $[N(CH_3)_2]_5(g)$ and NC_2H_7 (g) amine have been observed only in the low temperature below 623 K [26, 29]. The deviation vs. equilibrium could be analyzed experimentally by the use of various sizes of cracking cell as well as the deliberate and controlled introduction of $H_2(g)$. However, despite these limitations, these results indicate the main features of the precursor thermal behavior which can be very useful in the first stages of the development of any new ALD or CVD (for precursor transport) processes.

3.2. High power electronics: Stability of SiC in H_2 atmosphere

Silicon carbide (SiC) possesses many favorable properties making it interesting for a multitude of applications, from high temperature to high frequency and high power device.

Among them, its excellent physico-chemical and electronic properties such as wide band gap and high breakdown field, together with the degree of maturity of technology, makes SiC a good candidate for mass production of Schottky diodes [30].

SiC device processing is conditioned to the fabrication of large area single crystal wafers with the lowest defect density associated to deposition of epitaxial thin films which present good structural quality and controlled doping level [31]. The most common processes used to develop SiC wafers and SiC thin films are the seeded sublimation growth technique so called the "Modified Lely method" and the Chemical Vapor Deposition technique from propane and silane, respectively.

Huge improvements for both processes have been observed in the last decades. They come mainly from extensive experimental effort, all over different groups in the world. However, macroscopic modeling has given valuable information to understand the impact of some growth parameters and propose new design of experiment to enlarge wafer size and deposition area.

On both processes [32-35], some modeling trends were largely reported combined with experimental results obtained in our research's groups.

Special emphasis is given to chemical related results. To carry out modeling, it was followed the different levels of complexity procedure described in the earlier paragraphs. Owing to the similarity of the two systems, studies on species and material databases have been naturally used for both processes.

CVD- grown SiC films can be obtained from a variety of precursors which are generally part of the Si-C-H system. However, to obtain high crystal quality of 4H and 6H SiC layers, which are the most interesting polytypes for the power devices applications, experimental investigations have demonstrated that silane ($SiH_4(g)$) – propane ($C_3H_8(g)$) gave the most stable growth, in the typical conditions (temperature higher than 1700K, pressure between 10 kPa to 100kPa, hydrogen as carrier gas) [36]. Operations are separated in two steps, first an in situ etching step to prevent epitaxy-induced defects, then the deposition step.

A great body of literature dealing with both theoretical and experimental results has been devoted to understanding chemistries relevant to the separate Si-H and C-H systems.

However, it appears that most is unknown about the chemical reactions in which organosilicon species, that include the three elements, can be involved. This is related to the difficulty to measure thermochemical properties of such reactive, short life time, species. Most of the thermodynamic data that have been used for these species come from ab initio electronic structure calculations combined with empiric bond additivity corrections [37, 38]. Mass spectrometry measurements have been carried out to estimate the thermodynamic data of the gaseous species $Si_2C(g)$, $SiC_2(g)$, $SiC(g)$ and the condensed $SiC(s)$ phase [39, 40].

With a purely thermodynamic approach, it was examined the preliminary operation of in situ etching. It was found that the H_2(g) etching of SiC(s) at 1700 K under 100 kPa can lead to the formation of a condensed silicon phase, as shown on Figure 2. Thermodynamic study was made to understand the impact of the temperature, the pressure and the composition of the gas mixture [41].

Heterogeneous thermodynamic calculations show that the mixture (H_2(g) + condensed SiC(s)) ends in the formation of gaseous species such as CH_4(g) for the C-containing species and SiH_2(g), SiH_4(g), SiH(g), Si(g) and Si(s) in condensed phase for the Si-containing species. With the etching, the amount of the gaseous C-species formed (mainly CH_4(g)) is three times higher than the gaseous Si-species one. So, there is Si in excess which is condensed at the SiC(s) surface in a solid or liquid phase depending whether the etching temperature is higher or below the Si melting temperature. When the temperature is higher than 1800K, at atmospheric pressure, the quantity of formed gaseous Si-species becomes equal to the quantity of formed gaseous C-species. Consequently, the formation of liquid silicon is avoided.

3.2.1. *Conclusions*

Thermodynanic simulations have revealed the main phenomena and indicated some solutions. Reducing pressure would provide the same beneficial effect, though the etching rate decreases, as illustrated in Figure 3.

To compensate the formation of gaseous CH_4(g), the addition of an hydrocarbon species such as propane in the initial gaseous mixture would prevent the formation of condensed silicon.

All these effects have been confirmed with experimental studies (Figure 2).

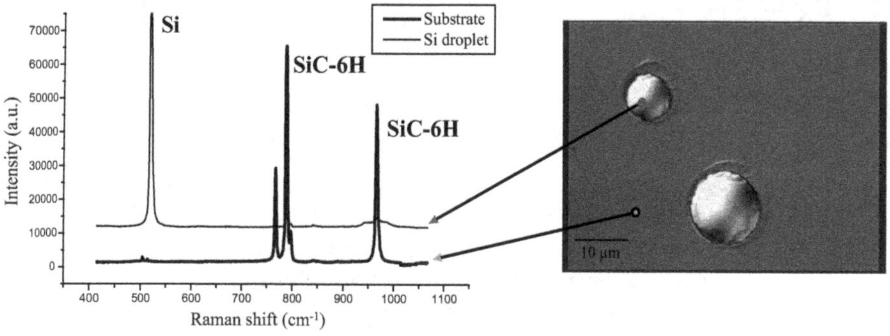

Figure 2. Optical Micrograph of SiC(s) surface etched with H_2(g) at 1700 K under 100 kPa, showing silicon droplets (right). The silicon phase is identified by Raman spectroscopy (left) [41].

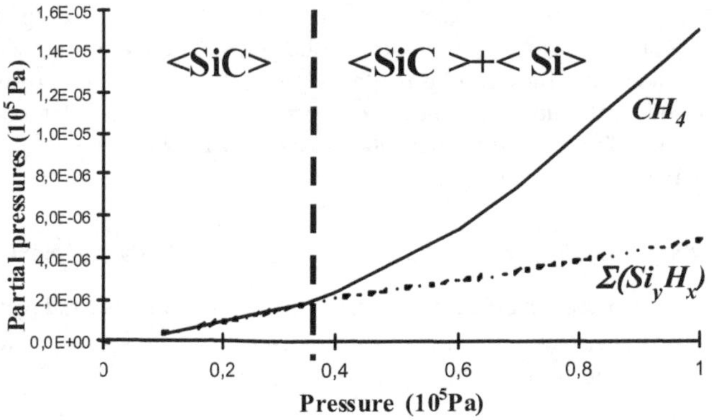

Figure 3. Heterogeneous equilibrium of the SiC(s) - H_2(g) system at 1685K. Gaseous species created by the etching are represented as a function of the pressure [41].

3.3. Thermodynamic analysis of plasma etching processes for microelectronics

With the constant downscaling of Complementary Metal-Oxide-Semiconductor (CMOS) devices and the consequent replacement of SiO_2 many high-*k* gate materials such as Al_2O_3, La_2O_3, Ta_2O_5, TiO_2, HfO_2, ZrO_2 and Y_2O_3 have been investigated. For each high-*k* material integration, the etch process has to be revisited.

In the case of the etching of HfO_2, one of the main issues is the low volatility of halogenated based etch by-products [42].

Compared to SiO_2 halide based etching process, thermodynamic data shows that, Hf based etch by-products ($HfCl_4$(g), $HfBr_4$(g), HfF_4(g)) are less volatile than Si etch by products ($SiCl_4$(g), $SiBr_4$(g), SiF_4(g)) [43]. Therefore, for HfO_2(s) etching, the choice of the halogenated based chemistry and substrate temperature are crucial parameters. In this work, thermodynamic studies have been carried out in the pressure (0.5 Pa) and temperature range (425 K to 625 K) conditions in order to select the most appropriate gas mixture and temperature leading to the formation of Hf and O based volatile products. Based on thermodynamic calculations in a closed system, the HfO_2(s) etching process has been simulated.

With this thermodynamic analyses, it is possible to determine an etch chemistry leading to volatile compounds and to estimate an etch rate under pure chemical etching conditions. It should be noted that the thermodynamic approach does not take into account the ion bombardment of the plasma.

3.3.1. Pure thermodynamic calculations of HfO_2 etching

For example, let's consider the etching of HfO_2(s) in CCl_4(g) plasma at 400 K and 0.5Pa. In such case, the thermodynamic system is composed by four elements Hf, O, C, and Cl. The thermodynamic calculation inputs are:each element of CCl_4(g) (C and Cl atoms) with

$HfO_2(s)$ as a solid phase. The main gaseous species are $CO_2(g)$ and $HfCl_4(g)$ and the main condensed species are $HfCl_4(s)$ on $HfO_2(s)$ in a solid phase. There are other gaseous species in very low amount so that they can be neglected as (CO, Cl_2, Cl). These results show that carbon and chlorine containing chemistries can lead to the etching of $HfO_2(s)$ by forming $CO_2(g)$ and $HfCl_4(g)$. Similar results have been obtained for the other halide chemistries $CCl_3F(g)$, $CCl_2F_2(g)$, $CCl_3F(g)$, $CCl_4(g)$.

3.3.2. *Thermodynamic analysis coupled to mass transport: evaluation of etching rate*

To point out the more promising chemistry among the usually adopted halogens precursors, the etch rate has been estimated from the flow calculations of each gaseous and condensed species under open conditions assuming molecular flow and the validity of the Hertz-Knudsen relation [44].

For these processes operating at low pressure (<10 Pa), it is possible to associate the incident and emitted flows from a given surface to the equilibrium partial pressures [45].

These calculations are based on the effusion calculations principles from the gas kinetic theory.

For a gaseous species e, the total flow Φ_e which is emitted from a vaporizing surface can be calculated according to the Hertz-Knudsen relation:

$$\Phi_e = \frac{p_e}{\sqrt{2\Pi M_e RT}} mol / s.m^2 \tag{25}$$

Where p_e et M_e are the partial pressure and molar mass of the species e, respectively.

For each etched or deposited element i, there is equality between the incident flow and the emitted or produced from reactions flows:

$$\Phi_i(incident) = \Phi_i(emitted) + \Phi_i(condensed) \tag{26}$$

on the deposited or etched surface

For example, in the case of $HfO_2(s)$ etching by $CHCl_3(g)$ with Ar(g), the system is Hf, O, C, H, Cl, Ar

For the previous thermodynamic calculations ($HfO_2(s)$ and $CHCl_3(g)$) the major species at equilibrium are:

- $CO_2(g)$, HCl(g), $HfCl_4(g)$, Ar(g) in the gaseous phase
- C(s), $HfO_2(s)$, $HfCl_4(s)$ in the solid phase

for a temperature of 300 K and a pressure of 5 Pa.

The flow equations are in this case (principle of mass conservation) :

- Flow of incident C − flow of evaporated C + flow of condensed C:

$$\frac{1}{\sqrt{2\pi.R.T}}\frac{P_{CO_2}}{\sqrt{M_{CO_2}}}+F(C(s))=\phi(CHCl_3(g)) \tag{27}$$

- Flow of incident H = flow of evaporated H :

$$\frac{1}{\sqrt{2\pi.R.T}}\frac{P_{HCl}}{\sqrt{M_{HCl}}}=\phi(CHCl_3(g)) \tag{28}$$

- Flow of incident Cl = flow of evaporated Cl + flow of condensed Cl :

$$\frac{1}{\sqrt{2\pi.R.T}}\left(\frac{P_{HCl}}{\sqrt{M_{HCl}}}+4.\frac{P_{HfCl_4}}{\sqrt{M_{HfCl_4}}}\right)+F(HfCl_4(s))=3.\phi(CHCl_3(g)) \tag{29}$$

with F(.(s)) the flow in solid phase species and $\phi(CHCl_3(g))=\frac{p^0_{CHCl_3}}{\sqrt{2\pi.M_{CHCl_3}.R.T_0}}$ (T_0 the temperature at the gases inlet and $p^0_{CHCl_3}$ the inlet partial pressure).

So five unknowns are obtained: $P_{CO_2}, F(C(s)), P_{HCl}, P_{HfCl_4}, F(HfCl_4(s))$

To be able to solve this system, five equations are needed. Already, three equations with these ones of flow exist. For the both missing, it's enough to consider:

The equation of total pressure: $P_{CO_2}+P_{HCl}+P_{HfCl_4}+P_{Ar}=P_{totale}$

And the value of the equilibrium constant for the assessment of mass equation of the system, the fifth global equation is supplied by:

$$HfO_2(s)+2.CHCl_3(g)\rightarrow C(s)+\frac{1}{2}.HfCl_4(s)+CO_2(g)+2.HCl(g)+\frac{1}{2}.HfCl_4(g) \tag{30}$$

the equilibrium constant of this equation is :

$$K_p=\frac{P_{CO_2}.P_{HfCl_4}^{\frac{1}{2}}.P_{HCl}^2}{P_{CHCl_3}^2} \tag{31}$$

$$\text{with}:\begin{cases}P_{CO_2}=x_1\\P_{HCl}=x_2\\P_{HfCl_4}=x_3\\F(C(s))=x_4\\F(HfCl_4(s))=x_5\end{cases} \tag{32}$$

The system of five equations and five unknowns is obtained:

$$\begin{cases} \dfrac{1}{\sqrt{2\pi.R.T}}\dfrac{x_1}{\sqrt{M_{CO_2}}}+x_4=\phi(CHCl_3) \\ \dfrac{1}{\sqrt{2\pi.R.T}}\dfrac{x_2}{\sqrt{M_{HCl}}}=\phi(CHCl_3) \\ \dfrac{1}{\sqrt{2\pi.R.T}}\left(\dfrac{x_2}{\sqrt{M_{HCl}}}+4.\dfrac{x_3}{\sqrt{M_{HfCl_4}}}\right)+x_5=3.\phi(CHCl_3) \\ x_1+x_2+x_3+P_{Ar}=P_{Totale} \\ \dfrac{x_1.x_3^{\frac{1}{2}}.x_2^2}{P_{CHCl_3}^2}=K_p \end{cases} \tag{33}$$

The partial pressures of the main species are obtained. To determine the etch rate, it is needed to use the calculated values for the pressures of the gases containing the elements of material to etch.

In our example, the following gases $CO_2(g)$ and $HfCl_4(g)$ are considered.

The theoretical etch rate ER is given by the lowest value between:

ER = flow $HfCl_4(g)$.molar volume $HfCl_4(g)$ or flow $CO_2(g)$.molar volume $CO_2(g)$ (in m/s)

Where:

$$Flow_{CO_2}=\frac{\mathrm{N}}{\sqrt{2\pi.R.T}}.\frac{P_{CO_2}}{\sqrt{M_{CO_2}}} \tag{34}$$

and

$$Flow_{HfCl_4}=\frac{\mathrm{N}}{\sqrt{2\pi.R.T}}.\frac{P_{HfCl_4}}{\sqrt{M_{HfCl_4}}}\ \left(\text{in molecules / s.m}^2\right) \tag{35}$$

with P_{CO_2} and P_{HfCl_4} determined by the resolution of the mathematical system and N the Avogadro number, P in Pa and M in kg.

Figure 4 shows the evolution of the calculated $HfO_2(s)$ etch rate as a function of temperature for different F/Cl ratios in $CCl_xF_y(g)$ based chemistries. The etch rate is lower when the F/Cl ratio increases in the gas mixture at temperature higher than 400 K. The decrease in the etch rate is explained by the non volatility of $HfF_4(g)$ in the investigated temperature range.

3.3.3. *Conclusions*

From these results, thermodynamic studies predict that a chlorocarbon gas mixture -such as CCl_4 -seems to be the most promising chemistry to etch HfO_2(s) under pure chemical etching conditions.

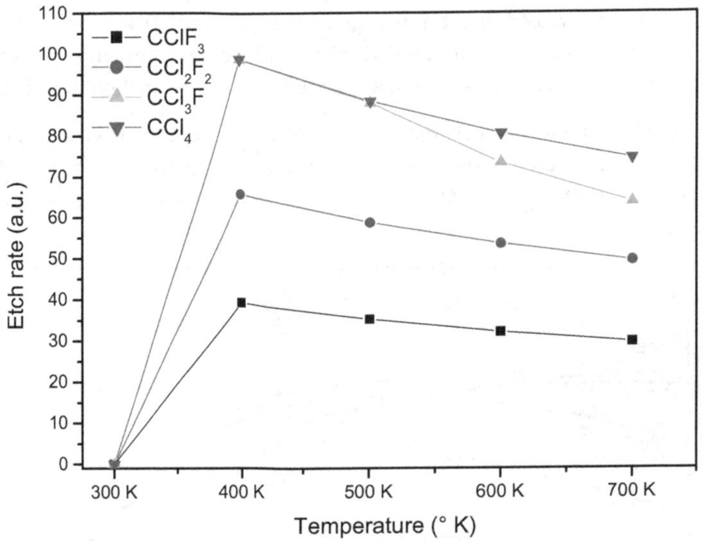

Figure 4. Evolution of the thermodynamically calculated HfO_2(s) etch rate as a function of F-Cl ratio in CCl_xF_y(g) based chemistries.

3.4. Optics: SiO_2 PVD deposition

From optics applications, the example of the evaporation/condensation process to obtain SiO_2 films is chosen. In that process, the surface of the evaporating source is heated by electronic bombardment, while the substrate is held at low temperature.

The control and the reproducibility of this type of process is based on the following points:

- Stability of the source with time (chemical composition, morphology of surface of the evaporating zone)
- Temperature and surface of the evaporated area (what is linked to the parameters of the electronic bombardment).

The object of this study is to simulate the evaporation of a source of glassy silica with the aim of depositing SiO_2. The heated zone is about 3-7 cm^2, the reactor has a volume about 1 m^3, the substrate is located at 1 m from the source.

The first paragraph is dedicated to the pure thermodynamic simulations to determine the major species originated from the evaporation. In the second one, the calculations of the flows of evaporation at equilibrium as well as exchanged flows between source and substrate surfaces are presented.

3.4.1. *Pure thermodynamic calculations of SiO_2 evaporation*

The thermodynamic simulations corresponding to the $SiO_2(s)$ evaporation are realized by considering an excess of solid $SiO_2(s)$ at a given temperature, in a constant volume. The range of tested temperature is 1600 - 2500 K. The results of the simulation indicate that the only solid present at equilibrium is $SiO_2(s)$ and that there is no formation of solid silicon.

In the range of selected temperature, the evaporation of the silica is thus congruent (the ratio of the quantity of silicon and oxygen produced in the gaseous form is equal to 2). Figure 5 presents the nature and the partial pressures of the gaseous species formed at equilibrium.

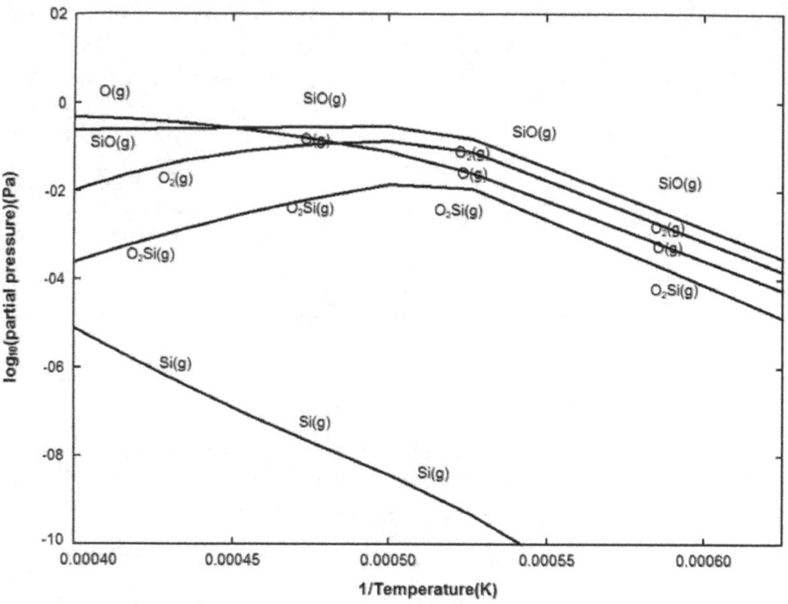

Figure 5. Calculated partial pressures for SiO_2 vaporization.

The major species in this range of temperature are $SiO(g)$, $O_2(g)$, $O(g)$ and $SiO_2(g)$, with trace of $Si(g)$. It can be noted that the mainly evaporated species is not $SiO_2(g)$ as it could believed to justify the stoichiometric composition of the deposits. From the results of the molar fractions calculated for various species, the gas phase reaction which takes place is globally the following:

$$SiO_2(s) = SiO(g) + 0.42\,O_2(g) + 0.15\,O(g) \tag{36}$$

These curves show that the evaporated material quantity and consequently the evaporation rate increases with temperature. That explains the best results obtained with sources carried beyond their melting point (besides the higher quality of the surface with regard to a solid source).

The calculated total pressure above silica is represented on the figure 5.

For the temperatures of evaporation above 1600 K, the total pressure over the silica is superior to 0.1 Pa. If the total pressure is fixed to a lower value, there is then complete consumption of the quantity of silica carried at the evaporation temperature from a thermodynamic point of view.

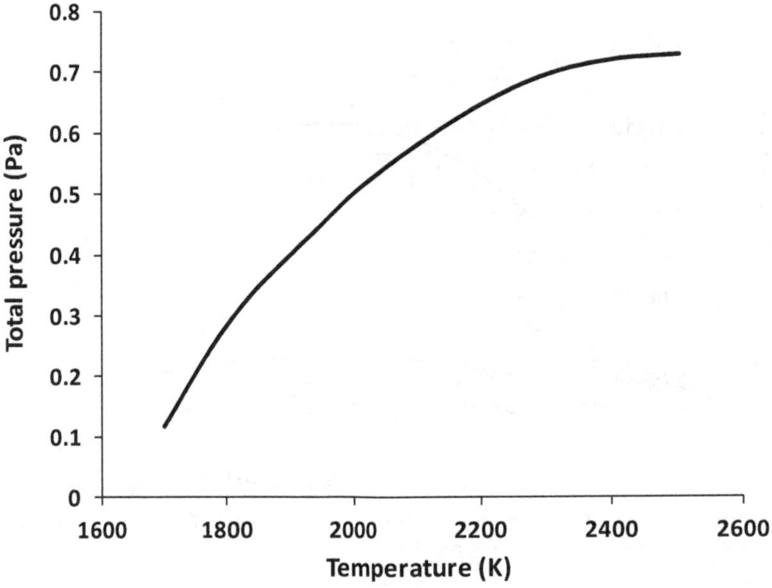

Figure 6. Calculated total pressure for $SiO_2(s)$ vaporization

3.4.2. Thermodynamic analysis coupled to mass transport: evaluation of deposition rate

The evaporation flows of the gaseous species originating from the $SiO_2(s)$ evaporation should respect the congruent vaporization relation (ratio Si/O in the gaseous phase = 2), demonstrated by the previous approach.

$$2*\Phi_{O_2}+\Phi_{SiO}+\Phi_{O}+2*\Phi_{Si_2O_2}+2*\Phi_{SiO_2}=\Phi_{Si}+2*\Phi_{SiO}+\Phi_{O}+4*\Phi_{Si_2O_2}+2*\Phi_{SiO_2} \quad (37)$$

where Φ_e is the molecular flow of the species e, according to equation (25).

As illustrated by the previous etching case, it is possible to calculate all the species partial pressures (p_e) which verify the equation (37).

From the calculated flows from equation (37), the molar volume of $SiO_2(s)$, the reactor geometry, and the temperature conditions on the surface, it is possible to estimate the growth rate and the deposition profile (Figure 7). The growth rate on the substrate is given from the exchanged flows between the two surfaces source and substrate, from basic assumptions of molecular flows.

With coaxial source and substrate, the exchanged flow between the r_0 radius source and r_1, radius substrate, separated by a distance h is given by the relation (38).

$$\Phi_e = \frac{\alpha_e p_e}{\sqrt{2\Pi M_e RT}} * \frac{\Pi}{2} * [h^2 + r_0^2 + r_1^2 - \sqrt{(h^2 + r_0^2 + r_1^2)^2 - 4 * r_0^2 r_1^2}] mol / s.m^2 \quad (38)$$

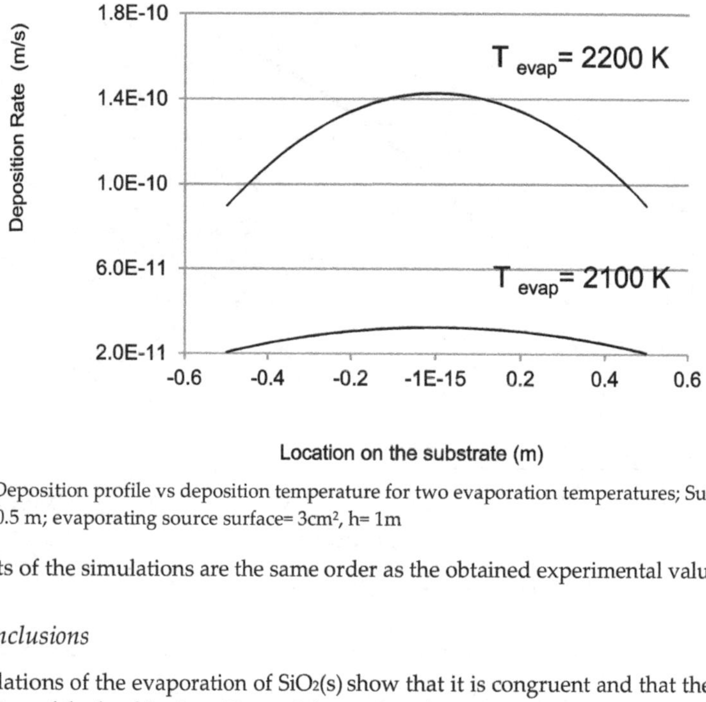

Figure 7. Deposition profile vs deposition temperature for two evaporation temperatures; Substrate radius r_0= 0.5 m; evaporating source surface= 3cm², h= 1m

The results of the simulations are the same order as the obtained experimental values.

3.4.3. Conclusions

The simulations of the evaporation of $SiO_2(s)$ show that it is congruent and that there is thus no evolution of the load in time. The mainly produced gaseous species are SiO(g), $O_2(g)$, and O(g). Their proportions remain constant but their quantities increase with the temperature of evaporation. Simulations of the evaporation/condensation process provide good estimations of the deposition rate.

4. Conclusions

This review illustrates the interest to operate a priori an thermodynamic approach to determine the feasibility and optimize a fabrication process, specially gas-solid fabrication process. Kinetic approaches will give rise to more realistic simulations but are often difficult to implement, for lack of reliable information. The classic pure thermodynamic can provide useful information. It can be the only approach in the case of complex chemical systems for which few kinetic data are available. To take into account the dynamic character of the processes, the approaches mixing thermodynamics simulations and calculations of exchanged flows are possible. In every case, the methodology has to contain continuous comparisons between experimental results and simulations.

Author details

Elisabeth Blanquet and Ioana Nuta
Laboratory "Science et Ingénierie des Matériaux et Procédés (SIMaP)"
Grenoble INP/CNRS/UJF, Saint Martin d'Hères, France

Acknowledgement

This paper was inspired by the collaborative works with the colleagues of the SIMaP (Laboratoire de Science et Ingénierie des Matériaux et Procédés): Jean-Noël Barbier, Claude Bernard, Christian Chatillon, Alexander Pisch, Arnaud Mantoux, Raphaël Boichot, Michel Pons.

5. References

[1] Cheynet, B., Chevalier, P.-Y. ,Fischer, E. (2002) Thermosuite. Calphad. 26: 167-174.
[2] Hack, K. (1996) The SGTE casebook: thermodynamics at work. Scientific Group Thermodata Europe and Institute of Materials.
[3] McBride, B. J., Zehe, M. J. ,Gordon, S. (2002) NASA Glenn Coefficients for Calculating Thermodynamic Properties of Individual Species NASA.
[4] Miedema, A., De Chatel, P. ,De Boer, F. R. (1980) Cohesion in alloys—fundamentals of a semi-empirical model. Physica B+ C. 100: 1-28.
[5] Hansen, M. ,Anderko, K. (1958) Constitution of Binary Alloys, Metallurgy and Metallurgical Engineering Series. New York: McGraw-Hill
[6] Elliott, R. (1965) Constitution of Binary Alloys, First Supplement. New York: McGraw-Hill.
[7] Massalski, T. B., Okamoto, H., Subramanian, P. R. ,Kacprazac, L. (1990) Binary alloy phase diagrams. Ohio: ASM International, Materials Park.
[8] Villars, P., Prince, A. ,Okamoto, H. (1995) Handbook of Ternary Alloy Phase Diagrams. Ohio: ASM International, Materials Park.
[9] Chase, M. (1998) NIST-JANAF thermochemical tables. Washington, D.C. and Woodbury, N.Y.: American Chemical Society.
[10] Barin, I. (1993) Thermochemical data of pure substances. New York: VCH.
[11] Gurvich, L. V., Veyts, I. ,Alcock, C. B. (1990) Thermodynamic Properties of Individual Substances. New York: Hemisphere Pub.

[12] SGTE, S. G. T. E. 38402 Saint Martin d'Heres, France.

[13] GTT-Technologies. 52134 Herzogenrath, Germany.

[14] Bale, C., Chartrand, P., Degterov, S., Eriksson, G., Hack, K., Ben Mahfoud, R., Melançon, J., Pelton, A. ,Petersen, S. (2002) FactSage thermochemical software and databases. Calphad. 26: 189-228.

[15] Belov, G. V., Iorish, V. S. ,Yungman, V. S. (1999) IVTANTHERMO for Windows--database on thermodynamic properties and related software. Calphad. 23: 173-180.

[16] Warnatz, J. (1984) Combustion chemistry. New York.

[17] (2005) NIST Chemistry WebBook, NIST Standard Reference Database. Gaithersburg MD.

[18] Smith, W. R. ,Missen, R. W. (1982) Chemical reaction equilibrium analysis: theory and algorithms. New York: Wiley

[19] Barbier, J. N. ,Bernard, C. (1986). In Calphad XV. Fulmer Grange, U.K.

[20] Davies, R., Dinsdale, A., Gisby, J., Robinson, J. ,Martin, S. (2002) MTDATA-thermodynamic and phase equilibrium software from the National Physical Laboratory. Calphad. 26: 229-271.

[21] Sundman, B., Jansson, B. ,Andersson, J. O. (1985) The thermo-calc databank system. Calphad. 9: 153-190.

[22] Thomas, N., Suryanarayana, P., Blanquet, E., Vahlas, C., Madar, R. ,Bernard, C. (1993) LPCVD WSi_2 Films Using Tungsten Chlorides and Silane. Journal of the Electrochemical Society. 140: 475-484.

[23] Bernard, C., Pons, M., Blanquet, E. ,Madar, R. (1999) Computer simulations from thermodynamic data: Materials productions and develeompment-Thermodynamic Calculations as the Basis for CVD Production of Silicide Coatings. MRS Bulletin-Materials Research Society. 24: 27-31.

[24] Pons, M., Bernard, C., Blanquet, E. ,Madar, R. (2000) Combined thermodynamic and mass transport modeling for material processing from the vapor phase. Thin Solid Films. 365: 264-274.

[25] Brize, V., Prieur, T., Violet, P., Artaud, L., Berthome, G., Blanquet, E., Boichot, R., Coindeau, S., Doisneau, B., Farcy, A., Mantoux, A., Nuta, I., Pons, M. ,Volpi, F. (2011) Developments of TaN ALD Process for 3D Conformal Coatings. Chem. Vapor Depos. 17: 284-295.

[26] Violet, P., Blanquet, E., Monnier, D., Nuta, I. ,Chatillon, C. (2009) Experimental thermodynamics for the evaluation of ALD growth processes. Surface and Coatings Technology. 204: 882-886.

[27] Violet, P., Nuta, I., Chatillon, C. ,Blanquet, E. (2007) Knudsen cell mass spectrometry applied to the investigation of organometallic precursors vapours. Surface and Coatings Technology. 201: 8813-8817.

[28] Dinsdale, A. (1991) SGTE data for pure elements. Calphad. 15: 317-425.

[29] Violet, P., Nuta, I., Artaud, L., Collas, H., Blanquet, E. ,Chatillon, C. (2009) A special reactor coupled with a high-temperature mass spectrometer for the investigation of the vaporization and cracking of organometallic compounds. Rapid Communications in Mass Spectrometry. 23: 793-800.

[30] Di Cioccio, L. ,Billon, T.(2002) Advances in SiC materials and technology for Schottky diode applications. In Silicon Carbide and Related Materials 2001, Pts 1 and 2, Proceedings, eds. Yoshida, S., Nishino, S., Harima, H. & Kimoto, T., 1119-1124.
[31] Matsunami, H. (2000) An overview of SiC growth. 125-130. Trans Tech Publ.
[32] Pons, M., Blanquet, E., Dedulle, J., Garcon, I., Madar, R. ,Bernard, C. (1996) Thermodynamic heat transfer and mass transport modeling of the sublimation growth of silicon carbide crystals. Journal of the Electrochemical Society. 143: 3727-3735.
[33] Pons, M., Anikin, M., Chourou, K., Dedulle, J., Madar, R., Blanquet, E., Pisch, A., Bernard, C., Grosse, P. ,Faure, C. (1999) State of the art in the modelling of SiC sublimation growth. Materials Science and Engineering: B. 61: 18-28.
[34] Pons, M., Baillet, F., Blanquet, E., Pernot, E., Madar, R., Chaussende, D., Mermoux, M., Di Coccio, L., Ferret, P., Feuillet, G., Faure, C. ,Billon, T. (2003) Vapor phase techniques for the fabrication of homoepitaxial layers of silicon carbide: process modeling and characterization. Applied Surface Science. 212–213: 177-183.
[35] Bernard, C., Blanquet, E. ,Pons, M. (2007) Chemical vapor deposition of thin films and coatings: Evaluation and process modeling. Surface and Coatings Technology. 202: 790-797.
[36] Hallin, C., Ivanov, I.G., Henry, A., Egilsson, T., Kordina, O., and Janzen, E. 1998, J. Crystal Growth, 183, 1-2, 163.
[37] Allendorf, M. D. ,Melius, C. F. (1992) Theoretical study of the thermochemistry of molecules in the silicon-carbon-hydrogen system. The Journal of Physical Chemistry. 96: 428-437.
[38] Allendorf, M. D. ,Melius, C. F. (1993) Theoretical study of thermochemistry of molecules in the silicon-carbon-chlorine-hydrogn system. The Journal of Physical Chemistry. 97: 720-728.
[39] Rocabois, P., Chatillon, C. ,Bernard, C. (1995) Thermodynamics of the Si-C system I. mass spectrometry studies of the condensed phases at high temperature. High Temperatures. High Pressures. 27: 3-23.
[40] Rocabois, P., Chatillon, C., Bernard, C. ,Genet, F. (1995) Thermodynamics of the Si-C system II. Mass spectrometric determination of the enthalpies of formation of molecules in the gaseous phase. High Temperatures. High Pressures. 27: 25-39.
[41] Neyret, E., Di Cioccio, L., Blanquet, E., Raffy, C., Pudda, C., Billon, T. ,J., C. (2000) SiC in situ pre-growth etching : a thermodynamic study. In Int. Conf. on Silicon Carbide and Related Materials, 1041-1044. Materials Science Forum.
[42] Helot, M., Chevolleau, T., Vallier, L., Joubert, O., Blanquet, E., Pisch, A., Mangiagalli, P. ,Lill, T. (2006) Plasma etching of HfO at elevated temperatures in chlorine-based chemistry. Journal of Vacuum Science & Technology A: Vacuum, Surfaces, and Films. 24: 30.
[43] Chen, J., Yoo, W. J., Tan, Z. Y. L., Wang, Y. ,Chan, D. S. H. (2004) Investigation of etching properties of HfO based high-K dielectrics using inductively coupled plasma. Journal of Vacuum Science & Technology A: Vacuum, Surfaces, and Films. 22: 1552-1558.

[44] Dumas, L., Chatillon, C. ,Quesnel, E. (2001) Thermodynamic calculations of congruent vaporization and interactions with residual water during magnesium fluoride vacuum deposition. Journal of Crystal Growth. 222: 215-234.

[45] Shen, J.-y. ,Chatillon, C. (1990) Thermodynamic calculations of congruent vaporization in III–V systems; Applications to the In-As, Ga-As and Ga-In-As systems. Journal of Crystal Growth. 106: 543-552.

Chapter 8

Group Contribution Methods for Estimation of Selected Physico-Chemical Properties of Organic Compounds

Zdeňka Kolská, Milan Zábranský and Alena Randová

Additional information is available at the end of the chapter

1. Introduction

Thermodynamic data play an important role in the understanding and design of chemical processes. To determine values of physico-chemical properties of compounds we can apply experimental or non-experimental techniques. Experimental techniques belong to the most correct, accurate and reliable. All experimental methods require relevant technical equipment, time necessary for experiment, sufficient amount of measured compounds of satisfactory purity. Compound must not affect technical apparatus and should not be decomposed during experiment. Other aspect is a valid legislation, which limits a usage of dangerous compounds by any users.

If due to any of these conditions mentioned above causes the experiment cannot be realized, some non-experimental approaches can be applied.

2. Non-experimental approaches to determine physico-chemical properties of compounds

If due to any of conditions results in that experimental determination cannot be realized and data on physico-chemical property are necessary, we have to employ some non-experimental approaches, either calculation methods or estimation ones. Due to the lack of experimental data for several industrially important compounds, different estimation methods have been developed to provide missing data. Estimation methods include those based on theory (e.g. statistical thermodynamics or quantum mechanics), various empirical relationships (correlations of required property with variable, experimentally determined compound characteristics, e.g. number of carbon atoms in their molecule, molecular weight, normal boiling temperature, etc.), and several classes of "additivity-principle" methods

(Baum, 1989; Pauling et al., 2001). Estimation methods can be divided into several groups from many aspects, e.g. into methods based on theoretical, semi-theoretical relations and the empirical ones. Books and papers of last decades divide estimation methods depending on the required input data into QPPR or QSPR approaches (Baum, 1989). QPPR methods (Quantity-Property-Property-Relationship) are input data-intensive. They require for calculation of searched value property knowledge of other experimental data. We can use them successfully only when we have input data. On the other hand QSPR (Baum, 1989) methods (Quantity-Structure-Property-Relationship) need only knowledge of the chemical structure of a compound to predict the estimated property. QSPR methods use some structural characteristics, such as number of fragments (atoms, bonds or group of atoms in a molecule), topological indices or other structural information, molecular descriptors, to express the relation between the property and molecular structure of compound (Baum, 1989; Pauling et al., 2001; Gonzáles et al., 2007a). Empirical and group contribution methods seem to be the most suitable (Pauling et al., 2001; Majer et al., 1989) due to their simplicity, universality and fast usage.

2.1. Group contribution methods

Group contribution methods are presented as empirical QSPR approaches. The easiest models were based on study of property on number of carbon atoms n_C or methylen groups n_{CH2} in molecules of homological series. In Fig. 1 is presented dependence of normal boiling temperature T_b on number of carbon atoms n_C (bottom axis) or methylen groups n_{CH2} (top axis) in molecules of homological series *n*-alkanols C_1-C_{12} (Majer & Svoboda, 1985; NIST database). As we can see, this dependence is clearly linear in some range of n_C=C_2-C_{10}. But increasing discrepancy is evident either for low number of carbon atoms C_1 or for higher one n_C>C_{10}. Due to these departures from linear behaviour some parameters covering structural effects on property were inclusive to these easy models (e.g. Chickos et al., 1996). From these approaches structural fragments and subsequently group contribution methods have been established.

Group contribution methods are based on the so called "additive principle". That means any compound can be divided into fragments, usually atoms, bonds or group of atoms, etc. All fragments have a partial value called a contribution. These contributions are calculated from known experimental data. Property of a compound is obtained by a summing up the values of all contributions presented in the molecule. Example of division of molecule of ethanol into atomic, bond and group fragments is presented in Fig. 2. When we divide this molecule into atomic fragments, the total value of property X of ethanol is given by summing up the values for two carbon atom contributions $X_{(C)}$, six hydrogen atom contributions $X_{(H)}$ and one oxygen atom contribution $X_{(O)}$. The second way is the division of ethanol molecules into the following bond fragments with their contribution: $X_{(C-C)}$, $X_{(C-O)}$, $X_{(C-H)}$ and $X_{(O-H)}$. Due to increasing quality and possibility of computer technique a fragmentation into more complex group structural fragments is applied in present papers (Baum, 1989; Pauling et al., 2001). Some of ways to divide the molecule ethanol into group structural fragments are presented in Fig. 2. Ethanol molecule can be divided either into: (i)

CH_3-, -CH_2- and -OH, or: (ii) CH_3- and –CH_2OH. More complex compounds are described by more complex structural fragments.

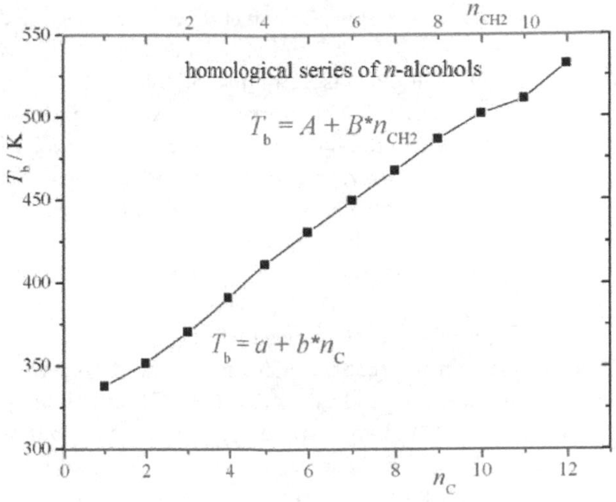

Figure 1. Dependence of normal boiling temperature of *n*-alkanols in homological series C_1-C_{12}

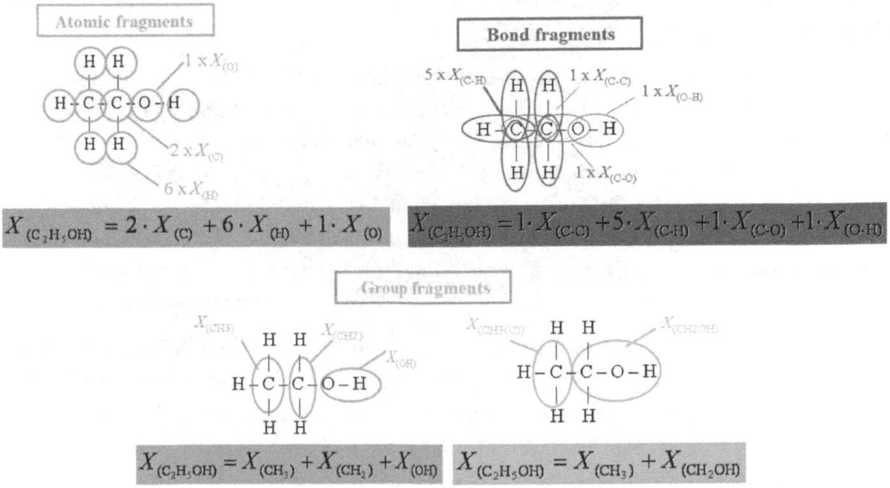

Figure 2. Example of division of ethanol molecule into atomic, bond and group structural fragments

Group contribution methods are essentially empirical estimation methods. A large variety of these models have been designed during last centuries, differing in a field of their applicability and in the set of experimental data. They were developed to estimate, e.g. critical properties (Lydersen, 1955; Ambrose, 1978; Ambrose, 1979; Joback & Reid, 1987;

Gani & Constantinou, 1996; Poling et al., 2001; Marrero & Gani, 2001; Brown et al., 2010; Monago & Otobrise, 2010; Sales-Cruz et al., 2010; Manohar & Udaya Sankar, 2011; Garcia et al., 2012), parameters of state equations (Pereda et al., 2010; Schmid & Gmehling, 2012), acentric factor (Constantinou & Gani, 1994; Brown et al., 2010; Monago & Otobrise, 2010), activity coefficients (Tochigi et al., 2005; Tochigi & Gmehling, 2011), vapour pressure (Poling et al., 2001; Miller, 1964), liquid viscosity (Joback & Reid, 1987; Conte et al., 2008; Sales-Cruz et al., 2010), gas viscosity (Reichenberg, 1975), heat capacity (Joback & Reid, 1987; Ruzicka & Domalski, 1993a; Ruzicka & Domalski, 1993b; Kolská et al., 2008), enthalpy of vaporization (e.g. Chickos et al., 1995; Chickos & Wilson, 1997; Marrero & Gani, 2001; Kolská et al. 2005, etc.), entropy of vaporization (Chiskos et al., 1998; Kolská et al. 2005), normal boiling temperature (Joback & Reid, 1987; Gani & Constantinou, 1996; Marrero & Gani, 2001), liquid thermal conductivity (Nagvekar & Daubert, 1987), gas thermal conductivity (Chung et al., 1984), gas permeability and diffusion coefficients (Yampolskii et al., 1998), liquid density (Campbell & Thodos, 1985; Sales-Cruz et al., 2010; Shahbaz et al., 2012), surface tension (Brock, 1955; Conte et al., 2008; Awasthi et al., 2010), solubility parameters of fatty acid methyl esters (Lu et al., 2011), flash temperatures (Liaw & Chiu, 2006; Liaw et al., 2011). Large surveys of group contribution methods for enthalpy of vaporization and liquid heat capacity have been presented in references (Zábranský et al., 2003; Kolská, 2004; Kolská et al., 2005; Kolská et. al, 2008; Zábranský et al, 2010a). Group-contribution-based property estimation methods ca be also used to predict the missing UNIFAC group-interaction parameters for the calculation of vapor-liquid equilibrium (Gonzáles at al., 2007b).

Group contribution methods can be used for pure compounds, even inorganic compounds (e.g. Williams, 1997; Briard et al., 2003), organometallic compounds (e.g. Nikitin et al., 2010) and also for mixtures (e.g. Awasthi et al., 2010; Papaioannou et al., 2010; Teixeira et al., 2011; Garcia et al., 2012). Also e.g. estimation of thermodynamic properties of polysacharides was presented (Lobanova et al., 2011). Discussion about determination of properties of polymers has been also published (Satyanarayana et al., 2007; Bogdanic, 2009; Oh & Bae, 2009). Property models based on the group contribution approach for lipid technology have been also presented (Díaz-Tovar et al., 2007).

During last years also models for ionic liquids and their variable properties were developed, e.g. for density, thermal expansion and viscosity of cholinium-derived ionic liquids (Costa et al., 2011; Costa et al., 2012), viscosity (Adamová et al., 2011), the glass-transition temperature and fragility (Gacino et al., 2011), experimental data of mixture with ionic liquid were compared with group contribution methods (Cehreli & Gmehling, 2010) or thermophysical properties were studied (Gardas et al., 2010).

Some of these group contribution methods were developed for only limited number of compounds, for some family of compounds, e.g. for fluorinated olefins (Brown et al., 2010), hydrocarbons (Chickos et al., 1995), fatty acid methyl esters (Lu et al., 2011), etc., most of approaches were established for a wide range of organic compounds.

2.1.1. Group contribution methods by Marrero-Gani

In this chapter for most of estimations the modified group contribution method by Marrero and Gani (Marrero & Gani, 2001; Kolská et al., 2005; Kolská et. al. 2008) was applied, which has been originally developed for estimation of different thermodynamic properties at one temperature only (Constantinou & Gani, 1996; Marrero & Gani, 2001). Determination of group contribution parameters is performed in three levels, primary, secondary and third. At first, all compounds are divided into the primary (first) order group contributions. This primary level uses contributions from simple groups that allow description of a wide variety of organic compounds. Criteria for their creation and calculation have been described (Marrero & Gani, 2001; Kolská et al., 2005; Kolská et al., 2008). The primary level groups, however, are insufficient to capture a proximity effect (they do not implicate an influence of their surroundings) and differences between isomers. Using primary level groups enables to estimate correctly properties of only simple and monofunctional compounds, but the estimation errors for more complex substances are higher. The primary level contributions provide an initial approximation that is improved at the second level and further refined at the third level, if that is possible and necessary. The higher levels (second and third) involve polyfunctional and structural groups that provide more information about a molecular structure of more complex compounds. These higher levels are able to describe more correctly polyfunctional compounds with at least one ring in a molecule, or non-ring chains including more than four carbon atoms in a molecule, and multi-ring compounds with a fused or non-fused aromatic or non-aromatic rings. The differences between some isomers are also able to distinguish by these higher levels. Complex polycyclic compounds or systems of fused aromatic or nonaromatic rings are described by the third order contributions. They are still bigger and more complex than the first, even the second order ones. The multilevel scheme enhances the accuracy, reliability and the range of application of group contribution method for an almost all classes of organic compounds.

After these all three levels the total value of predicted property X is obtained by the summing up of all group contributions, which occur in the molecule. First order groups, second and third order ones, if they are in.

$$X = x_0 + \sum_{i=1}^{n} N_i C_{xi} + \varpi \sum_{j=1}^{m} M_j D_{xj} + z \sum_{k=1}^{o} O_k E_{xk} \qquad (1)$$

where X stands the estimated property, x_0 is an adjustable parameter for the relevant property, C_{xi} is the first-order group contribution of type i, D_{xj} is the second-order group contribution of type j, E_{xk} is the third-order group contributions of the type k and N_i, M_j, O_k denote the number of occurrences of individual group contributions. The more detail description of parameters calculation is mentioned in original papers (Marrero & Gani, 2001; Kolská et al., 2005; Kolská et al., 2008)

To develop reliable and accurate group contribution model three important steps should be realized: (i) to collect input database, rather of critically assessed experimental data, from which parameters, group contributions, would be calculated; (ii) to design structural fragments for description of all chemical structures for compounds of input database; (iii) to divide all chemical structures into defined structural fragments correctly. It can be realized either manually, when databases of chemical structures and structural fragments inclusive several members only, either via computer program, when databases contain hundreds of compounds and structural fragments are more complex. To calculate group contribution parameters for thermophysical properties the ProPred program has been used (Marrero, 2002). Description and division chemical structures for other estimations have been made handy. Molecular structures for electronical splitting of all compounds from the basic data set were input in the Simplified Molecular Input Line Entry Specification, so-called the SMILES format (Weininger et al., 1986; Weininger, 1988; Weininger et al., 1989; Weininger, 1990).

For more universal usage of computer fragmentation a suitable computer program has been developed (Kolská & Petrus, 2010). The main goal of the newly developed program is to provide a powerful tool for authors using group contribution methods for automatic fragmentation of chemical structures.

2.2. Estimation of selected physico-chemical properties of compounds

The models for estimation of several physical or physico-chemical properties of pure organic compounds, such as enthalpy of vaporization, entropy of vaporization (Kolská et al., 2005), liquid heat capacity (Kolská et al., 2008) and a Nafion swelling (Randová et al, 2009) is presented below. Most of them are developed to estimate property at constant temperature 298.15 K and at normal boiling temperature (Kolská et al., 2005; Kolská et al., 2008; Randová et al., 2009), liquid heat capacity as a temperature dependent (Kolská et al., 2008). Hitherto unpublished results for estimation of a flash temperature or organic compounds and for determination of reactivation abbility of reactivators of acetylcholinesterase inhibited by inhibitor are presented in this chapter.

2.2.1. *Enthalpy of vaporization and entropy of vaporization*

Enthalpy of vaporization, $\Delta H\text{v}$, entropy of vaporization, $\Delta S\text{v}$ are important thermodynamic quantities of a pure compound, necessary for chemical engineers for modelling of many technological processes with evaporation, for extrapolation and prediction of vapour pressure data, or for estimation of the other thermodynamic properties, e.g. solubility parameters. It can be also used for extrapolation and prediction of vapour pressure data.

There are several methods to determine these properties, experiment-based and model-based. Experiment-based methods, such as calorimetry or gas chromatography, provide generally reliable data of good accuracy. In the case of model-based methods, we can distinguish several groups of methods on the basis of the input information they require. Methods based on the Clausius-Clapeyron equation and vapour pressure data, variable empirical correlations, methods based on the tools of statistical thermodynamics or quantum mechanics. During last decades the group contribution methods are widely used for their universality and simplicity. More rich survey of estimation methods for enthalpy of vaporization is presented in papers (Kolská, 2004; Kolská et al., 2005).

Large databases of critically assessed data have been used for group contribution calculations: data for 831 compounds have been used for estimations at 298.15 K, and data for 589 compounds have been used for estimations at the normal boiling temperature. Organic compounds were divided into several classes (aliphatic and acyclic saturated and unsaturated hydrocarbons, aromatic hydrocarbons, halogenated hydrocarbons, compounds containing oxygen, nitrogen or sulphur atoms and miscellaneous compounds). Especially calorimetrically measured experimental data from the compilation (Majer & Svoboda, 1985) and data from some other sources mentioned in original paper (Kolská et al., 2005) were employed.

Results for estimations of these three properties are presented in the following Tables, Table 1 for enthalpy of vaporization at 298.15 K, Table 2 for enthalpy of vaporization at normal boiling temperature and Table 3 for entropy of vaporization at normal boiling temperature, *NC* means a number of compounds used for development of model and contributions calculation, *NG* is number of applied structural fragments (groups), *AAE* is absolute average error and *ARE* is average relative error (Kolská et al., 2005).

Table 1 shows that values for 831 compounds were used for estimation of enthalpy of vaporization at 298.15 K. When only first level groups were used, the prediction was performed with the *AAE* and the *ARE* of 1.3 kJ/mol and 2.8%, resp. Values of 116 group contributions were calculated at this step. Then, 486 compounds were described by the second order groups. Prediction of these compounds improved after the use of these contributions from the value of 1.3 kJ/mol to 0.8 kJ/mol (from 2.8% to 1.8%) in comparison when using only the first level groups. At the end only 55 compounds were suitable for refining by the third order groups. The results were refined from the values of 1.4 kJ/mol to 1.1 kJ/mol (from 2.5% to 2.1%). The total prediction error was cut down from the value of 1.3 kJ/mol to 1.0 kJ/mol for *AAE* and from 2.8% to 2.2 % for *ARE* after usage of all three-level groups, as it is obvious from this table. A similar pattern of results for other predicted properties are presented in Tables 2 and 3.

Estimation level	*NC*	*NG*	*AAE* / kJ/mol	*ARE* / %
FIRST	**831**	**116**	**1.3**	**2.8**
SECOND	**486**	**91**	**0.8**	**1.8**
486 compounds	**after only the FIRST**		**(1.3)**	**(2.8)**
THIRD	**55**	**15**	**1.1**	**2.1**
55 compounds	**After FIRST + SECOND**		**(1.4)**	**(2.5)**
ALL LEVELS	**831**	**222**	**1.0**	**2.2**

Table 1. Results for estimation of enthalpy of vaporization at 298.15 K (Kolská et al., 2005)

Estimation level	*NC*	*NG*	*AAE* / kJ/mol	*ARE* / %
FIRST	**589**	**111**	**1.2**	**3.2**
SECOND	**377**	**100**	0.9	**2.5**
377 compounds	**after only the FIRST**		**(1.2)**	**(3.4)**
THIRD	**23**	**14**	**1.1**	**2.1**
23 compounds	**After FIRST + SECOND**		**(1.3)**	**(2.7)**
ALL LEVELS	**589**	**225**	**0.9**	**2.6**

Table 2. Results for estimation of enthalpy of vaporization at normal boiling temperature (Kolská et al., 2005)

Estimation level	*NC*	*NG*	*AAE* / J/(K·mol)	*ARE* / %
FIRST	**589**	**111**	**2.1**	**2.2**
SECOND	**377**	**100**	**1.8**	**1.9**
377 compounds	**after only the FIRST**		**(2.3)**	**(2.4)**
THIRD	**23**	**14**	**1.9**	**1.9**
23 compounds	**After FIRST + SECOND**		**(2.5)**	**(2.5)**
ALL LEVELS	**589**	**225**	**1.7**	**1.8**

Table 3. Results for estimation of entropy of vaporization at normal boiling temperature (Kolská et al., 2005)

As an example of the use of all three levels we have chosen the molecule of 1,1,4,7-tetramethylindane. Its chemical structure is shown in Fig. 3 and its division into individual first, second and third order groups with the result for vaporization enthalpy at 298.15 K is presented in Table 4. When we sum up all group contribution of the first level, we have got value of 64.48 kJ/mol. The first level provides an initial approximation with the relative error of estimated value exceeding 5 % in comparison with experimental value 61.37 kJ/mol. Estimated value of vaporization enthalpy at 298.15 K is then improved at the second level and further refined at the third level, after those the relative error reduced to 1.2 %.

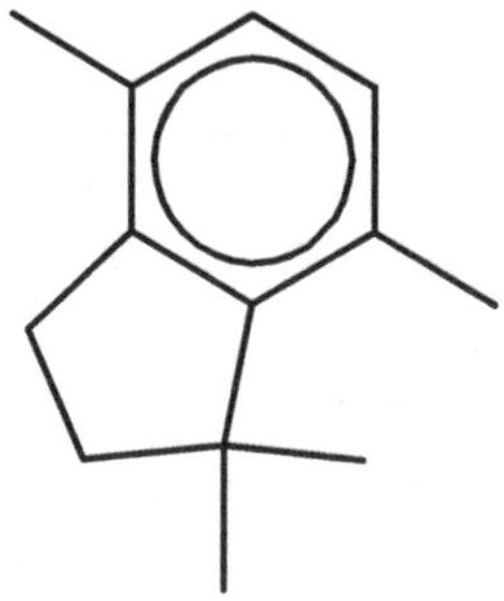

Figure 3. Chemical structure of 1,1,4,7-tetramethylindane

Estimation level	Group fragment no.	Group fragment definition	Its frequency	Group contribution value for ΔHv at 298.15 K / kJ/mol
FIRST	x_0	**Adjustable parameter**	**1**	**9.672**
	1	CH3	2	2.266
	13	aCH	2	4.297
	15	aC fused with nonaromatic subring	2	6.190
	18	aC-CH3	2	8.121
	107	CH2 (cyclic)	2	4.013
	109	C (cyclic)	1	3.667
Estimated value				64.48
SECOND	55	Ccyc-CH3	2	-1.355
Estimated value				63.59
THIRD	6	aC-CHncyc (fused rings) (n in 0..1)	2	0.279
	19	AROM.FUSED[2]s1s4	1	-0.615
Estimated value				62.13

Table 4. Results for estimation of enthalpy of vaporization at 298.15 K for 1,1,4,7-tetramethylindane, aC means carbon atom in aromatic ring, abbreviation cyc is used for cycle (Kolská et al., 2005)

Group contribution methods by Ducros (Ducros et al., 1980; Ducros et al., 1981; Ducros & Sannier, 1982; Ducros & Sannier, 1984), by Chickos (Chickos et al., 1996), the empirical method, equations nos. 6 and 7 by Vetere (Vetere, 1995) and method by Ma and Zhao (Ma & Zhao, 1993) were used for comparison of results obtained in this work for estimation at 298.15 K and at normal boiling temperature, resp. While the new approach (Kolská et al., 2005) was applied for enthalpy of vaporization at 298.15 K for 831 organic compounds with the *ARE* of 2.2 %, the Ducros´s method could be applied to only 526 substances with the *ARE* of 3.1 % and the Chickos's one for 800 compounds with the *ARE* of 4.7 %. For comparison of the results of estimation at the normal boiling temperature the new model provided for 589 compounds, the *ARE* was 2.6 % for enthalpy of vaporization and 1.8 % for entropy of vaporization (Kolská et al., 2005), the Vetere´s method was capable of estimating the values of for the same number of compounds with the following results: 4.6 % (Eq. 6, Vetere, 1995) and 3.4 % (Eq. 7, Vetere, 1995), model by Ma and Zhao (Ma & Zhao, 1993) for 549 compounds with the *ARE* of 2.5 %. The error for the enthalpy of vaporization, based on an independent set of various 74 compounds not used for correlation, has been determined to be 2.5%. Group contribution description and values for next usage of readers are presented in original paper (Kolská et al., 2005).

2.2.2. *Liquid heat capacity*

Isobaric heat capacity of liquid C^l_p is an important thermodynamic quantity of a pure compound. Its value must be known for the calculation of an enthalpy difference required for the evaluation of heating and cooling duties. Liquid heat capacity also serves as an input parameter for example in the calculation of temperature dependence of enthalpy of vaporization, for extrapolation of vapour pressure and the related thermal data by their simultaneous correlation, etc.

In work (Kolská et al., 2008) the three-level group contribution method by Marrero and Gani (Marrero & Gani, 2001) mentioned above, which is able to calculate liquid heat capacity at only one temperature 298.15 K, was applied, and this approach has been extended to estimate heat capacity of liquids as a function of temperature. Authors have employed the combination of equation for the temperature dependence of heat capacity and the model by Marrero and Gani to develop new model (Kolská et al., 2008).

For parameter calculation 549 organic compounds of variable families of compounds were taken. In Table 5 are presented results of this estimation. *NG* means number of applied structural groups and *ARE* is the average relative error. More detailed results are presented in original paper (Kolská et al., 2008).

Estimation level	*NG*	*ARE* / %
First	111	1.9
Second	88	1.6
Third	25	1.5

Table 5. Results for estimation of liquid heat capacity in temperature range of pure organic compounds (Kolská et al., 2008)

Also these estimated values were compared with results obtained by other estimation methods (Zábranský & Růžička, 2004; Chickos et al., 1993) for the basic dataset (compounds applied for parameter calculation) and also for 149 additional compounds not used in the parameter calculation (independent set). The first method (Zábranský & Růžička, 2004) was applied for all temperature range, the method proposed by Chickos (Chickos et al., 1993) was only used for temperature 298.15 K with the following results: new model was applied for 404 compounds with *ARE* of 1.5 %, the older method by Zábranský (Zábranský & Růžička, 2004) for the same number of compounds with the *ARE* of 1.8 % and the Chickos´s one for 399 compounds with the *ARE* of 3.9 %.

For the heat capacity of liquids authors used recommended data from the compilations by (Zábranský et al., 1996; Zábranský et al., 2001). Because the experimental data are presented permanently, it is necessary to update database of critically assessed and recommended data. Therefore authors´s work has been also aimed at updating and extending two publications prepared earlier within the framework of the IUPAC projects (Zábranský et al., 1996; Zábranský et al., 2001). These publications contain recommended data on liquid heat capacities for almost 2000 mostly organic compounds expressed in terms of parameters of correlating equations for temperature dependence of heat capacity. In new work (Zábranský et al., 2010b) authors collected experimental data on heat capacities of pure liquid organic and inorganic compounds that have melting temperature below 573 K published in the primary literature between 1999 and 2006. Data from more than 200 articles are included into the database. Compounds were divided into several families, such as hydrocarbons (saturated, cyclic, unsaturated, aromatic), halogenated hydrocarbons containing atoms of fluorine, chlorine, iodine, bromine, compounds containing oxygen (alcohols, phenols, ethers, ketones, aldehydes, acids, esters, heterocycles, other miscellaneous compounds), compounds containing nitrogen (amines, nitriles, heterocycles, other miscellaneous compounds), compounds containing sulphur (thioles, sulphides, heterocycles) and compounds containing silicon. Also data of organometallic compounds, compounds containing atoms of phosphorus and boron as well as some inorganic compounds were included. Also the list of families of compounds has been extended by a new group denoted as ionic liquids due to an increased interest in physical-chemical properties of these compounds in recent years. Data for approximately 40 ionic liquids were included. Altogether new data for almost 500 compounds, out of them about 250 compounds were not covered the in previous works (Zábranský et al., 1996; Zábranský et al., 2001), were compiled and critically evaluated.

2.2.3. *Nafion swelling*

Prediction of the physical and chemical properties of pure substances and mixtures is a serious problem in the chemical process industries. One of the possibilities for prediction of the properties is the group contribution method. The anisotropic swelling of Nafion 112 membrane in pure organic liquids (solvents) was monitored by an optical method. Nafion is a poly(tetrafluoroethylene) (PTFE) polymer with perfluorovinyl pendant side chains ended by sulfonic acid groups. The PTFE backbone guarantees a great chemical stability in both reducing and oxidizing environments. Nafion membrane is important in chemical industry. It is used in fuel cells, membrane reactors, gas dryers, production of NaOH, etc. (Randová et al., 2009). In many applications Nafion is immersed in liquid, which significantly affects the membrane properties, namely swelling and transport properties of permeates (Randová et al., 2009). The change in the size of the membrane sample is taken as a measure of swelling. All experimental data were presented (Randová et al., 2009) and these results were used as a basis for application of the group contribution method to the relative expansion in equilibrium. From a total of 38 organic liquids under study, 26 were selected as an evaluational set from which the group and structural group contributions were assigned. The remaining 12 compounds were used as the testing set.

Due to limited number of compounds the more complex and known group contribution methods could not been taken. Authors have to develop new group structural fragments. The proposed method utilizes the four kinds of the structural units: constants, C-backbone, functional groups, and molecular geometry (Randová et al., 2009). Constants were presented as alcohols, ketones, ethers, esters, carboxylic acids. As C-backbone were taken groups CH3, -CH2- and >CH-. Functional groups as hydroxyl OH-, carbonyl -C=O and ether –O- and fragments for molecular geometry for cycles and branched chains were taken. The relative expansions A_{exp} (for the drawing direction) and/or B_{exp} (for the perpendicular direction) were calculated from the side lengths of the dry membrane sample (a_{10}, a_{20}, b_{10}, b_{20}) and the side lengths of the swelled membrane sample in equilibrium (a_1, a_2, b_1, b_2) according to the eq. (3). Description of mentioned sizes is presented in Fig. 4.

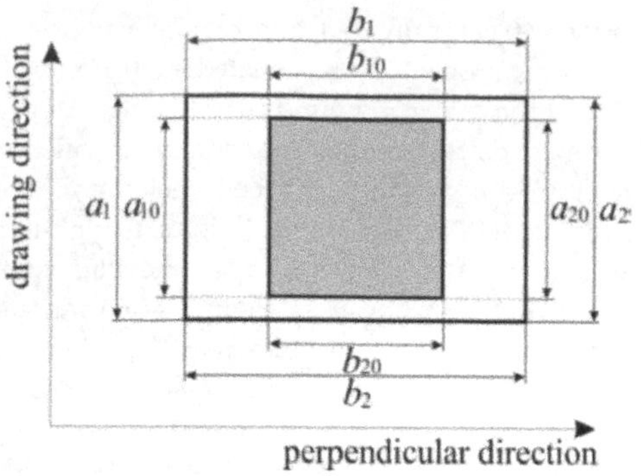

Figure 4. Description of membrane dimensions a_{10}, a_{20}, b_{10}, b_{20} side lengths of the dry membrane and a_1, a_2, b_1, b_2 side lengths of the swelled membrane in equilibrium (Randová et al., 2009)

$$u_{a1} = \frac{a_1 - a_{10}}{a_{10}} \quad , \quad u_{a2} = \frac{a_2 - a_{20}}{a_{20}} \quad , \quad u_{b1} = \frac{b_1 - b_{10}}{b_{10}} \quad , \quad u_{b2} = \frac{b_2 - b_{20}}{b_{20}} \tag{3}$$

Calculation approach is presented in original paper (Randová et al., 2009). Value of ±1.5% in relative expansions was determined to be the experimental error. Maximum differences between the experimental and calculated relative expansions in both sets did not exceed the value of ±3% (Randová et al., 2009).

The values of 13 contributions for individual membrane relative expansions were determined on the basis of experimental data on relative expansion of Nafion membrane. Obtained results are in good agreement with experimental data. Maximum differences between experimental and calculated values are nearly the same, only twice greater than the experimental error.

2.2.4. *Flash temperature of organic compounds*

The flash temperature T_f and lower flammability limit (LFL) are one of the most important variables to consider when designing chemical processes involving flammable substances. These characteristics are not fundamental physical points. Flash temperature is one of the most important variables used to characterize fire and explosion hazard of liquids. The flash temperature is defined as the lowest temperature at which vapour above liquid forms flammable mixture with air at a pressure 101 325 Pa. Usual approach for flash temperature estimation is linear relationship between flash temperature T_f and normal boiling temperature T_b (Dvořák, 1993). Some models for flash temperature were presented earlier (Liaw & Chiu, 2006; Liaw et al., 2011).

In this work to estimate flash temperature of organic compound authors applied the modified group contribution method (Kolská et al., 2005) and calculate group contribution values data for 186 compounds (Steinleitner, 1980) were used. The database for calculation of parameters contains data for aliphatic and acyclic saturated and unsaturated hydrocarbons, aromatic hydrocarbons, alcohols, halogenated hydrocarbons, compounds containing oxygen, nitrogen or sulphur atoms and miscellaneous compounds. To collect more data for development of reliable method was not able due to that all databases collect some values obtained via closed cup type measuring method and others measured by open cup one and data both of methods vary.

Flash temperature was calculated by relationship (4) similar to eq. (1):

$$T_f = T_f^\circ + \sum_{i=1}^{n} N_i C_i + \varpi \sum_{j=1}^{m} M_j D_j + z \sum_{k=1}^{o} O_k E_k \tag{4}$$

where T_f^o is an adjustable parameter, C_i is the first-order group contribution of type i, D_j is the second-order group contribution of type $_j$, E_k is the third-order group contribution of the type k and N_i, M_j, O_k denote the number of occurrences of individual group contributions. Determination of contributions and of adjustable parameters was performed by a three-step regression procedure (Marrero & Gani, 2001). To evaluate the method error the following statistical quantities for each compound, absolute error *AE* (eq. 5) and relative error *ARE* (eq. 6) were used:

$$AE[T_f] = \left| (T_f)_{\exp} - (T_f)_{est} \right| \tag{5}$$

$$RE[T_f] = \left(\frac{\left| (T_f)_{\exp} - (T_f)_{est} \right|}{(T_f)_{\exp}} \right) \cdot 100 \tag{6}$$

where subscripts "exp" and "est" mean experimental and estimated value of the flash temperature. 186 compounds from the basic data set were described by the first level group contributions (Kolská et al., 2005). From this large database only 114 compounds could be selected to be described by the original second level groups as defined earlier (Kolská et al., 2005). The total absolute and the relative average errors for all 186 compounds were equal to 6.3 K and 2.0 %. Results for individual estimation levels are presented in Table 6.

Estimation level	*NC*	*AAE* / K	*ARE* / %
FIRST	186	7.9	2.4
SECOND	105	5.7	1.8
THIRD	11	2.9	0.8
ALL LEVELS	186	6.5	2.0

Table 6. Results for Estimation of flash temperature, *NC* is number of compounds

Individual calculated structural fragments of the first, second and third estimation levels are presented in Tables 7-9, resp.

Structural fragment	Contribution / K	Structural fragment	Contribution / K	Structural fragment	Contribution / K
Tf^o	194.35	aCH	12.39	aC-OH	85.26
CH3	5.38	aC	21.15	CH2Cl	45.56
CH2	13.28	aC	26.84	CHCl	42.83
CH	15.77	aC	31.53	CCl	37.51
C	13.59	aN	25.51	CHCl2	67.42
CH2=CH	11.51	aC-CH3	28.32	CCl3	100.38
CH=CH	34.57	aC-CH2	37.11	aC-Cl	50.70
CH2=C	19.63	aC-CH	37.38	aC-F	53.47
CH=C	27.52	aC-C	19.88	aC-Br	64.97
C=C	29.27	aC-CH=CH2	50.66	-I	78.85
CH#C	14.94	OH	64.22	-Br	59.25
C#C	15.04	-SH	55.62	CH=CH	18.21
-F	2.99	CH2	10.89	CH=C	37.59
-Cl	29.32	CH	22.89	N	52.99
CH2SH	56.33	C	-9.50	O	-3.39

Table 7. Group contribution of the first level for estimation of flash temperature

Structural fragment	Contribution / K	Structural fragment	Contribution / K	Structural fragment	Contribution / K
(CH3)2CH	-1.45	CHm=CHn-Cl (m,n in 0..2)	-0.50	CHcyc-OH	-1.31
(CH3)3C	-3.98	aC-CHn-X (n in 1..2) X: Halogen	1.04	Ccyc-CH3	-0.23
CH(CH3)CH(CH3)	7.68	aC-CHn-OH (n in 1..2)	5.51	>Ncyc-CH3	-1.11E-17
CH(CH3)C(CH3)2	22.69	aC-CH(CH3)2	1.80	AROMRINGs1s2	-2.24
CHn=CHm-CHp=CHk (k,m,n,p in 0..2)	0.53	aC-CF3	0.13	AROMRINGs1s3	1.72
CH3-CHm=CHn (m,n in 0..2)	-2.01	(CHn=C)cyc-CH3 (n in 0..2)	0.46	AROMRINGs1s4	-0.84
CH2-CHm=CHn (m,n in 0..2)	2.23	CHcyc-CH3	-5.70	AROMRINGs1s2s4	-2.84
CHp-CHm=CHn (m,n in 0..2; p in 0..1)	3.78	CHcyc-CH2	17.80	AROMRINGs1s2s4s5	6.22
CHOH	-3.92	CHcyc-CH=CHn (n in 1..2)	8.52	PYRIDINEs3s5	9.98E-18
COH	-4.98	CHcyc-C=CHn (n in 1..2)	-1.20	(CH=CHOCH=CH)cyc	-4.24
CHm(OH)CHn(OH) (m,n in 0..2)	13.41	CHcyc-Cl	1.59	(3 F)	-0.13
				(perFlouro)	2.66E-17

Table 8. Group contribution of the second level for estimation of flash temperature

Structural fragment	Contribution / K	Structural fragment	Contribution / K	Structural fragment	Contribution / K
OH-(CHn)m-OH (m>2, n in 0..2)	-33.36	CHcyc-CHcyc (different rings)	-1.87	AROM.FUSED[2]	8.69
aC-aC (different rings)	-6.04	CH multiring	0.98	AROM.FUSED[4a]	-26.07
aC-CHncyc (fused rings) (n in 0..1)	-4.34	aC-CHm-aC (different rings) (m in 0..2)	12.92		

Table 9. Group contribution of the third level for estimation of flash temperature

2.2.5. Reactivation ability of some reactivators of acetylcholinesterase

In the last years regarding to valid legislation on dangerous compounds it is necessary to know many of important characteristics of chemical compounds. Due to this new models for their estimation were developed. New models for estimation of reactivation ability of reactivators for acetylcholinesterase inhibited by (i) chloropyrifos (*O,O*-diethyl *O*-3,5,6-trichloropyridin-2-yl phosphorothioate) as a representative of organophosphate insecticide and by (ii) sarin ((*RS*)-propan-2-yl methylphosphonofluoridate) as a representative of nerve agent is now presented. Both of these family compounds, organophosphate pesticide and nerve agent, are highly toxic and have the same effect to living organisms, which is based on an inhibition of acetylcholinesterase (AChE). New compounds able to reactivate the inhibited AChE, so-called reactivators of AChE, are synthesized. Reactivation ability of these reactivators is studied using standard reactivation *in vitro* test (Kuča & Kassa, 2003). Reactivation ability of reactivators means the percentage of original activity of AChE (Kuča & Patočka, 2004). New models for determination values of reactivation ability of reactivators AChE inhibited by (i) chloropyrifos and (ii) sarin have been developed. Concentration of reactivators was c=1·10^{-3} mol·dm^{-3}. In comparison with previous cases (estimations of thermophysical properties) authors have only less experimental data for development of model (about 20 for each of cases). Due to their long names and complex chemical structures these compounds in this chapter only are presented as their codes taken from original papers (Kuča & Kassa, 2003; Kuča et al., 2003a; Kuča et al., 2003b; Kuča et al., 2003c; Kuča & Patočka, 2004; Kuča & Cabal, 2004a; Kuča & Cabal, 2004b; Kuča et. al., 2006. Data of reactivation ability for these reactivators were given by the mentioned author team (Kuča et al.). Classical group contribution method includes groups describing some central atom, central atom with its bonds, or central atom with its nearest surrounding. However these models commonly used experimental data of hundreds or thousands compounds for parameters calculation. Due to for much small database in these cases it was necessary to design new fragments depending on the molecular structures available compounds. Structural fragments in this work cover larger and more complex part of molecules in comparison with other papers focused to group contribution methods. Reactivation potency is given in the group contribution method by the following relation, eq. 7:

$$R_{\mathrm{p}} = \sum_{i=1}^{n} x_i \cdot R_{\mathrm{p}i} \qquad (7)$$

where R_{pi} is value of individual fragment i presented in molecule by which it contributes to total value of R_p, x is number of frequency of this fragment i in molecule. Parameters R_{pi} were obtained by minimization function S_{Rp}, eq. 8:

$$S_{Rp} = \sum_{i=1}^{m} \left(R_{pi,calc} - R_{pi,exp} \right)^2 \tag{8}$$

where suffix exp presents experimental data and suffix calc the calculated values of R_p, m is number of compounds in dataset. The results obtained by this new approach were compared with experimental data using the following statistical quantities - an absolute error of individual compounds AE (eq. 9) and the average absolute error of dataset AAE (eq. 10):

$$AE_i = R_{pi,calc} - R_{pi,exp} \tag{9}$$

$$AAE = \sum_{i=1}^{m} \left(\frac{\left| R_{pi,calc} - R_{pi,exp} \right|}{m} \right) \tag{10}$$

Parameters of new model were calculated from the experimental data of the basic dataset. For model for reactivators AChE inhibited by chloropyrifos the input database included data of reactivation ability R_p for 24 reactivators (K 135, K 078, TO 096, TO 100, K 076, TO 094, TO 063, TO 097, TO 098, K 347, TO 231, K 117, K 074, K 033, K 106, K 107, K 110, K 114, HI-6, K 282, K 283, K 285, K 129, K 099) of concentration $c=1\cdot10^{-3}$ mol·dm^{-3}. Values for 17 groups with the AAE of 1.85 % of R_p were calculated. Designed groups with their calculated values of R_{pi} are presented in Table 10. These calculated parameters were tested on the test set of 5 independent compounds (TO 238, K 111, K 113, Methoxime, K 280) of which experimental data were not applied to group contributions determination. The AAE of R_p prediction for this test-set was 1.45 %. Table 11 presents experimental data and predicted values for these 5 independent compounds. Also illustration of usage of this method for two compounds from this test set is added below.

As it is clear from Table 10 the highest values of contributions are given for fragments P_3, P_7 for monoaromatic reactivators and P_{11}, P_{12} and P_{14} for two aromatic rings in reactivator molecule. On the other hand the smallest contribution (the negative ones) to total value of reactivation ability yields fragments P_5 a P_6 for monoaromatic compounds and P_{16} and P_{17} for two aromatic ring reactivators. These values resulted in fact that reactivation ability of new reactivators for reactivation AChE inhibited by chloropyrifos should be increased by presence of the following functional groups in molecules: another quarternary nitrogen atom in aliphatic ring bonded to aromatic quarternary nitrogen atom, the oxime groups in *para-* or *meta-* positions and presence of other aliphatic rings bonded to aromatic ring in other position than quarternary nitrogen and oxime groups. In all cases it is clear that reactivation ability decreases with presence of cycle ring, double bond and also in a less range with the presence of oxygen atoms presented in molecules. Also *ortho-* position of oxime group does not contribute positively.

no.	Fragment description	R_{pi} / %	no.	Fragment description	R_{pi} / %
P_1	Oxime group (=NOH) in position *o*- due to a quarternary nitrogen N^+ atom in aromatic ring	26.365	P_9	two oxime groups in positions *o*- due to a quarternary nitrogen atom in aromatic ring	26.580
P_2	Oxime group (=NOH) in position *p*- due to a quarternary nitrogen N^+ atom in aromatic ring	15.365	P_{10}	two oxime groups in positions *m*- due to a quarternary nitrogen atom N^+ in aromatic ring	15.737
P_3	Other quarternary nitrogen atom N^+ with 4 CH_x- groups in molecule, in aliphatic ring bonded to nitrogen atom N in aromatic ring	46.792	P_{11}	two oxime groups in positions *p*- due to a quarternary nitrogen atom N^+ in aromatic ring	47.105
P_4	Number of members bonded in aliphatic ring after the group P_3	-1.047	P_{12}	two oxime groups, one in position *o*-, other in position *m*- due to a quarternary nitrogen atom N^+ in aromatic ring	52.105
P_5	Cycle ring bonded to nitrogen atom N in aromatic ring	-10.365	P_{13}	two oxime groups, one in position *o*-, other in position *p*- due to a quarternary nitrogen atom N^+ in aromatic ring	25.842
P_6	Oxygen atom O bonded in aliphatic ring bonded to one aromatic ring	-32.437	P_{14}	two oxime groups, one in position *m*-, other in position *p*- due to a quarternary nitrogen atom N^+ in aromatic ring	56.105
P_7	Presence of other aliphatic ring bonded to aromatic one	88.073	P_{15}	Oxygen atom O bonded in aliphatic ring between two aromatic rings	-5.842
P_8	Number of members bonded in aliphatic ring following group N-CH_x- (nitrogen atom N is a part of aromatic ring), (which are not included in other groups)	-2.344	P_{16}	Cycle between two aromatic rings	-10.474
			P_{17}	Double bond between two aromatic rings	-22.105

Table 10. List of structural fragments and their values for estimation of reactivation ability of reactivators for acetylcholinesterase inhibited by chloropyrifos

Reactivator	$R_{p,exp}$ / %	$R_{p,calc}$ / %	Deviation / %
TO 238	48.00	49.55	1.55
K 111	8.00	5.26	-2.74
K 113	37.00	36.63	-0.37
Methoxime	45.00	47.11	2.11
K 280	4.00	4.48	0.48

Table 11. Results for estimation of reactivation ability of the test dataset of 5 reactivators of acetylcholinesterase inhibited by chloropyrifos

Illustration of new method for reactivation ability prediction of two reactivators (TO 238 and K 280) of which experimental data were not used for parameters calculation follows.

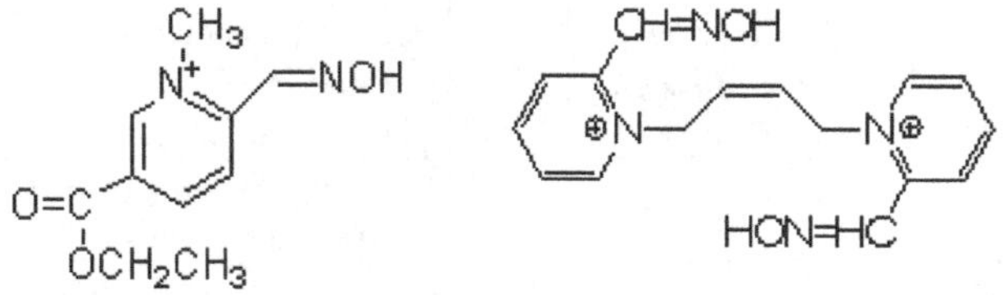

TO 238 K 280

Figure 5. Chemical structure of two reactivators of acetylcholinesterase signed as TO 238 and K 280

Example of usage of the new model for reactivation ability prediction for TO 280 reactivator:

$R_{p,calc}$(TO 238) = $P_1 + 2 \cdot P_6 + P_7 = 26.365 + 2 \cdot (-32.437) + 88.073 = 49.546$ %
$R_{p,exp}$(TO 238) = 48.00 %
$AE = R_{p,calc}$(TO 238) - $R_{p,exp}$(TO 238) = 1.55 %.

Example of usage of the new model for reactivation ability prediction for K 280 reactivator:

$R_{p,calc}$(K 280) = $P_9 + P_{17} = 26.580 + (-22.105) = 4.475$ %
$R_{p,exp}$(K 280) = 4.00 %
$AE = R_{p,calc}$(K 280) - $R_{p,exp}$(K 280) = 0.48 %.

For model development for reactivators AChE inhibited by sarin the input database included data of reactivation ability R_p for 18 reactivators (K 127, K 128, K 141, K 276, K 311, K 277, K 077, K 142, K 131, K 100, K 233, K 194, K 191, K 067, K 119, K 053, Pralidoxime, HI-6) of concentration $c=1 \cdot 10^{-3}$ mol·dm^{-3} were taken. Due to the smaller database in comparison with the chloropyrifos-inhibited case it was not possible to apply the same structural fragments. Values for 11 new structural different groups with the *AAE* of 3.39 % of R_p have been calculated. Designed groups with their calculated values of R_{pi} are presented in Table 12. These calculated parameters were tested on the test set of 4 independent compounds (TO 055, TO 058, K 197, Obidoxime) of which experimental data were not applied to group contributions determination. The *AAE* of R_p prediction for this test-set was 2.18 %. Table 13 presents experimental data and predicted values for 4 independent compounds.

no.	Fragment description	R_{pi} / %	no.	Fragment description	R_{pi} / %
P_1	Quarternary nitrogen atom N inclusive in aromatic ring	22.50	P_7	Other member of ring between two quarternary nitrogen atoms N^+ or/and bonded at the last quarternary nitrogen atom N^+ of molecule	-6.41
P_2	Presence of oxime group	-31.21	P_8	Presence of oxygen atom O in molecule other than mentioned in the following group	2.16
P_3	*ortho*- position of substituent on aromatic ring	46.03	P_9	Presence of group >C=O in molecule	7.88
P_4	*meta*- position of substituent on aromatic ring	14.49	P_{10}	Presence of group $-NH_x (x = 0, .., 2)$ in molecule	-12.20
P_5	*para*- position of substituent on aromatic ring	40.01	P_{11}	Presence of a double bond between two carbon atoms in a ring between two quarternary nitrogen atoms N^+ in molecule	1.66
P_6	Presence of cycle in a molecule	-10.03			

Table 12. List of structural fragments and their values for estimation of reactivation ability of reactivators for acetylcholinesterase inhibited by sarin

As it is shown in Table 12, the highest and the positive values of group contributions are given for fragments P_1, P_3-P_5, P_8 and P_9. On the other hand the smallest contribution (the negative ones) to the total value of reactivation ability yield fragments P_6, P_7 and P_{10}. Also the value of fragment P_2 for oxime group seems to have a negative effect to the total value but it should be said, that the oxime group has to be summed up with some group for its position on aromatic ring. It results in a fact that the oxime group in *meta*- position has the negative influence to the total value of reactivation ability, on the other hand the total value of R_p increases with oxime group in positions of *ortho*- or *para*-. These values resulted in fact that reactivation ability of new reactivators for reactivation AChE inhibited by sarin should be increased by the presence of the following function groups in molecules: another quarternary nitrogen atom in aromatic ring, the oxime groups in *ortho*- or *para*– positions, presence of oxygen atom or group >C=O in molecule. It is clear that reactivation ability decreases with presence of cycle ring and also with presence of the group NH_x $(x = 0, .., 2)$ in molecules. Also *meta*- position of oxime group, as same as the longer ring $(CH_x)_n$ $(x = 0, .., 2)$ bonded at quarternary nitrogen atoms, that means group P_7, do not contribute positively.

Reactivator	$R_{p,exp}$ / %	$R_{p,calc}$ / %	Deviation / %
TO 055	30.00	32.38	2.38
TO 058	25.00	27.63	2.63
K 197	4.00	4.08	0.08
Obidoxime	41.00	44.70	3.70

Table 13. Results for estimation of reactivation ability of the test dataset of 4 reactivators of acetylcholinesterase inhibited by sarin

Illustration of new method for reactivation ability prediction of two reactivators (TO 055 and TO 058) of which experimental data were not used for parameters calculation follows.

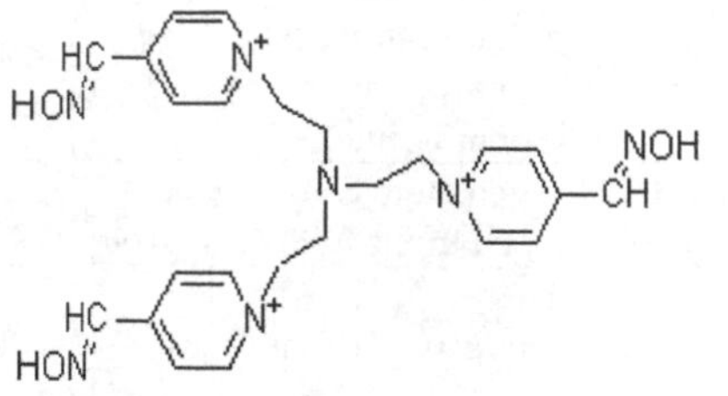

TO 055

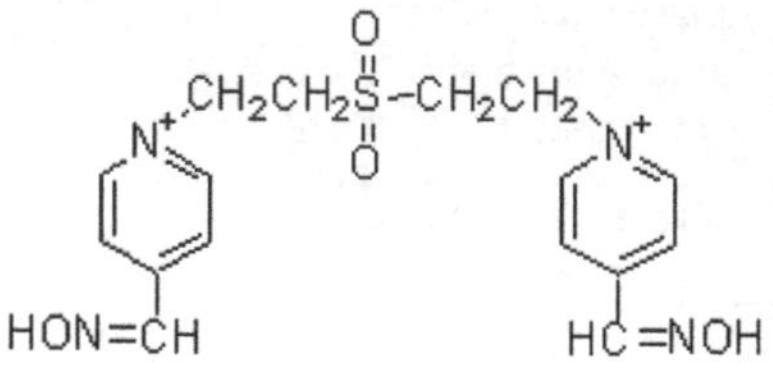

TO 058

Figure 6. Chemical structure of two reactivators of acetylcholinesterase signed as TO 055 and TO 058

Example of usage of the new model for reactivation ability prediction for TO 055 reactivator: $R_{p,calc}$(TO 055) = $3 \cdot P_1 + 3 \cdot P_2 + 3 \cdot P_5 + 3 \cdot P_6 + 3 \cdot P_7 + P_{10} = 3 \cdot (22.50) + 3 \cdot (-31.21) + 3 \cdot (40.01) + 3 \cdot (-10.03) + 3 \cdot (-6.41) + (-12.20) = 32.38$ %; $R_{p,exp}$(TO 055) = 30.00 %

$AE = R_{p,calc}$(TO 055) - $R_{p,exp}$(TO 055) = 2.38 %.

Example of usage of the new model for reactivation ability prediction for TO 055 reactivator: $R_{p,calc}$(TO 058) = $2 \cdot P_1 + 2 \cdot P_2 + 2 \cdot P_5 + 2 \cdot P_6 + 3 \cdot P_7 + 2 \cdot P_8 = 2 \cdot (22.50) + 2 \cdot (-31.21) + 2 \cdot (40.01) + 2 \cdot (-10.03) + 3 \cdot (-6.41) + 2 \cdot (2.16) = 27.63$ %; $R_{p,exp}$(TO 058) = 25.00 %

$AE = R_{p,calc}$(TO 058) - $R_{p,exp}$(TO 058) = 2.63 %.

As it is clear, in comparison with the previous cases, these models are applicable only for the same inhibitors but for new reactivators of ACHE inhibited by the same inhibitors (the first for chloropyrifos, the second one for sarin). But on the other hand, it can be also used as a tool for easy prediction of reactivation potency of some newly synthesized reactivators without any other *in vitro* standard tests.

3. Conclusion

Most of the industrial applications and products contain a mixture of many components and for the production it is important to know the properties of individual substance and the properties of aggregates. The accomplishments of all of these experiments are too expensive and time-consuming, so the calculation or estimation methods are good way to solve this problem. The group contribution methods are the important and favourible estimation method, because they permit to determine value of property of extant or hypothetic compound. Group contribution methods are the suitable tool for estimation of many physico-chemical quantities of pure compounds and mixtures too as it was showed and confirmed above for some cases. It can be used for estimation of pure compounds, as well as mixtures, for one temperature estimation, as well as for temperature range, etc. The biggest advantage of these methods is they need knowledge only chemical structure of compounds without any other input information. The presented models have been developed for estimation of many variable properties, enthalpy of vaporization, entropy of vaporization, liquid heat capacity, swelling of Nafion, flash temperature and reactivation ability of reactivators of acetylcholinesterase inhibited by organophosphate compounds. Proposed models and their structural fragments, accuracy and reliability depend mainly on frequency of input data and their accuracy, correctness and reliability. The most of presented models of group contribution methods, not only in the cases presented in this chapter, can be applied for the wide variety of organic compounds, when groups describing these molecules are presented. Some of models can be applied from only limited families of compounds due to their parameters were calculated only for limited database of compounds. Group contribution methods can be applied either for estimation or prediction of properties at one temperature or as a temperature function depending on their development. The accuracy of developed models is the higher, the input database is more reliable.

Author details

Zdeňka Kolská
Faculty of Science, J. E. Purkyně University, Usti nad Labem, Czech Republic

Milan Zábranský and Alena Randová
Institute of Chemical Technology, Prague, Czech Republic

Acknowledgement

This work was supported by the GA CR under the project P108/12/G108. Authors also thank to Ing. Michal Karlík from ICT Prague, Czech Republic, for data for flash temperature estimation and Prof. Kamil Kuča from Department of Toxicology, Faculty of Millitary Health Science Hradec Kralove, Czech Republic, for data on reactivation ability.

4. References

Baum, E. J. (1997). *Chemical Property Estimation: Theory and Application*, CRC Press LLC, ISBN 978-0873719384, Boca Raton, USA

Poling, B. E.; Prausnitz, J. M. & O´Connell, J. P. (2001). *The Properties of Gases and Liquids*, fifth edition. McGraw-Hill, ISBN 0-07-011682-2, New York, USA

González, H. E.; Abildskov, J.; & Gani, R. (2007). Computer-aided framework for pure component properties and phase equilibria prediction for organic systems. *Fluid Phase Equilibria*, Vol.261, No.1-2, pp. 199–204, ISSN 0378-3812

Majer, V.; Svoboda, V. & Pick, J. (1989). *Heats of Vaporization of Fluids*. Elsevier, ISBN 0-444-98920-X. Amsterdam, Netherlands

Majer, V. & Svoboda V. (1985). *Enthalpies of Vaporization of Organic Compounds, Critical Review and Data Compilation*. IUPAC. Chemical Data Series No. 32, Blackwell, Oxford, ISBN 0-632-01529-2

NIST database. http://webbook.nist.gov/chemistry/

Chickos, J. S.; Acree, W. E. Jr. & Liebman J. F. (1998). Phase Change Enthalpies and Entropies. In: *Computational Thermochemistry: Prediction and Estimation of Molecular* Thermodynamics. D. Frurip and K. Irikura, (Eds.), ACS Symp. Ser. 677, pp. 63-91, Washington, D. C.

Lydersen, A. L. (1955). *Estimation of Critical Properties of Organic Compounds*. Eng. Exp. Stn. rept. 3; University of Wisconsin College of Engineering: Madison, WI

Ambrose, D. (1978). *Correlation and Estimation of Vapour-Liquid Critical Properties. I. Critical Temperatures of Organic Compounds*. National Physical Laboratory, Teddington: NPL Rep. Chem. 92

Ambrose, D. (1979). *Correlation and Estimation of Vapour-Liquid Critical Properties. II. Critical Pressures and Volumes of Organic Compounds*. National Physical Laboratory, Teddington: NPL Rep. Chem. 98

Joback, K. G. & Reid, R. C. (1987). Estimation of pure-component properties from group-contributions. *Chemical Engineering Communications*, Vol.57, No.1-6, pp. 233-243, ISSN 0098-6445

Gani, R. & Constantinou, L. (1996). Molecular structure based estimation of properties for process design. *Fluid Phase Equilibria*, Vol.116, No.1-2, pp. 75-86, ISSN 0378-3812

Marrero J. & Gani R. (2001). Group-contribution based estimation of pure component properties. *Fluid Phase Equilibria*, Vol.183, Special Issue, pp. 183-208, ISSN 0378-3812

Brown, J. S.; Zilio, C. & Cavallini, A. (2010). Thermodynamic properties of eight fluorinated olefins. *International Journal of Refrigeration*, Vol.33, No.2, pp. 235-241, ISSN 0140-7007

Monago, K. O. & Otobrise, C. (2010). Estimation of pure-component properties of fatty acid sand esters from group contributions. *Journal of Chemical Society of Nigeria*, Vol.35, No.2, pp. 142-148

Sales-Cruz, M.; Aca-Aca, G.; Sanchez-Daza, O. & Lopez-Arenas, T. (2010). Predicting critical properties, density and viscosity of fatty acids, triacylglycerols and methyl esters by group contribution methods. *Computer-Aided Chemical Engineering*, Vol.28, pp. 1763-1768

Manohar, B. & Udaya Sankar, K. (2011). Prediction of solubility of Psoralea corylifolia L. Seed extract in supercritical carbon dioxide by equation of state models. *Theoretical Foundations of Chemical Engineering*, Vol.45, No.4, pp. 409-419, ISSN 0040-5795

Garcia, M.; Alba, J.; Gonzalo, A.; Sanchez, J. L. & Arauzo, J. (2012). Comparison of Methods for Estimating Critical Properties of Alkyl Esters and Its Mixtures. *Journal of Chemical & Engineering Data*, Vol.57, No.1, pp. 208-218, ISSN 0021-9568

Pereda, S.; Brignole, E. & Bottini, S. (2010). Equations of state in chemical reacting systems. In: *Applied Thermodynamics of Fluids*, Goodwin, A. R. H.; Sengers, J. V. & Peters, C. J. (Eds.), 433-459, Royal Society of Chemistry; 1st Ed., ISBN 978-1847558060, Cambridge, UK

Schmid, B. & Gmehling, J. (2012). Revised parameters and typical results of the VTPR group contribution equation of state. *Fluid Phase Equilibria*, Vol.317, pp. 110-126, ISSN 0378-3812

Constantinou, L. & Gani, R. (1994). New Group Contribution Method for Estimating Properties of Pure Compounds. *AIChE Journal*, Vol.40, No.10, pp. 1697-1710, ISSN ISSN 0001-1541

Tochigi, K.; Kurita, S.; Okitsu, Y.; Kurihara, K. & Ochi, K. (2005). Measurement and Prediction of Activity Coefficients of Solvents in Polymer Solutions Using Gas Chromatography and a Cubic-Perturbed Equation of State with Group Contribution. *Fluid Phase Equilibria*, Vol.228, No. Special Issue, pp. 527-533, ISSN 0378-3812

Tochigi, K. & Gmehling, J. (2011). Determination of ASOG Parameters-Extension and Revision. *Journal of Chemical Engineering of Japan*, Vol.44, No.4, pp. 304-306, ISSN 0021-9592

Miller, D. G. (1964). Estimating Vapor Pressures-Comparison of Equations. *Industrial and Engineering Chemistry*, Vol.56, No.3, pp. 46-&, ISSN 0019-7866

Conte, E.; Martinho, A.; Matos, H. A.; & Gani, R. (2008). Combined Group-Contribution and Atom Connectivity Index-Based Methods for Estimation of Surface Tension and Viscosity. *Industrial & Engineering Chemistry Research*, Vol.47, No.20, pp. 7940–7954, ISSN 0888-5885

Reichenberg, D. (1975). New Methods for Estimation of Viscosity Coefficients of Pure Gases at Moderate Pressures (With Particular Reference To Organic Vapors). *AIChE Journal*, Vol.21, No.1, 181-183, ISSN 0001-1541

Ruzicka, V. & Domalski, E. S. (1993a). Estimation of the Heat-Capacities of Organic Liquids as a Function of Temperature Using Group Additivity. 1. Hydrocarbon Compounds. *Journal of Physical and Chemical Reference Data*, Vol.22, No.3, pp. 597-618, ISSN 0047-2689.

Ruzicka, V. & Domalski, E. S. (1993b). Estimation of the Heat-Capacities of Organic Liquids as a Function of Temperature Using Group Additivity. 2. Compounds of Carbon, Hydrogen, Halogens, Nitrogen, Oxygen, and Sulfur. *Journal of Physical and Chemical Reference Data*, Vol.22, No.3, pp. 619-657, ISSN 0047-2689

Kolská, Z.; Kukal, J.; Zábranský, M. & Růžička, V. (2008). Estimation of the Heat Capacity of Organic Liquids as a Function of Temperature by a Three-Level Group Contribution Method. *Industrial & Engineering Chemistry Research*, Vol.47, No.6, pp. 2075-2085, ISSN 0888-5885

Chickos, J. S.; Hesse, D. G.; Hosseini, S.; Liebman, J. F.; Mendenhall, G. D.; Verevkin, S. P.; Rakus, K.; Beckhaus, H.-D. & Ruechardt, C. (1995). Enthalpies of vaporization of some highly branched hydrocarbons. *Journal of Chemical Thermodynamics*, Vol.27, No.6, pp. 693-705, ISSN 0021-9614

Chickos, J. S. & Wilson, J. A. (1997). Vaporization Enthalpies at 298.15K of the n-Alkanes from C21-C28 and C30. *Journal of Chemical and Engineering Data*, Vol.42, No.1, pp. 190-197, ISSN 0021-9568

Kolská, Z.; Růžička, V. & Gani, R. (2005). Estimation of the Enthalpy of Vaporization and the Entropy of Vaporization for Pure Organic Compounds at 298.15 K and at Normal Boiling Temperature by a Group Contribution Method. *Industrial & Engineering Chemistry Research*, Vol.44, No.22, pp. 8436-8454, ISSN 0888-5885

Nagvekar, M. & Daubert, T. E. (1987). A Group Contribution Method for Liquid Thermal-Conductivity. *Industrial & Engineering Chemistry Research*, Vol.26, No.7, pp. 1362-1365, ISSN 0888-5885

Chung, T. H.; Lee, L. L. & Starling, K. E. (1984). Applications of Kinetic Gas Theories and Multiparameter Correlation for Prediction of Dilute Gas Viscosity and Thermal-Conductivity. *Industrial & Engineering Chemistry Fundamentals*, Vol.23, No.1, pp. 8-13, ISSN 0196-4313

Yampolskii, Y.; Shishatskii, S.; Alentiev, A. & Loza, K. (1998). Group Contribution Method for Transport Property Predictions of Glassy Polymers: Focus on Polyimides and Polynorbornenes. *Journal of Membrane Science*, Vol.149, No.2, pp. 203-220, ISSN 0376-7388

Campbell, S. W. & Thodos, G. (1985). Prediction of Saturated-Liquid Densities and Critical Volumes for Polar and Nonpolar Substances. Journal of Chemical and Engineering Data, Vol.30, No.1, pp. 102-111, ISSN 0021-9568

Shahbaz, K.; Baroutian, S.; Mjalli, F. S.; Hashim, M. A. & AlNashef, I. M. (2012). Densities of ammonium and phosphonium based deep eutectic solvents: Prediction using artificial intelligence and group contribution techniques. *Thermochimica Acta*, Vol.527, pp. 59-66, ISSN 0040-6031

Brock, J. R. & Bird, R. B. (1955). Surface Tension and the Principle of Corresponding States. *AIChE Journal*, Vol.1, No.2, pp. 174-177, ISSN 0001-1541

Awasthi, A.; Tripathi, B. S. & Awasthi, A. (2010). Applicability of corresponding-states group-contribution methods for the estimation of surface tension of multicomponent liquid mixtures at 298.15 K. *Fluid Phase Equilibria*, Vol.287, No.2, pp. 151-154, ISSN 0378-3812

Lu, X.; Yang, Y. & Ji, J. (2011). Application of solubility parameter in solubility study of fatty acid methyl esters . *Zhongguo Liangyou Xuebao*, Vol.26, No.6, pp. 60-65

Liaw, H. & Chiu, Y. (2006). A general model for predicting the flash point of miscible mixtures. *Journal of Hazardous Materials*, Vol.137, No.1, pp. 38-46, ISSN 0304-3894

Liaw, H.; Gerbaud, V. & Li, Y. (2011). Prediction of miscible mixtures flash-point from UNIFAC group contribution methods. *Fluid Phase Equilibria*, Vol.300, No.1-2, pp. 70-82, ISSN 0378-3812

Zábranský, M.; Růžička, V. & Malijevský, A. (2003). Odhadové metody tepelných kapacit čistých kapalin. *Chemické Listy*, Vol.97, No.1, pp. 3-8, ISSN 0009-2770

Kolská, Z. (2004). Odhadové metody pro výparnou entalpii. *Chemické Listy*, Vol.98, No.6, pp. 328-334, ISSN 0009-2770

Zábranský, M.; Kolská, Z.; Růžička, V. & Malijevský, A. (2010a). The Estimation of Heat Capacities of Pure Liquids. In *Heat capacities: liquids, solutions and vapours*. T.M. Letcher & E. Wilhelm (Eds.), 421-435, The Royal Society of Chemistry, ISBN 978-0-85404-176-3, London

Gonzalez, H. E.; Abildskov, J.; Gani, R.; Rousseaux, P. & Le Bert, B. (2007). A Method for Prediction of UNIFAC Group Interaction Parameters. *AIChE Journal*, Vol. 53, No.6, pp. 1620-32, ISSN 0001-1541

Williams, J. D. (1997). *Prediction of melting and heat capacity of inorganic liquids by the method of group contributions.* Thesis, New Mexico State Univ., Las Cruces, NM, USA

Briard, A. J.; Bouroukba, M.; Petitjean, D. & Dirand, M. (2003). Models for estimation of pure n-alkanes' thermodynamic properties as a function of carbon chain length. *Journal of Chemical and Engineering Data,* Vol.48, No.6, pp. 1508-1516, ISSN 0021-9568

Nikitin, E. D.; Popov, A. P.; Yatluk, Y. G. & Simakina, V. A. (2010). Critical Temperatures and Pressures of Some Tetraalkoxytitaniums. *Journal of Chemical and Engineering Data,* Vol.55, No.1, pp. 178-183, ISSN 0021-9568

Papaioannou, V.; Adjiman, C. S.; Jackson, G. & Galindo, A. (2010). Group Contribution Methodologies for the Prediction of Thermodynamic Properties and Phase Behavior in Mixtures. In: *Molecular Systems Engineering.* Pistikopoulos E. N.; Georgiadis, M. C.; Due, V.; Adjiman, C. S.; Galindo, A. (Eds.), 135-172, Wiley-VCH, ISBN 978-3-527-31695-3, Weinheim

Teixeira, M. A.; Rodriguez, O.; Mota, F. L.; Macedo, E. A. & Rodrigues, A. E. (2010). Evaluation of Group-Contribution Methods To Predict VLE and Odor Intensity of Fragrances. *Industrial & Engineering Chemistry Research,* Vol.50, No.15, pp. 9390-9402, ISSN 0888-5885

Lobanova, O.; Mueller, K.; Mokrushina, L. & Arlt, W. (2011). Estimation of Thermodynamic Properties of Polysaccharides. *Chemical Engineering & Technology,* Vol.34, No.6, pp. 867-876, ISSN 0930-7516

Satyanarayana, K. C.; Gani, R.; & Abildskov, J. (2007). Polymer property modeling using grid technology for design of structured products. *Fluid Phase Equilibria,* Vol.261, No.1-2, pp. 58–63, ISSN 0378-3812

Bogdanic, G. (2009). Additive Group Contribution Methods for Predicting the Properties of Polymer Systems. In: *Polymeric Materials.* Nastasovic, A. B. & Jovanovic, S. M. (Eds.), 155-197, Transworld Research Network, ISSN 978-81-7895-398-4, Kerala

Oh, S. Y. & Bae, Y. C. (2009). Group contribution method for group contribution method for estimation of vapor liquid equilibria in polymer solutions. *Macromolecular Research,* Vol.17, No.11, pp. 829-841, ISSN 1598-5032

Díaz-Tovar, C.; Gani, R. & Sarup, B. (2011). Lipid technology: Property prediction and process design/analysis in the edible oil and biodiesel industries. *Fluid Phase Equilibria,* Vol.302, No.1-2, pp. 284–293, ISSN 0378-3812

Costa, A. J. L.; Esperanca, J. M. S. S.; Marrucho, I. M. & Rebelo, L. P. N. (2011). Densities and Viscosities of 1-Ethyl-3-methylimidazolium n-Alkyl Sulfates. *Journal of Chemical & Engineering Data,* Vol.56, No.8, pp. 3433-3441, ISSN 0021-9568

Costa, A. J. L.; Soromenho, M. R. C.; Shimizu, K.; Marrucho, I. M.; Esperanca, J. M. S. S.; Lopes, J. N. C. & Rebelo, L. P. N. (2012). Density, Thermal Expansion and Viscosity of Cholinium-Derived Ionic Liquids. *Chemphyschem: a European journal of chemical physics and physical chemistry,* Ahead of Print, ISSN 1439-7641

Adamova, G.; Gardas, Ramesh L.; Rebelo, L. P. N.; Robertson, A. J. & Seddon, K. R. (2011). Alkyltrioctylphosphonium Chloride Ionic Liquids: Synthesis and Physicochemical Properties. *Dalton Transactions,* Vol.40, No.47, pp. 12750-12764, ISSN 1477-9226

Gacino, F. M.; Regueira, T.; Lugo, L.; Comunas, M. J. P. & Fernandez, J. (2011). Influence of Molecular Structure on Densities and Viscosities of Several Ionic Liquids. *Journal of Chemical & Engineering Data*, Vol.56, No.12, pp. 4984-4999, ISSN 0021-9568

Cehreli, S. & Gmehling, J. (2010). Phase Equilibria for Benzene-Cyclohexene and Activity Coefficients at Infinite Dilution for the Ternary Systems with Ionic Liquids. *Fluid Phase Equilibria*, Vol.295, No.1, pp. 125-129, ISSN 0378-3812

Gardas, R. L.; Ge, R.; Goodrich, P.; Hardacre, C.; Hussain, A. & Rooney, D. W. (2010). Thermophysical Properties of Amino Acid-Based Ionic Liquids. *Journal of Chemical & Engineering Data*, Vol.55, No.4, pp. 1505-1515, ISSN 0021-9568

Marrero J. (2002). Programm *ProPred*, Version 3.5, Jorge Marrero, Department of Chemical Engineering, DTU Denmark, released date: May 15, 2002

Weininger, D.; Weininger, A. & Weininger, J. (1986). Smiles. A Modern Chemical Language and Information System. *Chemical Design Automation News*, Vol.1, No.8, pp. 2-15, ISSN 0886-6716

Weininger, D. (1988). Smiles. A Chemical Language and Information-System. 1. Introduction to Metodology and Encoding Rules. *Journal of Chemical Information and Computer Sciences*, Vol.28, No.1, pp. 31-36, ISSN 0095-2338

Weininger, D.; Weininger, A. & Weininger, J. (1989). Smiles. 2. Algorithm for Generation of Unique Smiles Notation. *Journal of Chemical Information and Computer Sciences*, Vol.29, No.2, pp. 97-101, ISSN 0095-2338

Weininger, D. (1990). Smiles. 3. Depict – Graphical Depiction of Chemical Structures. *Journal of Chemical Information and Computer Sciences*, Vol.30, No.3, pp. 237-243, ISSN 0095-2338

Kolská, Z. & Petrus P. (2010). Tool for group contribution methods – computational Fragmentation *Collection of Czechoslovak Chemical Communications*, Vol.75, No.4, pp. 393–404, ISSN 0010-0765

Randová, A.; Bartovská, L.; Hovorka, Š.; Poloncarzová, M.; Kolská, Z. & Izák, P. (2009). Application of the Group Contribution Approach to Nafion Swelling. *Journal of Applied Polymer Science*, Vol.111, No.4, pp. 1745-1750, ISSN 0021-8995

Ducros, M.; Gruson, J. F. & Sannier H. (1980). Estimation of enthalpies of vaporization for liquid organic compounds. Part 1. Applications to alkanes, cycloalkanes, alkenes, benzene hydrocarbons, alkanethiols, chloro- and bromoalkanes, nitriles, esters, acids, and aldehydes. *Termochimica Acta*, Vol.36, No.1, pp. 39-65, ISSN 0040-6031

Ducros, M.; Gruson, J. F. & Sannier H. (1981). Estimation of the enthalpies of vaporization of liquid organic compounds. Part 2. Ethers, thioalkanes, ketones and amines. *Termochimica Acta*, Vol.44, No.2, pp. 131-140, ISSN 0040-6031

Ducros, M. & Sannier, H. (1982). Estimation of the enthalpies of vaporization of liquid compounds. Part 3. Application to unsaturated hydrocarbons. *Termochimica Acta*, Vol.54, No.1-2, pp. 153-157, ISSN 0040-6031

Ducros, M. & Sannier, H. (1984). Determination of vaporization enthalpies of liquid organic compounds. Part 4. Application to organometallic compounds. *Termochimica Acta*, Vol.75, No.3, pp. 329-340, ISSN 0040-6031

Vetere A. (1995). Methods to predict the vaporization enthalpies at the normal boiling temperature of pure compounds revisited. *Fluid Phase Equilibria*, Vol.106, No.1-2, pp. 1-10 , ISSN 0378-3812

Ma, P. & Zhao, X. (1993). Modified Group Contribution Method for Predicting the Entropy of Vaporization at the Normal Boiling Point. *Industrial & Engineering Chemistry Research*, Vol.32, No.12, pp. 3180-3183, ISSN 0888-5885

Zábranský, M. & Růžička, V. (2004). Estimation of the Heat Capacities of Organic Liquids as a Function of Temperature Using Group Additivity. An Amendment. *J. Phys. Chem. Ref. Data*, Vol.33, No.4, pp. 1071-1081, ISSN 0047-2689

Chickos, J. S.; Hesse, D. G. & Liebman, J. F. (1993). A Group Additivity Approach for the Estimation of Heat-Capacities of Organic Liquids and Solids at 298 K. *Structural Chemistry*, Vol.4, No.4, pp. 261-269, ISSN 1040-0400

Zábranský, M.; Růžička, V.; Majer, V. & Domalski E. S. (1996). Heat capacity of liquids: Volume II. Critical review and recommended values. *Journal of Physical and Chemical Reference Data*, Monograph, 815-1596, American Chemical Society: Washington, D.C., ISSN 1063-0651

Zábranský, M.; Růžička, V. & Domalski E. S. (2001). Heat Capacity of Liquids: Critical Review and Recommended Values. Supplement I. *Journal of Physical and Chemical Reference Data*, Vol.30, No.5, pp. 1199-1689

Zábranský, M.; Kolská, Z.; Růžička, V. & Domalski, E. S. (2010b). Heat Capacity of Liquids: Critical Review and Recommended Values. Supplement II. *Journal of Physical and Chemical Reference Data*, Vol.39, No.1, pp. 013103/1-013103/404, ISSN 0047-2689

Dvořák, O. (1993). Estimation of the flash point of flammable liquids. *Chemický průmysl*, Vol.43, No.5, pp. 157-158., ISSN 0009-2789

Steinleitner, H. G. (1980). *Tabulky hořlavých a nebezpečných látek*, Transl. Novotný, V.; Benda, E. Svaz požární ochrany ČSSR , 1st ed. Praha

Kuča, K. & Kassa, J. (2003). A Comparison of the Ability of a New Bisperidinium Oxime-1,4-(hydroxyiminomethylpyridinium)-4-(4-carbamoylpyridinium)butane Dibromide and Currently used Oximes to Reactivate Nerve Agent-Inhibited Rat Brain Acetylcholinesterase by In Vitro Methods. *Journal of Enzyme Inhibition*, Vol.18, No.6, pp. 529-533, ISSN 8755-5093

Kuča, K.; Bielavský, J.; Cabal, J. & Kassa, J. (2003a). Synthesis of a New Reactivator of Tabun-Inhibited Acetylcholnesterase. *Bioorganic & Medicinal Chemistry Letters*, Vol.13, No.20, pp. 3545-3547, ISSN 0960-894X

Kuča, K.; Bielavský, J.; Cabal, J. & Bielavská, M. (2003b). Synthesis of a Potential Reactivator of Acetylcholinesterase-(1-(4-hydroxyiminomethylpyridinium)-3-(carbamoylpyridinium)-propane dibromide. Tetrahedron Letters, Vol.44, No.15, pp. 3123-3125, ISSN 0040-4039

Kuča, K.; Patočka, J. & Cabal, J. (2003c). Reactivation of Organophosphate Inhibited Acetyylcholinesterase Activity by *α,ω*-bis-(4-hydroxyiminomethylpyridinium)alkanes in vitro. *Journal of Applied Biomedicine*, Vol.1, No.4, pp. 207-211. ISSN 1214-0287

Kuča K. & Patočka J. (2004). Reactivation of Cyclosarine-Inhibited Rad Brain Acetylcholinesterase by Pyridinium-Oximes. *Journal of Enzyme Inhibition and Medicinal Chemistry*. Vol.19, No.1, pp. 39-43, ISSN 1475-6366

Kuča K. & Cabal, J. (2004a). In Vitro Reaktivace Acetylcholinesterázy Inhibované O-Isopropylmethylfluorofofonátem užitím biskvarterního oximu HS-6. *Česká a Slovenská Farmacie*, Vol.53, No.2, pp. 93-95, ISSN 1210-7816

Kuča K. & Cabal, J. (2004b). In Vitro Reactivation of Tabun-Inhibited Acetylcholinesterase Using New Oximes - K027, K005, K033 and K048. *Central European Journal of Public Health*, Vol.12, No.Suppl, pp. S59-S61, ISSN 1210-7778

Kuča K.; Jun, D. & Musílek, K. (2006). Structural Requirements of Acetylcholinesterase Reactivators. *Mini-Reviews in Medicinal Chemistry*, Vol.6, No.3, pp. 269-277, ISSN 1389-5575/06

Permissions

The contributors of this book come from diverse backgrounds, making this book a truly international effort. This book will bring forth new frontiers with its revolutionizing research information and detailed analysis of the nascent developments around the world.

We would like to thank Ricardo Morales-Rodriguez, for lending his expertise to make the book truly unique. He has played a crucial role in the development of this book. Without his invaluable contribution this book wouldn't have been possible. He has made vital efforts to compile up to date information on the varied aspects of this subject to make this book a valuable addition to the collection of many professionals and students.

This book was conceptualized with the vision of imparting up-to-date information and advanced data in this field. To ensure the same, a matchless editorial board was set up. Every individual on the board went through rigorous rounds of assessment to prove their worth. After which they invested a large part of their time researching and compiling the most relevant data for our readers. Conferences and sessions were held from time to time between the editorial board and the contributing authors to present the data in the most comprehensible form. The editorial team has worked tirelessly to provide valuable and valid information to help people across the globe.

Every chapter published in this book has been scrutinized by our experts. Their significance has been extensively debated. The topics covered herein carry significant findings which will fuel the growth of the discipline. They may even be implemented as practical applications or may be referred to as a beginning point for another development. Chapters in this book were first published by InTech; hereby published with permission under the Creative Commons Attribution License or equivalent.

The editorial board has been involved in producing this book since its inception. They have spent rigorous hours researching and exploring the diverse topics which have resulted in the successful publishing of this book. They have passed on their knowledge of decades through this book. To expedite this challenging task, the publisher supported the team at every step. A small team of assistant editors was also appointed to further simplify the editing procedure and attain best results for the readers.

Our editorial team has been hand-picked from every corner of the world. Their multi-ethnicity adds dynamic inputs to the discussions which result in innovative outcomes. These outcomes are then further discussed with the researchers and contributors who give their valuable feedback and opinion regarding the same. The feedback is then collaborated with the researches and they are edited in a comprehensive manner to aid the understanding of the subject.

Apart from the editorial board, the designing team has also invested a significant amount of their time in understanding the subject and creating the most relevant covers. They scrutinized every image to scout for the most suitable representation of the subject and create an appropriate cover for the book.

The publishing team has been involved in this book since its early stages. They were actively engaged in every process, be it collecting the data, connecting with the contributors or procuring relevant information. The team has been an ardent support to the editorial, designing and production team. Their endless efforts to recruit the best for this project, has resulted in the accomplishment of this book. They are a veteran in the field of academics and their pool of knowledge is as vast as their experience in printing. Their expertise and guidance has proved useful at every step. Their uncompromising quality standards have made this book an exceptional effort. Their encouragement from time to time has been an inspiration for everyone.

The publisher and the editorial board hope that this book will prove to be a valuable piece of knowledge for researchers, students, practitioners and scholars across the globe.

List of Contributors

Nikolai Bazhin
Institute of Chemical Kinetics and Combustion, Novosibirsk State University, Institutskaya 3, Novosibirsk, Russia

Ahmet Gürses
Ataturk University, K.K. Education Faculty, Department of Chemistry, Erzurum, Turkey

Mehtap Ejder-Korucu
Kafkas University, Faculty of Science and Literature, Department of Chemistry, Kars, Turkey

Yi Fang
Department of Mathematics, Nanchang University, 999 Xuefu Road, Honggutan New District, Nanchang, 330031, China

A. Plastino
Universidad Nacional de La Plata, Instituto de Física (IFLP-CCT-CONICET), C.C. 727, 1900 La Plata, Argentina
Physics Departament and IFISC-CSIC, University of Balearic Islands, 07122 Palma de Mallorca, Spain

Evaldo M. F. Curado
Centro Brasileiro de Pesquisas Fisicas, Rio de Janeiro, Brazil

M. Casas
Physics Departament and IFISC-CSIC, University of Balearic Islands, 07122 Palma de Mallorca, Spain

Bohdan Hejna
Institute of Chemical Technology Prague, Department of Mathematics, Studentská 6, 166 28 Prague 6, Czech Republic

Ronald J. Bakker
Department of Applied Geosciences and Geophysics, Resource Mineralogy, Montanuniversitaet, Leoben, Austria

Elisabeth Blanquet and Ioana Nuta
Laboratory "Science et Ingénierie des Matériaux et Procédés (SIMaP)" Grenoble INP/CNRS/UJF, Saint Martin d'Hères, France

Zdeňka Kolská
Faculty of Science, J. E. Purkyně University, Usti nad Labem, Czech Republic

Milan Zábranský and Alena Randová
Institute of Chemical Technology, Prague, Czech Republic

Printed in the USA
CPSIA information can be obtained
at www.ICGtesting.com
JSHW011414221024
72173JS00004B/541

9 781632 381729